CROP RESIDUES IN SUSTAINABLE
MIXED CROP/LIVESTOCK FARMING SYSTEMS

Crop Residues in Sustainable Mixed Crop/Livestock Farming Systems

Edited by

C. Renard
International Crops Research Institute for the Semi-Arid Tropics (ICRISAT)
India

CAB INTERNATIONAL
in association with the
International Crops Research Institute for the Semi-Arid Tropics
and the
International Livestock Research Institute

CAB INTERNATIONAL
Wallingford
Oxon OX10 8DE
UK

Tel: +44 (0)1491 832111
Fax: +44 (0)1491 833508
E-mail: cabi@cabi.org

CAB INTERNATIONAL
198 Madison Avenue
New York, NY 10016-4341
USA

Tel: +1 212 726 6490
Fax: +1 212 686 7993
E-mail: cabi-nao@cabi.org

Published in association with:
International Crops Research Institute for the Semi-Arid Tropics
Patancheru 502 324
Andhra Pradesh
India

and

International Livestock Research Institute
PO Box 30709
Nairobi
Kenya

A catalogue record for this book is available from the British Library, London, UK.
A catalogue record for this book is available from the Library of Congress,
Washington DC, USA.

ISBN 0 85199 177 7

Printed and bound in the UK by Biddles Ltd, Guildford and King's Lynn

Contents

List of Participants

M.M. Anders (Agronomist) International Crops Research Institute for the Semi-Arid Tropics (ICRISAT), Patancheru 502 324, Andhra Pradesh, India.

G. Ashiono (Agronomist) Kenya Agricultural Research Institute (KARI), PO Box 1275, Nakuru, Kenya.

F.R. Bidinger (Agronomist) International Crops Research Institute for the Semi-Arid Tropics (ICRISAT), Patancheru 502 324, Andhra Pradesh, India.

D.E. Byth (Breeder and Associate Director General, Research) International Crops Research Institute for the Semi-Arid Tropics (ICRISAT), Patancheru 502 324, Andhra Pradesh, India.

S. Chater (Editor) Hawson Farm, Buckfastleigh, Devon TQ11 OHX, UK.

M. Cissé (Animal Nutritionist) Institut Sénégalais de Recherches Agricoles (ISRA), BP 2057, Dakar, Senegal.

B. de Groot (Animal Productionist), Indo-Swiss Project in Andhra Pradesh (ISPA), Plot No. 1227, Road No. 62, Jubilee Hills, Hyderabad 500 034, India.

P.N. de Leeuw (Agroecologist) International Livestock Research Institute (ILRI), PO Box 30709, Nairobi, Kenya.

C. Devendra (Consultant Animal Productionist) International Livestock Research Institute (ILRI), 8 Jalan 9/5, 46000, Petaling Jaya, Selangor, Malaysia.

A. Douiyssi (Breeder) Institut National de la Recherche Agronomique (INRA), BP 589, Settat, Morocco.

S.L. Dwivedi (Breeder) International Crops Research Institute for the Semi-Arid Tropics (ICRISAT), Patancheru 502 324, Andhra Pradesh, India.

A.R. Egan (Nutritionist) Department of Agriculture and Resource Management, University of Melbourne, Parkville, Victoria 3052, Australia.

S. Fernández-Rivera (Animal Nutritionist) International Livestock Research Institute (ILRI), ICRISAT Sahelian Center, BP 12404, Niamey, Niger.

D.J. Flower (Agronomist), International Crops Research Institute for the Semi-Arid Tropics (ICRISAT), Patancheru 502 324, Andhra Pradesh, India.

R.N. Gates (Agronomist) United States Department of Agriculture (USDA), Coastal Plain Experiment Station, PO Box 748, Tifton, GA 31793, USA.

I. Gomaa (Animal Nutritionist) Animal Production Research Institute, PO Box 443, Nadi Ebid Street, El Dokki Giza, Cairo, Egypt.

A.V. Goodchild (Animal Nutritionist) International Center for Agricultural Research in the Dry Areas (ICARDA), PO Box 5466, Aleppo, Syria.

C.L.L. Gowda (Breeder) CLAN Coordinator, International Crops Research Institute for the Semi-Arid Tropics (ICRISAT), Patancheru 502 324, Andhra Pradesh, India.

S.D. Hall (Editor) International Crops Research Institute for the Semi-Arid Tropics (ICRISAT), Patancheru 502 324, Andhra Pradesh, India.

L. Harrington (Economist) Centro Internacional de Mejoramiento de Maiz y Trigo (CIMMYT), Lisboa 27, Apdo. Postal 6-641, CP 06600 Mexico, D F Mexico.

P. Hofs (Consultant Animal Nutritionist) Road No. 12, Banjara Hills, Hyderabad 500 034, Andhra Pradesh, India.

C. Johansen (Agronomist and Director, Agronomy Division) International Crops Research Institute for the Semi-Arid Tropics (ICRISAT), Patancheru 502 324, Andhra Pradesh, India.

Kiran Singh (Animal Nutritionist and Director) Indian Veterinary Research Institute, Izatnagar, Bareilly 243 122, Uttar Pradesh, India.

Z. Kouyate (Agronomist) Institut d'Economie Rurale (IER), Station de Recherche Agronomique de Cinzana, BP 214, Ségou, Mali.

K.A. Kumar (Breeder and Director, Genetic Enhancement Division) International Crops Research Institute for the Semi-Arid Tropics (ICRISAT), ICRISAT Sahelian Center, BP 12404, Niamey, Niger.

M. Latham (Soil Scientist) International Board for Soil Research and Management (IBSRAM), PO Box 9-109, Bangkok 10900, Thailand.

Laxman Singh (Breeder) International Crops Research Institute for the Semi-Arid Tropics (ICRISAT), Patancheru 502 324, Andhra Pradesh, India.

G.P. Lodhi (Forage Breeder) Chaudhary Charan Singh, Haryana Agricultural University, Hissar 125 004, Haryana, India.

J.H.H. Maehl (Consultant Animal Productionist) Gustav-Leo Strasse 4, D-20249, Hamburg, Germany.

V.M. Manyong (Economist) International Institute of Tropical Agriculture (IITA), PMB 5320, Ibadan, Nigeria.

R.J.K. Myers (Soil Scientist) International Crops Research Institute for the Semi-Arid Tropics (ICRISAT), Patancheru 502 324, Andhra Pradesh, India.

A. Nefzaoui (Animal Nutritionist) Institut National de la Recherche Agronomique de Tunisie, Département des Productions Animales et Fourragères, Laboratoire de Nutrition Animale, Rue Hédi Karray, 2049 Ariana, Tunisia.

Y.L. Nene (Pathologist and Deputy Director General) International Crops Research Institute for the Semi-Arid Tropics (ICRISAT), Patancheru 502 324, Andhra Pradesh, India.

A.J. Nianogo (Animal Nutritionist and Chef du Programme Production Animale) Institut National d'Etudes et de Recherches Agricoles (INERA), BP 7192, Ouagadougou, Burkina Faso.

N.E Nyirenda (Agronomist) Ministry of Agriculture, Chitedze Agricultural Research Station, PO Box 158, Lilongwe, Malawi.

O.L. Oludimu (Economist) Ogun State University, PMB 2002, Ago-Iwoye, Ogun State, Nigeria.

E. Owen (Animal Nutritionist) Department of Agriculture, University of Reading, Earley Gate, PO Box 236, Reading RG6 1DD, Berkshire, UK.

S.B. Panday (Animal Nutritionist) Nepal Agricultural Research Council (NARC),

Khumaltar, Lalitpur, Nepal.

P. Parthasarathy Rao (Economist) International Crops Research Institute for the Semi-Arid Tropics (ICRISAT), Patancheru 502 324, Andhra Pradesh, India.

D.A. Pezo (Agronomist) Universidad de Costa Rica, PO Box 235-2300, San José, Costa Rica.

J.M. Powell (Agroecologist), USDA-ARS Dairy Forage Research Center, 1925 Linden Drive West, Madison, Wisconsin 53706, USA.

R.A. Quiroz (Farming Systems Specialist) International Potato Center (CIP), PO Box 1558, Lima 12, Peru.

H.F.W. Rattunde (Breeder) International Crops Research Institute for the Semi-Arid Tropics (ICRISAT), Patancheru 502 324, Andhra Pradesh, India.

D.H. Rearte (Animal Productionist) Instituto Nacional de Technología Agropecuaria (INTA), CC 276-7620 Balcarce, Buenos Aires, Argentina.

M.R. Reddy (Agronomist, Professor and University Head) Andhra Pradesh Agricultural University (APAU), Department of Feed and Fodder Technology, College of Veterinary Science, Rajendra Nagar, Hyderabad 500 030, Andhra Pradesh, India.

C. Renard (Agronomist and Executive Director) International Crops Research Institute for the Semi-Arid Tropics (ICRISAT), Patancheru 502 324, Andhra Pradesh, India.

D.B. Roxas (Animal Nutritionist) Institute of Animal Science, University of The Philippines, Los Baños College, Laguna 4031, The Philippines.

A.D. Salman (Animal Nutritionist) IPA Agricultural Research Center, PO Box 39094, Baghdad, Iraq.

F. San Martín (Veterinarian) Centro de Investigacion, Instituto Veterinario de Investigaciones Tropicales y de Altura (IVITA), PO Box 41-0068, San Borja, Lima, Peru.

R. Sansoucy (Animal Productionist) Food and Agriculture Organization of the United Nations (FAO), Animal Production and Health Division, Via delle Terme di Caracalla, 00100 Rome, Italy.

J.B. Schiere (Animal Productionist) Department of Animal Production Systems,

Wageningen Agricultural University, PO Box 338, Wageningen 6700 AH, The Netherlands.

S.V.R. Shetty (Agronomist) International Crops Research Institute for the Semi-Arid Tropics (ICRISAT), ICRISAT Sahelian Center, BP 12404, Niamey, Niger.

B.B. Singh (Breeder) International Institute of Tropical Agriculture (IITA), IITA Kano Station, PMB 3112, Kano, Nigeria.

J. Steinbach (Animal Nutritionist) Department of Livestock Ecology, Tropical Sciences Centre, Justus-Liebig University, Ludwigstrasse 21, D-35390 Giessen, Germany.

J.W. Stenhouse (Breeder) International Crops Research Institute for the Semi-Arid Tropics (ICRISAT), Patancheru 502 324, Andhra Pradesh, India.

D.B. Tamang (Agronomist) Research Extension and Irrigation Division, Ministry of Agriculture, PO Box 119, Thimphu, Bhutan.

A. Tukue (Animal Nutritionist) Awassa College of Agriculture, PO Box 5, Awassa, Sidamo, Ethiopia.

S.M. Virmani (Agroclimatologist) International Crops Research Institute for the Semi-Arid Tropics (ICRISAT), Patancheru 502 324, Andhra Pradesh, India.

M. Wanapat (Animal Nutritionist) Department of Animal Science, Faculty of Agriculture, Khon Kaen University, Khon Kaen 4002, Thailand.

Wang Jiaqi (Animal Nutritionist) Animal Sciences Institute, Chinese Academy of Agricultural Sciences (CAAS), Haidian, Beijing 100094, China.

E. Weltzien-Rattunde (Breeder) International Crops Research Institute for the Semi-Arid Tropics (ICRISAT), Patancheru 502 324, Andhra Pradesh, India.

T.O. Williams (Economist) International Livestock Research Institute (ILRI), ICRISAT Sahelian Center, BP 12404, Niamey, Niger.

M. Winugroho (Animal Productionist) Research Institute for Animal Production, PO Box 221, Bogor 16002, Indonesia.

E. Zerbini (Animal Scientist) International Livestock Research Institute (ILRI), PO Box 5689, Addis Ababa, Ethiopia.

Organizing Committee

C. Renard (Convenor)	ICRISAT
R.J. Carsky	IITA
G.L. Denning	IRRI
S. Fernández-Rivera	ILRI
A.V. Goodchild	ICARDA

National agricultural research systems

Kiran Singh	India
A.J. Nianogo	Burkina Faso
D.B. Roxas	Philippines

ICRISAT

M.M. Anders
D.E. Byth
S.L. Dwivedi
C.L.L. Gowda
S.D. Hall
T.G. Kelley
P. Parthasarathy Rao
E. Weltzien-Rattunde

Preface

Since the 1970s, livestock production in developing countries has been growing strongly. In fact, growth rates in recent years appear to be as spectacular as the cereals growth rates achieved during the Green Revolution. For the period 1970-1990, meat production rose on average by 4.6% per annum, while rice rose by 3.0%. These trends seem likely to continue into the next century.

As incomes rise, so does the demand for foods of animal origin, such as meat, milk and dairy products. Such foods, which have higher income elasticities of demand than food grains, constitute an increasing proportion of total food intake, at the expense of traditional cereals and pulses. Over at least the next 25 years, rapid population growth and urbanization will fuel further increases in demand. To respond to this demand, livestock numbers will expand and production will have to be intensified.

In tropical rainfed agriculture in Africa, cattle obtain between 16 and 40% of their dry-matter intake from crop residues. The proportion may rise to 50-80% during the dry season, when few alternative feeds are available. For producers in mixed crop/livestock farming systems, the most widespread technical constraint to increased output of livestock products is their inability to feed animals adequately throughout the year. Yet significant opportunities exist in most tropical agroecological zones to increase ruminant livestock feed supplies by using crop residues and other feed resources more efficiently.

Until recently, crop breeding research in the developing world has focused mainly on direct human consumption, concentrating efforts on improving food grain yields and quality. With the increased demand for cereal straw and legume haulms for livestock feed, greater emphasis is now needed on vegetative production. It is important to improve both the quality and quantity of crop residues, and to develop improved, dual-purpose cultivars with high biomass and minimal trade-offs in grain yields.

While livestock are the most obvious users of crop residues, we should not forget that these resources may also play important alternative or complementary roles in sustaining future crop production by conserving soil moisture, preventing erosion and increasing soil organic matter content. They may also provide a range of products important in rural areas, such as fuel and thatch.

In this workshop a multidisciplinary systems approach was used to investigate the use of crop residues to improve the productivity and sustainability of mixed crop/livestock farming systems. Seventy-two scientists from 33 countries, six continents and a wide range of disciplines came together to review past work, report new results and strengthen links at both personal and institutional levels.

These proceedings are an important source of information for animal and crop scientists working on the challenges of feeding the developing world's rapidly rising livestock population and improving the productivity of its agriculture. It is hoped that, by collaborating more closely, these scientists will be able to develop the innovative approaches and new technologies needed in the next century.

James G. Ryan

Director General
International Crops Research Institute for the Semi-Arid Tropics

Welcome Address

Y.L. Nene

International Crops Research Institute for the Semi-Arid Tropics, Patancheru 502 324, Andhra Pradesh, India

Chairman Dr Byth, distinguished participants from all over the world and my ICRISAT colleagues:

Let me extend a hearty welcome to all of you on behalf of Dr J.G. Ryan, ICRISAT's Director General, and on my own behalf to this International Workshop on Crop Residues in Sustainable Mixed Crop/Livestock Farming Systems, organized by the International Crops Research Institute for the Semi-Arid Tropics (ICRISAT) and sponsored by the International Livestock Research Institute (ILRI).

A few years ago such a workshop could not have been thought of, simply because the research foci of the Consultative Group on International Agricultural Research (CGIAR), to which ICRISAT and its sister centres belong, were different from those of today. In the past 5 years the CGIAR research agenda has undergone considerable change because of the unprecedented changes—political, social, environmental and economic—that have occurred and are still occurring worldwide. The vigorous exchange of ideas that has recently taken place among all actors in the CGIAR has led to new, more realistic global initiatives. All of us here are aware of the new ecoregional initiatives, including the Systemwide Livestock Initiative. This initiative became possible for several reasons, most important being the establishment of the International Livestock Research Institute (ILRI), with a global agenda, by way of a merger of the former International Livestock Centre for Africa (ILCA) and the International Laboratory for Research on Animal Diseases (ILRAD).

Increases in population and incomes, and especially the rise of the so-called "middle class" in developing countries, will undoubtedly lead to increases in the demand for animal products. At the same time, because the demand for food crops will also increase, human beings and herbivorous animals will compete for certain commodities. Providing adequate good-quality feed to livestock to raise and maintain their productivity is and will be a major challenge to agricultural scientists and policy makers all over the world.

This international workshop is very timely. The agenda, which covers topics such as the dynamics of feed resource in different regions, crop residues in mixed farming systems, and the use of residues to sustain systems, is most appropriate. The participation of scientists from Asia, Africa and Latin America will enrich the discussions.

Let me once again extend to you a very hearty welcome, and wish you a pleasant stay at ICRISAT Asia Center, despite the summer heat.

Opening Remarks

Hank Fitzhugh

International Livestock Research Institute, PO Box 30709, Nairobi, Kenya

Thank you for this opportunity to join with the International Crops Research Institute for the Semi-Arid Tropics (ICRISAT) in welcoming participants to this workshop. Funding for this workshop was provided by the Consultative Group on International Agricultural Research (CGIAR), through its Systemwide Livestock Programme (SLP). The International Livestock Research Institute (ILRI), one of 16 international agricultural research centres supported by the CGIAR, is the convening centre for the SLP.

A new institute, ILRI, was established in 1995 by amalgamating the International Laboratory for Research on Animal Diseases (ILRAD) and the International Livestock Centre for Africa (ILCA). ILRI has a global mandate for livestock research to: (i) improve animal performance through technological research and conservation of animal genetic diversity in developing regions; (ii) improve and sustain the productivity of major livestock and crop/livestock systems; (iii) improve the technical and economic performance of the livestock sector; and (iv) improve the development, transfer and utilization of research-based technology by national programmes and their client farmers.

ILCA and ILRAD had focused their research on sub-Saharan Africa. In its first year, ILRI therefore initiated a series of consultations with livestock specialists and research and development leaders in Asia, Latin America and the Caribbean, and West Asia-North Africa. Through these consultations development priorities and research needs were identified for each region. Consistently across regions, improving livestock feeds from sustainable mixed crop/livestock farming systems was given high priority.

The SLP was established by the CGIAR in 1995 to develop integrated research programmes on livestock feed and production systems. In developing the SLP, these recommendations from the CGIAR's Technical Advisory Committee (TAC) were followed:

1. Priority is given to research to improve the production and utilization of feed resources in crop/livestock systems.

2. Priority is given to research, including natural resource management research, at the ecoregional level, but with the emphasis on objectives which have cross-regional, global relevance.

3. Priority is given to research by consortia involving plant-oriented centres and national agricultural research systems (NARSs).

The SLP is managed by an Inter-Center Livestock Programme Group (IC-LPG), comprising senior managers from nine CGIAR centres and chaired by ILRI's Director General.

Annual funding of US$ 4 million has been allocated for the SLP. These funds support research to improve the production and utilization of feed resources for livestock. This support is provided as annual grants for up to 3 years to consortia of CGIAR centres, NARSs and advanced research institutes (ARIs). The grants are awarded on a competitive basis. Research proposals from consortia are evaluated by an independent external panel.

The first call for proposals, in 1995, brought eight research proposals from the consortia. Based on evaluations by the external panel, the IC-LPG recommended 3-year funding for the following three proposals, starting in 1996:

1. CIAT-led consortium: improved legume-based feeding systems for smallholder dual-purpose cattle production in tropical Latin America (first-year grant of US$ 0.77 million).

2. ICARDA-led consortium: production and utilization of multi-purpose fodder shrubs in West Asia-North Africa (first-year grant of US$ 0.84 million).

3. ICRAF-led consortium: utilization of forage legume biodiversity for dairy production and natural resource management in the Eastern and Central African highlands (first-year grant of US$ 0.85 million).

Matching funds, approximately equal to the SLP grants, will be provided by each consortium from other sources (e.g. bilateral funds for participating NARSs).

The IC-LPG also decided that more attention should be given to research to improve the utilization of crop residues. It therefore allocated funds to support this research planning workshop, which ICRISAT agreed to convene.

Expectations as to the outcomes of this workshop include:

1. A comprehensive overview of the current state of knowledge on crop residues in sustainable mixed farming systems.

2. The identification of priorities for crop residues research at the agroecological and regional levels.

3. The identification of opportunities for coordinated research which has trans-regional importance (e.g. crop residues research with general relevance across the semiarid zones of Asia, Africa and Latin America). Trans-regional research could involve the use of similar methodologies in different locations, so that data can be pooled for broader trans-regional analysis.

The CGIAR members will meet 21-24 May 1996 in Jakarta, Indonesia. During this meeting, they will consider funding for the SLP. Although there has been a delay in SLP funding for 1996, we are encouraged by the strong endorsements from the CGIAR Chairman and from the Group's Technical Advisory Committee (TAC).

Annual grants to the CIAT-, ICARDA- and ICRAF-led consortia, together with SLP administrative expenses, will total about US\$ 2.6 million. Therefore, assuming full funding of US\$ 4 million, approximately US\$ 1.4 million of grant funds will be available for the next call for proposals. We anticipate that these grants will start in 1997. The amount of each grant will depend on the research proposed and the availability and amount of matching funds. However, on the basis of our initial experience, annual grants will probably be in the range of US\$ 0.4 to 0.8 million.

The decision as to whether there will be a call for proposals in 1996 will be made in June, following the CGIAR meeting in Jakarta. Regardless, the proceedings from this workshop will be an important contribution to knowledge about crop residues in sustainable crop/livestock farming systems.

1. Technological Constraints and Opportunities in Relation to Class of Livestock and Production Objectives

Adrian R. Egan

Department of Agriculture and Resource Management, University of Melbourne, Parkville, Victoria 3052, Australia

Abstract

Feed-year strategies involve matching the cycles of animal production with the changing availabilities of all sources of nutrients over time. The strategy must be consistent with the diverse production objectives of farmers and with the feasibility of achieving the nutritional support required. These in turn vary with farmers' socioeconomic circumstances, which in developing countries range from fully commercial to small-scale, "subsistence-plus" farmers.

The technical constraints to the development of a suitable strategy concern the animals' nutritional demands in relation to the desired type, level and timing of product outputs, the nutritional characteristics of the feeds available, and possible mismatches in timing or location. The issues of storage, treatment and transport of feedstuffs will be important in overcoming mismatches. Developing options that will optimize crop residue use within the labour and capital constraints of individual farms requires an effective but flexible model.

To improve the use of crop residues in a feed-year system, both the intake and the balance of nutrients may need to be adjusted for one or more phases in the animal's annual cycle of work, production and reproduction. The options for achieving desirable change include: (i) changing the animal production system or the timing of events in it; (ii) introducing crops with different agronomic features, giving more and better residues; (iii) adopting different grazing, feeding and storage strategies and methods; (iv) modifying crop residues by physical and chemical treatments; (v) changing feed availability, feed combinations, supplements and feeding procedures; and (vi) employing other technologies that enhance digestive and physiological performance. Animal behaviour, rumen function and physiology are critical in determining the impact of changes on performance.

Introduction

Improved livestock production provides an avenue for the rational use of herbage on uncropped land and of crop residues on farmland. Via this avenue, standards of living can be raised. The aim of this paper is: (i) to outline ways of improving livestock production systems in which crop residues form, or can form, an important feed base; and (ii) to promote a systematic approach to the evaluation of options for managing nutrition—either options now existing or those that can be sought through research and development.

The first step is to identify the socioeconomic conditions which must be fulfilled if any option is to find favour with producers; that is, to identify the product and the production conditions consistent with the livestock owner's goals. Given the enormous range of mixed farming systems—from highly developed, vertically integrated commercial systems to small farmer "subsistence-plus" systems—the improved utilization of crop residues and by-products poses complex business management, socioeconomic and educational as well as technical questions (e.g. Petheram *et al.*, 1985). Many influences, from the policies of governments to the attitudes and asset limitations of farmers and the perceptions of the wider community, act to limit opportunities, either preventing their recognition or constraining the adoption of technological innovations. It is not the purpose of this paper to address these issues in full, but I will nevertheless try to provide the basis for identifying those constraints that are truly technical and for seeking solutions that recognize these boundary conditions.

The keys to success lie in selecting objectives that are achievable within the constraints and capabilities of small production units and within the time and place at which feed resources are available. Strategies based on these principles will avoid dependence on: (i) expensive imports; (ii) resources for which competition is high and increasing; (iii) cash outlays against long-delayed product output; (iv) time and labour for which there are other high-priority demands; and (v) complex equipment and highly specialized or complicated technical skills.

The aim should be to optimize overall agricultural and livestock productivity from available resources through an integrated set of practices, considering crops as multipurpose and often also involving multipurpose animals. It should be to make effective use of residues and by-products as feeds but also for soil amendments, fuels and/or other non-feed purposes, and to recycle non-polluting wastes to the land.

Ruminant livestock production systems can meet these criteria, particularly where the ruminant animal walks to and harvests its own major feed resource. Stubbles, together with any grain scattered at harvest, can be "managed" simply by grazing them in situ. The grain harvesting method and the stage of maturity at harvest determine what fraction of the residue is left as stubble, what is left cut in the field, what is transported before threshing and what are the nutritional characteristics of the residues. Where residues are collected at a central site, transfer into storage or treatment systems is facilitated. The crop residues of most concern as feed resources are the fibrous harvest residues of cereals and the residues of other staple grain and oilseed crops. Grain legumes usually provide residues that have some advantage over those of cereal crops in terms of their nitrogen content, digestibility or intake. Crop residues are one component of the changing year-round feed supply, and the role they play depends on the production goals.

Livestock Management Systems

A livestock management system follows a year-round strategy incorporating a coordinated set of practices undertaken to meet specified objectives in terms of output and to maintain a level of efficiency that is consistent with genetic capabilities in the face of environmental variables. In mixed crop/livestock systems it is a part— and not always the high-priority part—of a whole farm system. The system often involves competition for capital and labour. Figure 1.1 shows the inputs and outputs of a livestock production system within a whole farm context.

The purposes of livestock production may variously be to accumulate capital and insurance, to meet domestic needs for a wide range of products and services, or to make a regular income from the sale of animals or of animal products. The

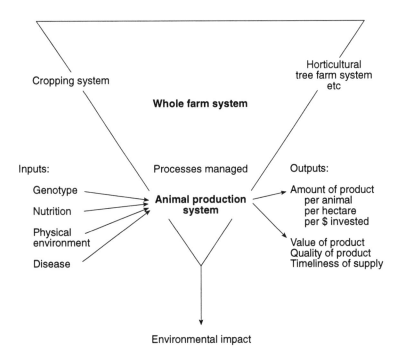

Fig.1.1. Inputs and outputs in a livestock production system within a whole-farm context.

product outputs may be: (i) more animals within the herd; (ii) live animals for sale at times of necessity or of opportunity (or regularly at set times, ages or weights); (iii) meat and meat products, and hides or skins; (iv) milk and milk products; (v) work; and (vi) manure.

The purposes of attempts to improve livestock systems, therefore, would be to meet increased household needs and/or to provide more output or more diversified outputs for sale. In the widest sense, the agenda for improvement is set both by farmers' attitudes and by the policies and strategies of national or local authorities, which may encourage or discourage change. Tradition, risk aversion, land tenure and the relationships between the producer and the market are all important considerations. The willingness to innovate is generally related to: (i) accuracy of identification and effective communication of the need/opportunity; (ii) the demonstrable effectiveness of the practice or technology involved; and (iii) the technical ease with which it can be implemented.

Managing a livestock system is a complex business, and any proposed change in practices must be sufficiently flexible to allow different expression in the circumstances of individual producers. Likewise, going from the general strategy (i.e. more effective use of crop residues) to the combination of specific practices and their timing to suit individual circumstances defies direct research. Somehow we need to build unique systems-based solutions from component knowledge— and that implies using a model.

Feed-year Strategies

A feed-year strategy is an essential basis for the efficient use of crop residues in any livestock system. It involves matching the cycles of animal production to the changing availabilities of sources of nutrients over time. The result must be consistent with the objectives of a livestock owner and the levels of nutritional support required to achieve those objectives. Crop residues are just one set of contributing feed resources, whose nutritional characteristics and times of availability will cover part of the cycle. Developing a strategy will require compromise. The process may demonstrate that the proposed production targets are unrealistic and unachievable. Alternatively, it will indicate that targets are feasible, but only with certain changes in management and resource use. In the latter case, adoption may be possible for some but not for all farmers in what appear to be similar circumstances.

The feed-year strategy model is thus a tool for recognizing the best combinations from available feeds to support a shift towards specified, achievable animal production goals. In setting a feed-year strategy in the context of a whole farm system, we confront a range of broader issues, as indicated in Box 1.1.

Technical constraints encountered in the development of a feed-year strategy relate to: (i) the product(s) required of the animals; (ii) the physiological capabilities of the animals; (iii) the characteristics of the feeds available; (iv) mismatches in timing or location; and (v) the storage, treatment and transport facilities available.

Box 1.1. Relating the feed-year strategy to the whole farm plan

Annual cyclic phenomena to be matched
- Feed supply and quality
    - Plant growing season; pasture species; crop harvest residues; browse
    - Grazing management; conserved, stored and treated materials; supplements
- Animal/herd cycles
    - Species; breed; sex; physiological state; age groups
    - Stocking rate and grazing intensity
    - Effects of preceding conditions or performance on current capabilities

Land resource impacts to be evaluated
- Soil fertility
    - Plant-water relationships; growth rate; mineral nutrient content; fertilizer use
    - Nutrient export; nutrient recycling
    - Soil structure; soil depth
- Slope, aspect, erodability, drainage

Risk factors to be determined
- Variability in cycles
    - Late break of season; low rainfall; early end to season; drought
- Diseases affecting mortality rates, production rates, quality of product
- Market movements

Productivity and product quality effects to be assessed
- Skills and knowledge level of operators
- Resource requirements and availability
- Meeting market specifications
    - Quality criteria; reliability and timeliness
- Achieving premium prices
    - Out-of-season production

Often all five of these sets of technical constraints intervene in hampering the use of crop residues, though partial solutions can be achieved by action on any one. Of prime importance is that any model should allow the cumulative and multiplicative effects of several partial solutions to be evaluated.

The events and processes in the animal production cycle that have to be accommodated are set by mating time, which determines the periods of pregnancy and births, the period for which offspring are suckled, and the time at which they will be weaned. In annual calving systems a cow will be mated in early- to mid-lactation, so recovery of body condition must be matched with weaning of the calf and mid- to late pregnancy. Thus the physiological traits of consequence are: (i) fertility and seasonality of oestrus (see Jainudeen, 1985); (ii) fecundity, ovulation rate and seasonality of ovulation (Teleni *et al.,* 1988); (iii) gestation length and nutritional support of the foetus, particularly in the final third of pregnancy (Hall *et al.,* 1992); (iv) lactation capability and nutritional support for lactation; (v) time and intensity of any period of draught work (Lawrence, 1987); (vi) time to weaning and rate of growth to weaning; (vii) weight at weaning and growth rate post-weaning; (viii) mature body size and body composition at age and weight; (ix) effect of previous performance demands; and (x) effect of all these factors on flock or herd structure.

The principal area of application of a feed-year strategy model in regard to smallholder cattle, buffalo or small ruminant production is in allocating available feed components for utilization in a "best-fit" seasonal animal production system (Egan, 1989, 1992; Egan and Dixon, 1992). The system must be flexible, in that the very best fit may not be possible owing to external constraints such as labour.

Ongoing work at the University of Melbourne, Australia, in developing such a model has involved networked research in a number of developing countries, including Nigeria, Thailand, Indonesia, Malaysia, The Philippines, Sri Lanka and India.

Nutritional Specifications

In allocating feed resources for a feed-year strategy, there needs to be a clear and quantifiable understanding of: (i) the nutritional needs of a given class of animal in the specific environment; (ii) the inherent nutritional qualities of each type of feed material available; (iii) the associative effects on intake and on the amounts of nutrients absorbed when available feeds are used in combinations; (iv) the nature and magnitude of any insufficiency in nutrients restricting performance below that set as the target; and (v) the consequences for production if the nutritional support is inadequate at any given stage of the animals' cycle of work, production and reproduction.

The precision of any prediction is, in the first instance, less important than a rough understanding of what any particular allocation is likely to achieve (Dixon and Egan, 1987). Where the nutritional demands associated with the production goal are high and exacting in terms of balance of nutrients (e.g. late pregnancy, peak lactation and rapid growth, or work), it must be certain that highly nutritional feed will be available (Ffoulkes *et al.,* 1987; Teleni and Hogan, 1989).

The nutrients the animal receives are not those chemically determined in the feed but are the products of a variable fermentation process in the rumen and of intestinal digestion of the material flowing on from there. The prediction of nutrient yield is imprecise, but it is possible to identify probable nutrient balances. Among the organic substrates necessary to support maintenance, growth, reproduction and lactation, the balance between glucose, amino acids and the remaining metaboliz-able compounds is critical (MacRae and Egan, 1980). The nutrient needs of the animal expressed in these terms are recognized as variable, depending on physi-ological state, body condition and the level of production being achieved (Pethick, 1984; Graham, 1985; Baumalim *et al.*, 1987). Both level and balance of nutrients interact with the capacity to draw on body reserves of fat and, less readily, protein. These aspects of metabolic efficiency and product/substrate quotients are addressed in complex mechanistic models (e.g. Baldwin, 1995) or, more simply, in classifica-tion and specification statements (e.g. Egan and Walker, 1975; ARC, 1980). The latter are being built up gradually for tropical forages and crop residues. From here we can proceed to a process of "diagnosis and recommended treatment". If the performance of the animal is predicted to be limited by amount of feed and/or nutrient balance, the model will deduce the physiological basis and relate this to the properties of the diet. It can then be used to identify, among existing feed options, the one that will most increase intake and/or improve nutrient balance. The improvement is made by altering the components of the diet so as to bring about a shift in such parameters as the rate of fermentation, the balance of fermentation products, the digesta flow rate, and the provision of unfermented feed residues for intestinal digestion.

Various methods and approaches exist for improving the quality of the feed ingested by the animal. For example, a change in the availability of feed components can be used either to permit greater selection by the animal or to adjust the proportional intakes of different feeds. For crop residues, this sort of intervention can go a step further, with modification of the crop residue components through changes in harvesting height, harvesting method, plant breeding (Pearce, 1986), physical and chemical treatments (Doyle *et al.*, 1991) and supplementation with a cheap source of nutrients limiting microbial fermentative activity and growth (Egan *et al.*, 1993).

The Feed Resources

Crop residues are included in most relevant surveys and statistical reports as contributing to the basic bulk feed resources of a country or region (e.g. Butterworth and Mosi, 1985). However, they are not a single class of feedstuff and their different feeding values are often lost in broad generalizations (Pearce, 1986). The extent and effectiveness of their utilization depends on traditional practices at village or farm level, where the total set of feed resources will vary considerably in both amount and seasonal availability. Practices differ between dryland and irrigated cropping systems. In some irrigated rice production systems, crop residues and stubbles are used only for short periods between the harvest and the preparation of the land for the next crop. Alternatively, they may even be burnt, whereas in other systems the straw is highly valued, often being traded or stored and treated for later use.

Native Pasture, Crop Residues and Fibrous Agroindustrial By-products

On their own, these feed components are usually regarded as being of limited nutritional value and of low to very low digestibility, supporting low intake by the animal (Egan, 1989). The products of digestion are considered to be poorly balanced for all productive purposes (Leng, 1987). However, a database being developed by network researchers indicates a wide range of nutritional values, and identifies some of the species and cultivar differences and the agronomic and harvesting conditions that contribute to this variability (Moog, 1986). In some cases, sufficient experimental work has been done to allow guidelines to be developed on the use of supplements or combinations of feeds to enhance animal performance on these basal roughages (Nordblom *et al.*, this proceedings, p. 131).

Where crop residues are collected and stored, treatment may improve their acceptability and nutritional value. The database mentioned above carries information on the probable effect of treatment method on magnitude of increase in intake, digestibility and nutrient balance. At present, precision is poor, owing to inadequate descriptions of the starting properties of the crop residues and to the variable effectiveness of the treatment process. It is clear that, even where physical and chemical treatments can be safely and effectively implemented at village level (Jayasuriya, 1983), animal production will remain limited by poor balance of nutrients (Egan, 1992). Education programmes are an essential component of any package involving treatment (Wanapat, 1985).

Supplements are needed that improve the pattern of fermentation. The direction of change can be reliably predicted, but its magnitude (in total yield of nutrients) less so. The extent to which supplements result in improved efficiency of overall feed use depends on changes in the rates of fermentation of fibrous carbohydrate, synthesis of microbial protein, degradation of dietary protein and outflow of particulate matter from the rumen. The tactics in supplying supplements are all related to timing, so as to best modify these rates, which both determine and are determined by the level of intake.

Special-purpose Pasture

When used strategically, feed grown on land set aside from cropping can provide special grazing for animals currently producing or being prepared to produce outputs of high value. The ability to predict intake and the contribution of such feed to nutrient status is being examined in the context of current research at Melbourne. While technically unproblematic, the growing of special-purpose pasture is, in a developing country context, frequently associated with various socioeconomic problems, including the value set on the survival of all animals in the herd rather than the high production of relatively few animals, and the difficulties of finding suitable land that is not required for food crops. The attitudes of livestock owners have to be carefully taken into account when designing interventions of this kind.

Trees and Shrubs

Trees and shrubs provide green biomass of moderate to high digestibility and protein content when other feed reserves are scarce and low in nitrogen. There are several commonly available options for making effective use of shrub and tree foliage in the tropics. One of the most successful is the cultivation of *Leucaena leucocephala* as a palatable, high-protein browse or cut-and-carry feed component, often used with crop residues or native grasses as the basal roughage. However, many other species are used, most of them locally. The feed-year model as yet has a poor database and predictive ability for most of these species, but this problem is being addressed through current research.

Nitrogen Supplements

Nitrogen (N) is frequently a principal limiting factor in the small-scale farmer's attempts to increase animal production (Doyle *et al.*, 1991). Urea has different effects when provided in different forms and at different times in the feeding cycle, but the model does not yet do other than indicate a probable period of ruminal ammonia inadequacy. Other N supplements may provide, to differing degrees also requiring quantification, N and protein for digestion in the small intestine (so-called bypass protein) (Egan, 1981).

Grain Products and Purchased Supplements

These components are used to obtain substantial gains in production and are introduced into the model on the basis of dose response data, which allow a separate round of "What if?" predictions. In general, the dose response curves we possess at present (e.g. Preston and Leng, 1987) are derived from insufficient data or data of unknown reliability. Often, clear and unambiguous descriptions of the history and condition of the animals before and at the time of the feeding trials are not available. This is important, as compensatory growth or a high level of milk production due to unusually high existing body condition can bias the evaluation of the contribution to performance being made by the supplement.

However, where such factors can be ruled out, these components can be used cost-effectively to improve the intake of other dietary components and the balance of nutrients. Supplements that contain slowly degraded proteins can improve the flow of protein to the small intestine and have been widely examined against other options. In many cases, improved animal performance can be attributed to both the timely release of N in the rumen and the amounts of bypass protein digested in the intestines. Such supplements can provide nutritional support for more demanding steps in the production cycle, permitting, for example, a strategic shift in the timing of mating. This can then allow a better fit of all subsequent production processes to the feed available throughout the year. At present we are examining the use of oilseed and legume grains at specific physiological moments. It is possible that very short periods of feeding may, for example, permit improved conception rates, or improved colostrum production and calf survival at low birthweight. These effects will be useful if the advantage can be subsequently sustained.

Figure 1.2 provides an example of the matching process in feed allocation for female draught cattle through pregnancy and lactation.

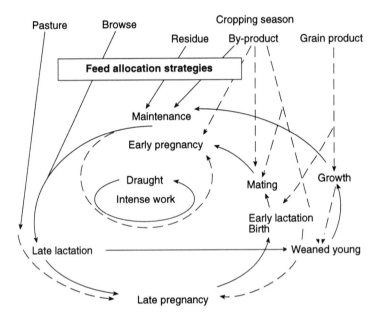

Fig. 1.2. Allocation of feed resources in the year-round reproductive cycle of a draught cow.

Improved Use of Crop Residues

The general principles relevant to any effort to improve the use of crop residues in a whole farm context have been outlined by such authors as Parra and Escobar (1985), Pearce (1986), Dixon and Egan (1987) and Preston and Leng (1987). They are as follows:

1. Understand the present production system and identify the critical different features of the potential new system.

2. Identify feed availability patterns and broadly classify all the feeds.

3. Identify all product outputs, existing and desired.
4. Consider the target path for liveweight and condition score required for current or future animal performance.
5. Develop a conceptual framework for the existing and potential feed-year model.
6. Consider the management constraints.
7. Consider the remaining nutritional constraints and risks.
8. Consider whether the above can be dealt with by minor supplementation.
9. Establish general "rules" that help to identify and adapt a useful model.
10. Recognize any special local circumstances.
11. Adopt, adapt or generate a specific model.
12. Repeat the process, considering and comparing alternative strategies.

Remember that the model is simply a way of organizing ideas and exploring the relationships between inputs and outputs from relevant experimental databases. It is a tool to aid what is a complex decision-making process. It will help to identify areas of uncertainty requiring further research, and to decide which measurements to make in on-station or on-farm research and development.

The Model

Our model will consist of an applied animal production model integrated with a number of modules, including:

1. A feed/nutrient specification simulation module with a supporting series of databases.
2. A linear programme that assembles the resource options against a specific production goal expressed as amounts of different products. This should carry a prediction of marginal return to the input of the limiting resource, even if, in reality, this return would not be directly visible in dollar terms.
3. An expert system that interprets the predictions from the simulation model. This will identify probable areas of biological inefficiency, together with areas of inaccuracy in the description of the conditions being simulated. It will also suggest indicators useful in testing the possible causes of inefficiency.

As already explained, the purpose of our model is to explore possible feed allocation patterns against a given herd structure and to identify realistic options that will improve productivity. This raises several important issues:

1. Intake from grazing is well nigh impossible to predict. Consequently, classes of conditions are used to identify probable liveweight changes and nutrient deficiencies. As discussed later, measurement and diagnostic approaches can be used to substitute the need to predict intake.
2. The outcome of treating crop residues and using particular supplements,

including timed access to fodder banks, can be modelled at an array of different levels and combinations. The model's first "best fit" will predict intake and balance of nutrients, but will warn of the variability of possible outcomes, expressed as probabilities. The model identifies indicators which can be used to reinforce the confidence attached to a preferred value.

3. Some proposed allocation patterns will be "illegal" because they are not consistent with the feed reserves held or because there is a higher-priority need to hold stocks against a later allocation. Some feed management practices can be varied, so as to identify potential improvements in the efficiency of resource use. These tactics are discussed below.

4. The production objectives and conditions can be reset by selecting altered levels of production, stocking rates, within-herd priorities or timing of major events determining production, such as mating time.

5. Production outcomes include liveweight, current growth rate, change in body condition score, level of milk production, conception rate, calving rate, calf survival to weaning, weaning weight, calving interval and period of intensive work. At least for some important production criteria, the model attempts to predict the future consequences of current performance.

A mechanistic approach is used to determine the outcomes of digestive and metabolic processes, but these outcomes are kept in line with reliable data available on feed conversion, nutrient balance, metabolite entry, liveweight and body composition, milk output and draught work.

Scope and Level

The principal features of the feed-year model relevant to the annual period of utilization of pasture and crop residues are as follows:

1. The model assumes that the basal roughages available to the animals are of low digestibility, low intake and poorly balanced in nutrients. They may be used alone to support bare survival, with varying rates of loss of liveweight depending on the initial condition score of the animal. The imbalanced supply of nutrients, if sustained through early development and growth, may result in permanently impaired performance.

2. The types of supplement and the level and timing of their use should be viewed in terms of their direct and indirect effects on rates of digestion, intake and balance of nutrients.

3. Supplements can be used to support increased liveweight gain or to reduce rate of loss of liveweight and condition score, but because of the potential for compensatory gain under better feeding conditions, they may not have long-term benefits for the overall efficiency of production. Also, liveweight changes may have consequences for later events (e.g. disease susceptibility).

4. Supplements can be used to trigger improved conception rates throughout the herd. However, the effect in individual cases needs to be considered in relation to the condition score of the cow, its past calving record and stage of lactation, and the conditions 9 months ahead for calving and lactation.

5. Supplements can be used to sustain milk production in cows when there is a transfer from green feed to higher-quality crop residues, but their effectiveness will depend on the cow's stage of lactation and its body condition at the time of the change.

With regard to point five, at Melbourne we have been exploring the following concepts and criteria relating to digesta behaviour when supplements are fed intermittently.

Ruminal Fermentation

The optimum conditions for ruminal fermentation are: (i) that the balance between rapidly fermented and slowly fermented materials should be kept relatively constant; (ii) that pH should not fall below 6.0; and (iii) that ruminal ammonia concentration should not fall below 120 mg N l^{-1}. In the model these are default conditions that can be reset if there is good reason to believe that other circumstances will prevail. Deviations from these conditions will result in periods of reduced microbial fermentation and growth efficiency. The rate of fermentation of fibrous constituents from past meals may be limited more by the accessibility of hydrolysable carbohydrates than by the availability of other nutrients. Discontinuities in substrate availability will have consequences for the relative abundance of specific groups of microorganisms, including amylolytic or cellulolytic bacteria, protozoans with different feeding strategies, and anaerobic fungi. The ecological patterns established and the consequences for the rate and extent of fermentation and growth of microbial biomass remain to be quantified.

Unfermented Dietary Nutrients

Some feed constituents are relatively inaccessible to microbial enzymic attack. Low solubility, linkages through bonds which are non-hydrolysable at near neutral pH, and physical encapsulation within resistant materials all contribute to this phenomenon. Colonization and localized development of populations of microorganisms with effective hydrolytic capabilities may be a function of time. Thus residence time in the rumen for particulate material of low degradability is an important aspect to consider. Some, but not all, material escaping fermentation is altered during its passage through the abomasum to become digestible in the small intestine.

Ruminal Digesta Load and Outflow

As feed particles are reduced in size through chewing and other forces, the probability that they will enter the flow paths through the reticulo-omasal orifice is increased. Particles of 1-2 mm or less pass relatively efficiently once presented during a cycle of fluid passage through the orifice (Kennedy *et al.*, 1987). The amount of fluid passing at each contraction, the numbers of contractions and the loading of fluid with solids of appropriate dimensions and specific gravity are variable. However, the rumen always contains a pool of particles having appropriate properties. This pool contributes between 40 and 80% of the total ruminal digesta load.

When feed is ingested, two new features are introduced. For roughages, few new particles immediately join the "flow-ready" pool, but added bulk and fluid alter the conditions for fluid outflow. For grain and readily fermented supplements, the rate at which new particles join the "flow-ready" pool is a property of the supplement, but again rumen load and flow characteristics will change. In both cases, the small particle pool will exhibit accelerated removal rates and so will be temporarily reduced. Less frequent large meals of supplements of either roughage or grain may purge the refractory material in the small particle pool and beneficially alter the total intake and balance of nutrients. Thus the timing of access to supplements or to fodder banks may substantially affect microbial fermentation patterns, the long-term balance of nutrients and the total intake.

Substitution

When animals fed diets of low digestibility are given supplements, the objective is to catalyse improved use of the basal feed. At the very least, a complementary set of nutrients should be added to the overall nutrient yield without adversely affecting roughage use. When a supplemented animal reduces its intake of the more abundant roughage, this substitution effect lowers the value of the combination. Substitution has variously been ascribed to intra-ruminal effects such as pH changes affecting fibre digestion rates, greater transient rumen load, and improved nutrient intake resulting in reduced intake drive for the least acceptable dietary component. The causes probably have both physiological and behavioural components. The classification of feeds into those with high or low substitution rates needs to take into account the physiological state and experience of the animal and the relative acceptability of the basal feed.

We are therefore exploring the frequency and timing of feeding, since this could add another management device, either for reducing labour input without adverse nutritional consequences or for improving utilization if substitution is reduced and/or there is a net gain in the amounts and balances of nutrients through the processes discussed above.

Management Practices and Technologies

The management practices and technologies that exist now or may exist in the future deal ultimately with microbial or physiological processes. Knowing the desired change in intake or nutrient yield, the manager can choose among the following options to achieve the change:

1. Alter the feed allocation level to allow increased selection.
2. Separate plant morphological parts in order to use them differently.
3. Treat plant parts to increase: (i) sites of microbial access; (ii) ease of physical particle size reduction; (iii) access of microbial enzymes to fibre constituents at molecular level.
4. Provide supplements of rate-limiting microbial nutrients, e.g. nitrogen, silicon or phosphorus.
5. Devise strategies to promote more efficient and effective microbial populations using flow rates, intermittent feeding tactics and special nutrient supplies.
6. Consider modifying the chemical and morphological characteristics of the residue by breeding.
7. Consider altering the harvest method.
8. Consider the scope for genetic engineering of microbes.
9. Consider the scope for genetic engineering of crop plants.
10. Select and breed higher-performing animals within the existing herd/system.
11. Consider the scope for genetic engineering of animals.

For the individual manager, options 8, 9 and 11 are promises for the future rather than real possibilities at present. But we should bear in mind that strategic research does have the potential to bring about major changes in the opportunities for using crop residues in the longer term. In addition, models can help to project the mind into the future to aid in determining where research priorities might lie.

Conclusions

Crop residues are and will continue to be a valuable feed resource for ruminant livestock production. Crop/livestock production systems and the socioeconomic conditions in which they operate will change. We need not only to take advantage of the opportunities within existing systems but also to prepare for new scenarios that will require us to rethink our management strategies and devise new technical tools to help us form and apply them.

Research will continue to improve the array of tools available. However, new tools will only be adopted if they are relatively simple to use, can be relied upon to bring tangible benefits to the farmer without undue levels of risk, and fit the farmer's resources and goals.

Year-round feed models allow managers to plan the best possible use of their limited feed resources. They also allow researchers to explore the potential impact of different interventions and to identify areas in which further strategic research is needed. Lastly, they allow the effects of successful interventions to be tested on a broader array of production systems.

References

ARC (1980) *The Nutrient Requirements of Ruminant Livestock.* Commonwealth Agricultural Bureaux, Slough, UK.

Baldwin, R.L. (1995) *Modelling Ruminant Digestion and Metabolism.* Chapman & Hall, London, UK.

Baumalim, A., Ffoulkes, D. and Fletcher, I.C. (1987) Preliminary observations on the effect of work on intake, digestibility, growth and ovarian activity of swamp buffalo cows. *DAP Project Bulletin* 3: 6-10.

Butterworth, M.H. and Mosi, A.K. (1985) Practical evaluation of crop residues and agroindustrial by-products for ruminants in developing countries with emphasis on East Africa. In: Preston, T.R., Kossila, V.L., Goodwin, J. and Reed, R.J.S. (eds), *Better Utilization of Crop Residues and By-Products in Animal Feeding.* Food and Agriculture Organization of the United Nations, Rome, Italy, pp. 15-20.

Dixon, R.M. and Egan, A.R. (1987) Strategies for optimizing use of fibrous crop residues as animal feeds. In: Dixon, R.M. (ed.), *Ruminant Feeding Systems Utilizing Fibrous Agricultural Residues.* International Development Program of Australian Universities and Colleges (IDP), Canberra, pp. 11-26.

Doyle, P.T., Dixon, R.M. and Egan, A.R. (1991) Treatment of roughages and their relevance to animal production in the tropics. In: Jalaludin, S. (ed.), *Proceedings of the Third International Symposium on the Nutrition of Herbivores*, 25-30 August 1991, Malaysian Society of Animal Production, Serdang, pp. 45-54.

Egan, A.R. (1981) Bypass protein: A review. In: Farrell, D.J. (ed.), *Recent Advances in Animal Nutrition.* University of New England, Armidale, Australia, pp. 42-50.

Egan, A.R. (1989) Living with and overcoming limits to feeding value of high-fibre roughages. In: Hoffman, D., Nari, J. and Petheram, R.J. (eds), *Draught Animals in Rural Development.* Australian Centre for International Agricultural Research (ACIAR), Canberra, pp. 176-180.

Egan, A.R. (1992) Strategies of feeding and nutrition of dairy cattle based on crop residues. In: Bunyvejchevin, P., Sangdid, S. and Hangsanet, K. (eds), *Proceedings of the Sixth Animal Sciences Congress of the Asian-Australian Association of Animal Production Societies,* 23-28 November 1992, Bangkok, Thailand, pp. 167-176.

Egan, A.R. and Dixon, R.M. (1992) Feed resource needs for draught animal power.

In: Pryor, W. (ed.), *Draught Animal Power in the Asian-Australasian Region*. Proceedings No. 46, Australian Centre for International Agricultural Research (ACIAR), Canberra, pp. 87-96.

Egan, A.R. and Walker, D.J. (1975) Resource allocation and efficiency of N use in ruminants. In: Reid, R.L. (ed.), *Proceedings of the Third World Conference on Animal Production*, 20-30 May 1973, Melbourne. University of Sydney Press, Sydney, Australia, pp. 551-563.

Egan, A.R., Hogan, J.P. and Leng, R.A. (1993) Effects of manipulation of rumen fermentation on ruminant production. In: El Shazly, K. (ed.), *Manipulation of Rumen Microorganisms*. Proceedings of the International Conference on Manipulation of Rumen Microorganisms to Improve Efficiency of Fermentation and Ruminant Production, 20-23 September 1993, Alphagraph, Alexandria, Egypt, pp. 70-83.

Ffoulkes, D., Baumalim, A. and Panggabean, T. (1987) Utilization of fibrous feeds by working buffaloes. In: Dixon, R.M. (ed.), *Ruminant Feeding Systems Utilizing Fibrous Agricultural Residues*. International Development Program of Australian Universities and Colleges (IDP), Canberra, pp. 161-169.

Graham, N. McC. (1985) Relevance of the British metabolism energy system to the feeding of draught animals. In: Copland, J.W. (ed.), *Draught Animal Power for Production*. Australian Centre for International Agricultural Research (ACIAR), Canberra, pp. 96-98.

Hall, D.G., Piper, L.R., Egan, A.R. and Bindon, B.R. (1992) Lamb and milk production from Booroola ewes supplemented in late pregnancy. *Australian Journal of Experimental Agriculture* 32: 587-593.

Jainudeen, M.R. (1985) Reproduction in draught animals: Does work affect female fertility? In: Copland, J.W. (ed.), *Draught Animal Power for Production*. Australian Centre for Interntional Agricultural Research (ACIAR), Canberra, pp. 130-133.

Jayasuriya, M.C.N. (1983) Problems of treating crop residues at the village level. In: Pearce, G.R. (ed.), *The Utilization of Fibrous Agricultural Residues*. Australian International Development Assistance Burea (AIDAB), Canberra, pp. 62-67.

Kennedy, P.M., John, A., McSweeney, C.S., Murphy, M.R. and Schlink, A.C. (1987) Comparative nutrition of cattle and swamp buffaloes given a rice straw-based diet, 2: Rumination and passage from the rumen. In: Rose, M. (ed.), *Second International Symposium on the Nutrition of Herbivores*, 18-23 July 1987, Australian Society of Animal Production, Brisbane, pp. 167-168.

Lawrence, P. (1987) Nutrition of working animals. In: Falvey, J.L. (ed.), *An Introduction to Working Animals*. MPW, Melbourne, Australia.

Leng, R.A. (1987) Determining the nutritive value of forage. In: Blair, G.J., Ivory, D.A. and Evans, T.R. (eds), *Forages in Southeast Asian and South Pacific Agriculture*. Australian Centre for International Agricultural Research (ACIAR), Canberra, pp. 111-123.

MacRae, J.C. and Egan, A.R. (1980) Measurement of glucose kinetics in sheep.

In: Mount, L.E. (ed.), *Energy Metabolism*. Butterworth, London, UK, pp. 421-426.

Moog, F.A. (1986) Forages in integrated food cropping systems. In: Blair, G.J., Ivory, D.A. and Evans, T.R. (eds), *Forages in Southeast Asian and South Pacific Agriculture*. Australian Centre for International Agricultural Research (ACIAR), Canberra, pp. 152-156.

Parra, R. and Escobar, A. (1985) Use of fibrous agricultural residues in ruminant feeding in Latin America. In: Preston, T.R., Kossila, V.L., Goodwin, J. and Reed, R.J.S. (eds), *Better Utilization of Crop Residues and By-Products in Animal Feeding*. Food and Agriculture Organization of the United Nations, Rome, Italy, pp. 81-98.

Pearce, G.R. (1986) Possibilities for improving the nutritive value of rice straw without pretreatment. In: Dixon, R.M. (ed.), *Ruminant Feeding Systems Utilizing Fibrous Agricultural Residues*. International Development Program of Australian Universities and Colleges (IDP), Canberra, pp. 101-105.

Petheram, R.J., Thahar, A. and Bernsten, R.H. (1985) Socioeconomic aspects of draught animal power in Southeast Asia with special reference to Java. In: Copland, J.W. (ed.), *Draught Animal Power for Production*. Australian Centre for International Agricultural Research (ACIAR), Canberra, pp. 13-19.

Pethick, D. (1984) Energy metabolism of skeletal muscle. In: Purser, D.B., Baker, S.K., Mackintosh, B.J. and Gawthorn, J.G. (eds), *Ruminant Physiology: Concepts and Consequences*. University of Western Australia, Perth, pp. 227-287.

Preston, T.R. and Leng, R.A. (1987) *Matching Ruminant Production Systems with Available Resources in the Tropics and Subtropics*. Penambul Books, Armidale, Australia.

Teleni, E. and Hogan, J.P. (1989) Nutrition of draught animals. In: Hoffman, D., Nari, J. and Petheram, R.J. (eds), *Draught Animals in Rural Development*. Australian Centre for International Agricultural Research (ACIAR), Canberra, pp. 118-133.

Teleni, E., Boniface, A.N., Sutherland, S. and Entwistle, K.W. (1988) The effect of liveweight loss on ovarian activity in *Bos indicus* cattle. *Proceedings of the Nutrition Society of Australia* 13: 126.

Wanapat, M. (1985) Nutritional status of draught buffaloes and cattle in northeast Thailand. In: Copland, J.W. (ed.), *Draught Animal Power for Production*. Australian Centre for International Agricultural Research (ACIAR), Canberra, pp. 90-95.

2. The Influence of Socioeconomic Factors on the Availability and Utilization of Crop Residues as Animal Feeds

Timothy O. Williams[1], Salvador Fernández-Rivera[1] and Timothy G. Kelley[2]

[1] *International Livestock Research Institute, ICRISAT Sahelian Center, BP 12404, Niamey, Niger*
[2] *International Crops Research Institute for the Semi-Arid Tropics, Patancheru 502 324, Andhra Pradesh, India*

Abstract

The fibrous by-products resulting from crop cultivation constitute a major source of nutrients for animal production in developing countries. On small farms, they form the principal feed of ruminant livestock during the dry seasons. Concerns about inadequate utilization of available feeds have led to the establishment of research programmes to improve the nutritive value and utilization of crop residues as ruminant feed. Despite this, farmer uptake of research findings has been limited. This paper explains why. It argues that the importance of crop residues as feed differs between production systems. Differences in production goals, resource endowments and socioeconomic conditions create different opportunities for the use of crop residues. Consequently, in designing research and extension projects that seek to improve use as livestock feed, it is pertinent to identify the main livestock production systems, farmers' production objectives and resource endowments, and determine the appropriate crop-residue-based diet for each system.

Introduction

Among the constraints facing livestock production in developing countries, poor animal nutrition and productivity arising from inadequate feed supplies stands as the most important. Growing concerns about this problem have prompted researchers and development planners to search for ways to promote the more efficient utilization of available feed resources. Crop residues—the fibrous by-products which result from the cultivation of cereals, pulses, oil plants, roots and tubers—represent an important feed resource for animal production in developing countries. These residues provide fodder at low cost since they are by-products of existing crop production activities. They are important adjuncts to natural pastures and planted forages and are often used to fill feed gaps during periods of acute shortage of other feed resources. In most parts of Africa, Asia, Latin America and the Caribbean, ruminants are dependent for at least part of the year on diets based on crop residues. Sandford (1989) reported that in various parts of semiarid sub-Saharan Africa, cattle derive up to 45% of their total annual feed intake from crop residues, and up to 80% during critical periods. Thole *et al.* (1988), in a village survey carried out in western Maharashtra, India, found that sorghum stover contributes between 20 and 45% of the total dry-matter (DM) feed provided to dairy animals by small-scale farmers. Similarly, McDowell and Hildebrand (1980) indicated that, in three livestock production systems found in Latin America, crop residues contribute 30-90% of ruminant feed. Parra and Escobar (1985) highlighted the importance of crop residues as feeds in different livestock production systems of Latin America and the Caribbean, from the integrated crop/livestock systems of Central America and Caribbean countries to the ranching systems of Mexico and Venezuela and the feedlot operations of Peru.

Nonetheless, there is still a perception that the potential of crop residues as livestock feed has not been fully exploited, particularly given the expansion in arable land that has taken place in many countries. This is partly due to the fact that crop residues have low nutritive value, i.e. low contents of metabolizable energy and crude protein. Consequently, many governments in both developed and developing countries have launched research programmes to improve the nutritive value and utilization of crop residues. Emphasis in much of this work has been on improving crop residue intake and digestibility in ruminants through upgrading and/or supplementation (Sundstøl and Owen, 1984; Doyle *et al.*, 1986; Little and Said, 1987; Kiran Singh and Schiere, 1993). Despite impressive animal production responses on the experiment station, farmer uptake of research findings has been minimal, partly because much less effort has been put into identifying the socio-economic factors limiting greater utilization of crop residues and adoption of new feeding systems by smallholder farmers (Owen, 1985; Owen and Jayasuriya, 1989; Devendra, 1991, this proceedings, p. 241).

The purpose of this paper is to review and summarize the information currently available on the utilization and management of crop residues at farm level in order to identify the constraints that have hindered the practical application of research results and to identify opportunities for greater utilization of crop residues as livestock feed. It is argued that the potential for such utilization differs from one production system to another. The substantial diversity that exists between production systems in resource endowments, availability of different feeds and type and level of animal production creates different opportunities for the use of crop residues as feeds. Consequently, it is important to identify the major livestock production systems and farmers' production goals and to determine what type of crop residue-based diet is appropriate for each system.

In what follows, the relative importance of crop residues in different ruminant production systems is examined. The social and economic factors that can influence crop residue availability at the farm level are discussed, as well as the opportunities and constraints to the adoption and utilization of technologies designed to improve the feed value of crop residues. A description of emerging socioeconomic trends and their likely influence on crop residue use as livestock feed concludes the paper.

Importance in Different Production Systems

Numerous livestock production systems can be found in developing countries. Their multiplicity necessitates a system of classification to order and group similar systems for the purpose of identifying opportunities and constraints to livestock development. In a recent multi-donor initiative coordinated by the Food and Agriculture Organization of the United Nations (FAO), a classification system based on three criteria was developed: (i) the agroecological environment; (ii) the level of integration of livestock with crop production; and (iii) the availability and type of land used for livestock production (Seré *et al.*, 1995). The study identified 11 livestock production systems involving monogastrics and ruminants, but the emphasis in this paper is on the nine ruminant production systems relevant to developing countries. This emphasis is justified because ruminants, with their ability to digest low-quality feeds and roughages, can utilize crop residues much more effectively than monogastrics.

Table 2.1 considers ruminant production systems in the context of the feed resources they use. These systems can be found in varying proportions in different developing regions of the world. The table indicates that the role of crop residues as livestock feed differs considerably between systems. Crop residues usually play a minor role in grass/rangeland systems, but are very important in mixed crop/ livestock systems.

For analytical purposes, and to obtain a better understanding of the relative availability of other productive resources within these systems, they have been regrouped into three management systems as shown in Table 2.2. The grass/ rangeland systems constitute the extensive grazing systems. Land is the principal resource in such systems, but its quality and productivity vary greatly between regions. Mixed crop/livestock systems are found in some of the most populous locations in the world. Potential labour availability in these systems is relatively high, but land and capital are less plentiful. The specialized production systems are made up of intensive enterprises, including the beef feedlots and dairy farms found in some developing countries, particularly near large urban centres. These systems are characterized by high capital inputs.

Table 2.1. Ruminant production systems and feed resources in selected countries and regions.

System	Region [1]	G/R [2]	FCs	CRs	CCs
Livestock/grassland (temperate zones, tropical highlands)	Mongolia, Parts of China S America E Africa	xxx			
Livestock/grassland (humid/subhumid tropics)	LAC (lowlands)	xxx			
Livestock/grassland (arid/semiarid tropics)	Parts of SSA WANA	xxx		x	
Mixed crop/livestock (rainfed; temperate zones, tropical highlands)	NE Asia Parts of E Africa Andean LAC (Ecuador, Mexico)	x	xx	xx	x
Mixed crop/livestock (rainfed; humid/subhumid tropics)	SE Asia LAC SSA	x	x	xxx	x
Mixed crop/livestock (rainfed arid/semiarid tropics)	WANA W Africa S Asia NE Brazil	xx	x	xxx	x
Mixed crop/livestock (irrigated; temperate zones, tropical highlands)	E Africa Parts of China	x	xx	xx	x
Mixed crop/livestock (irrigated; humid/ subhumid tropics)	Parts of SE Asia (Philippines, Vietnam)		x	xxx	
Mixed crop/livestock (irrigated; arid/semiarid tropics)	WANA S Asia Mexico		xx	xxx	

[1] SSA = sub-Saharan Africa; LAC = Latin America and the Caribbean; WANA = West Asia-North Africa.
[2] G/R = grassland/rangeland; FCs = forage crops; CRs = crop residues; CCs = concentrates. The number of crosses under each feed resource indicates its relative importance.
Source: Based on Seré *et al.* (1995).

Classified in this way, it is clear that livestock production systems differ in terms of their resource availability and potential to adopt and use technologies that demand different resource inputs. The relative availability of labour and capital compared with land, the existence of other feeds and the alternative uses to which crop residues can be put are some of the factors that determine the usefulness of crop residues as feeds within a production system (Giæver, 1984). Thus, in designing research and extension projects that will seek to improve the utilization of crop residues as livestock feed, it is important to identify what sort of system is being targeted and so to determine the resources available to support the changes to be introduced.

Table 2.2. Relative resource availability in ruminant production systems of developing countries.

Management system	Relative availability of:		
	Land	Labour	Capital
Extensive grazing	High	Low	Low
Mixed crop/livestock	Low	High	Low
Specialized production	Low	Low	High

Factors Influencing Availability

Having identified the production systems where crop residues can be utilized as livestock feed, it is important to determine the amounts of different residues available within the various systems. The availability of crop residues at the farm level depends not just on production levels but also on a variety of social and economic factors. Land, crop and animal ownership patterns, cultural practices, the use of modern crop varieties and the opportunities for market and non-market exchanges all influence a farmer's access to the residues that are locally produced. Seasonal and inter-year variations in crop residue production can also have a marked effect on availability at particular times of the year. These factors are usually ignored in national and regional assessments of crop residue production. Yet they are important not only for arriving at a realistic estimate of what is available but also for identifying the constraints that might limit improved utilization.

Land Tenure and Access Arrangements

A farmer's access to crop residues depends on the amount of land he or she owns and the institutional arrangements that exist for sharing or exchanging residues. Customary arrangements for sharing vary from country to country. In parts of sub-Saharan Africa where population pressure is low, privately cultivated fields revert to communal use after harvest so that crop residues can be freely grazed by village animals or collected by those who need them. This evens out inequality in land ownership and crop production within a village. But in areas where population density is high, access to crop residues is tightly controlled. Where farmers rely on animal manure to maintain soil fertility, they often enter into exchange relationships with pastoralists whereby the latters' animals graze the crop residues on their fields in exchange for the manure they deposit. In Asia, where share-cropping is widely practised, different arrangements exist for sharing the grain and fodder outputs. For example, in India the share-cropper takes the entire fodder output if he or she provides the bullocks used in field work, but shares the output with the landowner on an agreed basis if the latter supplied the draught animals (Parthasarathy Rao, personal communication).

Seasonality of Production

Crop residues are seasonally produced. Most become available only after grain harvest. If they are not used immediately they have to be conserved until needed. The difficulties of handling and storing crop residues have not been given adequate attention by researchers (Hilmersen *et al.,* 1984; Owen and Aboud, 1988). Devendra (1982) showed that exposure to weather decreased the nutritive value of rice straw. Other problems include pest infestation, moulding and fire risk. The combination of seasonality and storage problems creates an annual cycle of brief peaks in crop residue availability followed by long periods of scarcity. This is the case in many parts of semiarid sub-Saharan Africa, where crops are planted with the first rains in May or early June and harvesting occurs from September to November. Crop residues are plentiful for just 3 months, between December and February. By the end of the dry season, in April, they are often extremely scarce. Where the rainy season is longer or irrigation is available, farmers have much more flexibility in choosing when to grow crops. Planting is often phased to even out labour demands. As a result, a variety of crop residues may be available throughout the year.

Inter-annual fluctuations in rainfall can also affect crop residue yield, which may in turn affect the ratio between edible and non-edible fractions within residues. In northern Nigeria, van Raay and de Leeuw (1971) found a negative correlation ($r = -0.92$) between total crop residue yield and the edible fraction of sorghum. This was because high residue yields were associated with thicker and less edible stalks. Thus, inter-year variations in crop residue yields caused by rainfall may be partially offset by changes in the edible/non-edible ratio (Sandford, 1989).

Cultural Practices

The management practices used in the production, harvesting and processing of crops can affect the quantity and quality of crop residues. Shortage of labour often prevents farmers from harvesting crop residues. Where harvesting is done it is mainly carried out after grain harvest. Delayed harvesting can lead to greater loss of leaves and leaf sheaths—the most digestible parts of cereal straws—with a consequent reduction in nutritive value.

These factors taken together affect not only the quantity and quality of crop residues available to the farmer, but also the uses to which he or she decides to put those residues.

Alternative Uses

Besides serving as animal feed, crop residues have several other uses. In South Asia they are used as compost and mulch for crop production, bedding for live-stock, a substrate for growing mushrooms, fibre for paper manufacture and as fuel. In semiarid sub-Saharan Africa, they are used to control wind erosion and in the construction of roofs, fences, granaries, beds and doormats. Research has also shown that annual incorporation of millet stover into the sandy soils of this region increases soil pH, organic matter content and exchangeable cations, and—most important—crop yields (Pichot *et al.*, 1974; Bationo *et al.*, 1995). In discussing the alternative uses of crop residues, the quantities used for these other purposes and the relative returns to different uses are important issues that need to be considered.

Few estimates are available of the amounts of crop residues used for these other purposes. Data from village surveys carried out in the dry and wet zones of western Niger showed that farmers collect on average 33 and 90 kg ha^{-1} of millet stover for non-feed use. These amounts represent approximately 2.5 and 3.5% of the average millet stover yields in the two zones respectively (Baidu-Forson, 1995). In South Korea, Im and Park (1983) reported that about 66% of the rice straw produced annually (about 7 million t year^{-1}) was used as a soil amendment or as fuel.

What are the relative returns to these different uses of crop residues? McIntire *et al.* (1992) modelled the "competition" between animals and crops for crop residues in Niger, Nigeria and Ethiopia, representing the semiarid, subhumid and highland tropics respectively, with varying soils, population densities and degrees of mecha-nization. Cattle were used in the model in Niger and Nigeria, while cattle, draught oxen and sheep were used in Ethiopia. The results indicated that, in Niger, shifting from grazing crop residues to using them as mulch would reduce income because the yield response of pearl millet to crop residues is weak. In Ethiopia too, shifting to mulching would reduce income. In Nigeria, in contrast, it would increase in-come, because the yield of crop residues is high and crop response to them appears stronger. Field visits by the authors confirmed that feeding crop residues to live-stock is the predominant use.

These few observations indicate that large quantities of crop residues are available for feeding to animals, but that there are other competing uses which need to be considered when promoting increased use of this resource as livestock feed.

Feeding Methods

Farmers use various methods to feed crop residues to their animals. Arranged in increasing order of labour requirements, these methods include: (i) open access to whole residues on harvested fields; (ii) harvest and removal of stalks, with subsequent open access to stubble on harvested fields; (iii) harvest and removal of stalks, with subsequent restricted access to stubble on harvested fields; (iv) transport and storage for feed or sale; and (v) harvest of thinnings from cultivated fields for selective feeding before the main harvest (McIntire *et al.*, 1992).

The pattern of residue use is often dictated by population density, herd management practices and level of transport and marketing infrastructure. Open access to residues occurs in areas with low population densities and where animals are herded communally. In densely populated and heavily stocked areas, farmers restrict access to crop residues. The availability of labour, large livestock populations and easy access to markets encourage the removal of crop residues from fields. Direct grazing, through either open or restricted access, allows farmers to use residues as feed without incurring storage and processing costs. This method of feeding results in low utilization rates due to trampling and spoilage, but allows for the consumption of most nutritious plant parts and the return of nutrients to the soil (Klopfenstein *et al.*, 1987). Methods of residue feeding that involve harvesting (i.e. cut-and-carry systems) are more demanding in terms of labour, transport and storage facilities. The returns to these methods have to be reasonably high before they appeal to farmers.

Socioeconomic and Technological Constraints

Apart from the socioeconomic factors discussed above, there are other limitations to the use of crop residues as livestock feed. In many sub-Saharan African countries, inappropriate pricing policies and overvalued exchange rates have encouraged the use of imported grains and concentrates at the expense of locally produced feeds such as crop residues (Williams, 1989). Another major limitation is that crop residues have low contents of metabolizable energy and crude protein, so their intake by ruminants is low. This nutritional constraint has led to the development of various technologies to improve the quality and utilization of crop residues as feed. These include: (i) crop management practices (Bartle and Klopfenstein, 1988); (ii) variety selection (Ørskov, 1991); (iii) chemical (Sundstøl and Owen, 1984), physical (Riquelme-Villagrán, 1988) and biological (Burrows *et al.*, 1979; Latham, 1979; Kamra and Zadrazil, 1988) treatment; (iv) supplementation (Preston and Leng, 1987; Dixon and Egan, 1988); (v) feeding strategies, such as excess feeding or selective grazing (Owen and Aboud, 1988); (vi) genetic manipulation of rumen microbes (Armstrong and Gilbert, 1985; Ørskov and Flint, 1991; Wallace, 1994); and (vii) selection of animals better suited to the utilization of fibrous by-products

(Coombe, 1981; Ørskov, 1991). Reviewing the scientific principles underlying these technologies is not within the scope of this paper. However, their potential impacts are listed in Table 2.3.

Despite positive results achieved on the experiment station, the adoption of these technologies by farmers in developing countries remains low. Table 2.4 shows some of the constraints that have hindered adoption. In many cases, high labour and/or capital requirements have been the main deterrent. The difficulty of accurately predicting animal responses to the treatment and/or supplementation of crop residues has also been cited (Owen and Jayasuriya, 1989; Jayasuriya, 1993). In general, it appears that considerable efforts have been devoted to developing technologies and not enough to investigating the resource requirements of these technologies at farm level—a major determinant of their profitability and adoption. In most cases, the economic benefits of the new technologies have not been convincingly demonstrated to farmers (Devendra, this proceedings, p. 241). Economic

Table 2.3. Technologies available for improving the yield, quality and use of crop residues, and their potential impacts.

Technology	Impact
Crop management:	
Fertilizer application	Higher yield and quality
Planting density	Higher yield
Timely harvesting	Lower losses, higher quality
Selective harvesting	Improved availability, quality
Plant breeding	Improved yield, quality
Physical treatment (chopping, grinding, etc)	Higher intake, slight decrease in digestibility, less waste
Chemical treatment	Increased digestibility, intake
Biological treatment	Increased digestibility
Supplementation	Increased digestibility, intake, nutrient supply
Residue management:	
Excess feeding	Greater selectivity (quality)
Selective grazing	Greater selectivity (quality)
Livestock selection (species and breeds)	Animals better adapted to low-quality feeds
Genetic engineering of rumen microbes	Uncertain (technology under development)

Table 2.4. Potential constraints to the adoption of technologies for improving crop residues.

Technology	Labour	Capital	Other
Crop management:			
Fertilizer application	xx	xxx	Availability
Planting density	x		More nutrients needed
Timely harvesting	xx		
Selective harvesting	xxx		
Plant breeding		x	Potential trade-offs
Physical treatment	xxx	xxx	Uncertainty of animal response
Chemical treatment	xxx	xxx	Uncertainty of animal response, safety concerns, environmental pollution
Biological treatment	xxx	xxx	Variability in animal response, high technical skills needed
Supplementation	x	xxx	Availability, variability in animal response
Residue management:			
Excess feeding	xx		Residue availability
Selective grazing	xx		Higher management skills needed
Livestock selection		xx	Multiplication of superior animals

evaluation has too often been regarded as a secondary activity to be carried out after the technical experiments have been conducted (Potts, 1982).

Early economic evaluation can identify problems that might prevent adoption. Such evaluation needs to go beyond simple cost/benefit analysis of new technologies. It must identify farmers' production objectives and resource endowments, including available quantities of different feeds. And it must determine whether farmers are reluctant to adopt new technologies because these are too labour demanding, too risky, or simply not profitable enough compared with other investment opportunities.

Future Trends

Livestock production systems evolve in response to population and income growth, urbanization, and improvements in transport and marketing infrastructure. Projections for human population growth up to the year 2010 indicate an average annual growth rate of 1.4% in Asia, 1.6% in Latin America and the Caribbean, 2.2% in West Asia-North Africa and 3.1% in sub-Saharan Africa (Alexandratos, 1995). As population pressure increases on a finite resource base, extensive grazing will become less feasible. Rapid population growth and increased demand for food will provide strong incentives for the expansion of mixed crop/livestock systems.

Most of the population increase in developing countries will occur in cities. According to projections by the World Bank (1992), urban population growth over the period 1990-2030 will average 1.6% in Latin America, 3% in Asia and 4.6% in sub-Saharan Africa. Over the same period, average annual per capita income in real terms for the developing countries as a whole is expected to more than triple, from US$ 750 to US$ 2500. However, substantial regional differences will persist, with the figure ranging from a mere US$ 400 in sub-Saharan Africa to US$ 5000 in Latin America (World Bank, 1992). Where they occur, rapid urbanization and higher incomes will encourage the consumption of livestock products and stimulate demand for higher-value products. This should spur a general increase in animal production, together with the development of more specialized production units.

In countries where inappropriate sectoral and macroeconomic policies have inhibited the growth of the livestock sector in the past, structural adjustment and macroeconomic reform can be expected to provide a boost to the sector soon, especially as price distortions and trade barriers are removed and misaligned exchange rates are corrected. The resulting increase in livestock production will increase the demand for feeds, including crop residues. The removal of government subsidies on imported feeds should stimulate demand still further, by promoting the use of local feed resources. At the same time, similar policy reforms in the crop subsector are likely to discourage food imports and encourage domestic production, increasing the availability of crop residues.

The rate at which livestock production systems intensify will vary between countries, just as it will between regions. Specific trends are difficult to predict. What is clear is that the demographic and economic changes predicted for the developing countries generally will ensure the continuing utilization of crop residues as livestock feed. The shift towards mixed crop/livestock systems that these changes are expected to create will stimulate the demand for different types of crop residues, both treated and untreated. It will then become essential to identify what type of crop residues will fit best into a particular system.

In the longer term, as economic growth progresses and transport and marketing improve, mixed crop/livestock systems will gradually evolve into more specialized production systems (McIntire *et al.*, 1992). These specialized systems will require animals of higher genetic potential and feeds of better quality to achieve higher milk yields or animal growth rates. Under these conditions, the use of low-quality crop residues will diminish. Nevertheless, there is evidence that crop residues can be a suitable feed in specialized beef (Ward, 1978; Klopfenstein *et al.*, 1987) and dairy (Klopfenstein and Owen, 1981) enterprises, particularly during phases when animal nutritional demands are lowest.

Conclusions

The greatest potential for the use of crop residues as animal feeds exists in the mixed crop/livestock systems of the semiarid and subhumid tropics. The demographic and economic changes expected in these regions will generally reinforce the importance of crop residues as feeds. However, as production systems become more specialized, crop residues are likely to be included in ruminant diets in lower proportions or only at phases of production with lower nutritional requirements. This implies varying opportunities for the use of crop residues as feeds.

As a result, recommendations on feeding crop residues, and especially on upgrading their quality, should be developed for each production system only after careful consideration of the opportunities and constraints inherent in that system.

The potential of crop residues as a feed resource will only be fully realized if their use proves to be economically beneficial and compatible with the resource endowments and production goals of farmers.

References

Alexandratos, N. (1995) *World Agriculture Towards 2010: An FAO Study.* John Wiley & Sons, Chichester, UK.

Armstrong, D.G. and Gilbert, H.J. (1985) Biotechnology and the rumen: A mini-review. *Journal of Science, Food and Agriculture* 36: 1039-1046.

Baidu-Forson, J. (1995) Determinants of the availability of adequate millet stover for mulching in the Sahel. *Journal of Sustainable Agriculture* 5: 101-116.

Bartle, S.J. and Klopfenstein T.J. (1988) Nonchemical opportunities for improving crop residue feed quality: A review. *Journal of Production Agriculture* 1: 356-362.

Bationo, A., Buerkert, A., Sedogo, M.P., Christianson, B.C. and Mokwunye, A.U. (1995) A critical review of crop-residue use as soil amendment in the West African semiarid tropics. In: Powell, J.M., Fernández-Rivera, S., Williams, T.O. and Renard, C. (eds), *Livestock and Sustainable Nutrient Cycling in Mixed Farming Systems of sub-Saharan Africa,* vol. 2: Technical Papers. Proceedings of an International Conference, 22-26 November 1993, International Livestock Centre for Africa (ILCA), Addis Ababa, Ethiopia, pp. 305-322.

Burrows, I., Seal, K.J. and Eggins, H.O.W. (1979) The biodegradation of barley straw by *Coprinus cinereus* for the production of ruminant feed. In: Grossbard, E. (ed.), *Straw Decay and its Effect on Disposal and Utilization.* John Wiley & Sons, Chichester, UK, pp 147-154.

Coombe, J.B. (1981) Utilization of low-quality residues. In: Morley, F.W.H. (ed.), *Grazing Animals.* World Animal Science, B1. Elsevier, Amsterdam, The Netherlands, pp. 319-334.

Devendra, C. (1982) Perspectives in the utilization of untreated rice straw by ruminants in Asia. In: Doyle, P.T. (ed.), *The Utilization of Fibrous Agricultural Residues as Animal Feeds*. University of Melbourne, Parkville, Victoria, Australia, pp. 7-26.

Devendra, C. (1991) Technologies currently used for the improvement of straw utilization in ruminant feeding systems in Asia. In: Romney, D.L., Ørskov, E.R. and Gill, M. (eds), *Utilization of Straw in Ruminant Production Systems*. Proceedings of a Workshop, 7-11 October 1991, Natural Resources Institute and Malaysian Agriculture Research and Development Institute, Kuala Lumpur, pp. 1-19.

Dixon, R.M. and Egan, A.R. (1988) Strategies for optimizing use of fibrous crop residues as animal feeds. In: Dixon, R.M. (ed.), *Ruminant Feeding Systems Utilizing Fibrous Agricultural Residues*. Proceedings of the Seventh Annual Workshop of the Australian-Asian Fibrous Agricultural Residues Research Network. International Development Program of Australian Universities and Colleges (IDP), Canberra, pp. 11-26.

Doyle, P.T., Devendra, C. and Pearce, G.R. (eds) (1986) *Rice Straw as a Feed for Ruminants*. International Development Program of Australian Universities and Colleges (IDP), Canberra.

Giæver, H. (1984) The economics of using straw as feed. In: Sundstøl, F. and Owen, E. (eds), *Straw and Other Fibrous By-products as Feed*. Elsevier, Amsterdam, The Netherlands, pp. 558-574.

Hilmersen, A., Dolberg, F. and Kjus, O. (1984) Handling and storing. In: Sundstøl, F. and Owen, E. (eds), *Straw and Other Fibrous By-products as Feed*. Elsevier, Amsterdam, The Netherlands, pp. 25-44.

Im, K.S. and Park, Y.I. (1983) Animal Agriculture in Korea. Mimeo, the Asian-Australasian Association of Animal Production Societies, Department of Animal Sciences, College of Agriculture, Seoul National University, Korea.

Jayasuriya, M.C.N. (1993) Use of crop residues and agroindustrial by-products in ruminant production systems in developing countries. In: Gill, M., Owen, E., Pollott, G.E. and Lawrence, T.L.J. (eds), *Animal Production in Developing Countries*. Occasional Publication No. 16, British Society of Animal Production, pp. 47-55.

Kamra, D.N. and Zadrazil, F. (1988) Microbiological improvement of ligno-cellulosics in animal feed production: A review. In: Zadrazil, F. and Reiniger, P. (eds), *Treatment of Lignocellulosics with White Rot Fungi*. Elsevier Applied Science, London, UK, pp. 56-63.

Kiran Singh and Schiere, J.B. (eds) (1993) *Feeding of Ruminants on Fibrous Crop Residues: Aspects of Treatment, Feeding, Nutrient Evaluation, Research and Extension*. Proceedings of a Workshop, 4-8 February 1991, National Dairy Research Institute (NDRI)-Karnal. Indian Council of Agricultural Research (ICAR), New Delhi, India.

Klopfenstein, T. and Owen, F.G. (1981) Value and potential use of crop residues and by-products in dairy rations. *Journal of Dairy Science* 64: 1250-1268.

Klopfenstein, T., Roth, L., Fernández-Rivera, S. and Lewis, M. (1987) Corn residues in beef production systems. *Journal of Animal Science* 65: 1139-1148.

Latham, M.J. (1979) Pre-treatment of barley straw with white-rot fungi to improve digestion in the rumen. In: Grossbard, E. (ed.), *Straw Decay and its Effect on Disposal and Utilization.* John Wiley & Sons, Chichester, UK, pp. 131-137.

Little, D.A. and Said, A.N. (eds) (1987) *Utilization of Agricultural By-products as Livestock Feeds in Africa.* Proceedings of the African Research Network for Agricultural Byproducts (ARNAB) Workshop, September 1986, Blantyre, Malawi. International Livestock Centre for Africa (ILCA), Addis Ababa, Ethiopia.

McDowell, R.E. and Hildebrand, P.E. (1980) Integrated Crop and Animal Production: Making the Most of Resources Available to Small Farms in Developing Countries. Working Paper, The Rockefeller Foundation, New York, USA.

McIntire, J., Bourzat, D. and Pingali, P. (1992) *Crop/Livestock Interaction in sub-Saharan Africa.* World Bank Regional and Sectoral Studies, Washington DC, USA.

Ørskov, E.R. (1991) Manipulation of fibre digestion in the rumen. *Proceedings of the Nutrition Society* 50: 187-196.

Ørskov, E.R. and Flint H.J. (1991) Manipulation of rumen microbes or feed resources as methods of improving feed utilization. In: Hunter, A.G. (ed.), *Biotechnology in Livestock in Developing Countries.* University of Edinburgh, pp. 123-138.

Owen, E. (1985) Crop residues as animal feeds in developing countries: Use and potential use. In: Wanapat, M. and Devendra, C. (eds), *Relevance of Crop Residues as Animal Feeds in Developing Countries.* Funny Press, Bangkok, Thailand, pp. 25-42.

Owen, E. and Aboud, A. (1988) Practical problems of feeding crop residues. In: Reed, J.D., Capper, B.S. and Neate, P.J.H. (eds), *Plant Breeding and the Nutritive Value of Crop Residues.* Proceedings of a Workshop, 7-10 December 1987, International Livestock Centre for Africa (ILCA), Addis Ababa, Ethiopia, pp. 133-155.

Owen, E. and Jayasuriya, M.C.N. (1989) Use of residues as animal feeds in developing countries. *Research and Development in Agriculture* 6: 129-138.

Parra, R. and Escobar, A. (1985) Use of fibrous agricultural residues in ruminant feeding in Latin America. In: Preston, T.R., Kossila, V.L., Goodwin, J. and Reed, S.B. (eds), *Better Utilization of Crop Residues and By-products in Animal Feeding.* Proceedings of the FAO/ILCA Expert Consultation, 5-9 March 1984, International Livestock Centre for Africa (ILCA), Addis Ababa, Ethiopia, pp. 81-98.

Pichot, J., Burdin, S., Charoy, J. and Nabos, J. (1974) L'enfouissement des pailles de mil *Pennisetum* dans les sols sableux dunaires: Son influence sur les rendements et la nutrition minérale du mil; son action sur les caractéristiques chimiques du sol et la dynamique de l'azote minéral. *Agronomie Tropicale* (Paris) 29: 995-1005.

Potts, G.R. (1982) Application of research results on by-product utilization: Economic aspects to be considered. In: Kiflewahid, B., Potts, G.R. and Drysdale, R.M. (eds), *By-product Utilization for Animal Production.* Proceedings of a Workshop on Applied Research, 26-30 September 1982, Nairobi, Kenya. International Development Research Centre (IDRC), Ottawa, Canada, pp. 116-127.

Preston, T.R. and Leng, R.A. (1987) *Matching Livestock Production Systems to Available Resources.* Penambul Press, Armidale, New South Wales, Australia.

Riquelme-Villagrán, E. (1988) Feed processing technologies to increase nutritive values. In: *Proceedings of the Sixth World Conference on Animal Production,* 27 June-1 July 1988, Helsinki, Finland, pp. 72-84.

Sandford, S. (1989) Crop residues/livestock relationships. In: Renard, C., Vandenbeldt, R.C. and Parr, J.F. (eds), *Soil, Crop and Water Management Systems for Rainfed Agriculture in the Sudano-Sahelian Zone.* Proceedings of an International Workshop, 11-16 January 1987, ICRISAT Sahelian Centre, Niamey, Niger. International Crops Research Institute for the Semi-Arid Tropics, Patancheru, India, pp. 169-182.

Seré, C., Steinfeld, H. and Groenewold, J. (1995) World livestock systems: Current status, issues and trends. In: Gardiner, P. and Devendra, C. (eds), *Global Agenda for Livestock Research.* Proceedings of a Consultation, 18-20 January 1995, International Livestock Research Institute (ILRI), Nairobi, Kenya, pp. 11-38.

Sundstøl, F. and Owen, E. (eds) (1984) *Straw and Other fibrous By-products as Feed.* Developments in Animal and Veterinary Sciences 14, Elsevier, Amsterdam, The Netherlands.

Thole, N.S., Joshi, A.L. and Rangnekar, D.V. (1988) Feed availability and nutritional status of dairy animals in western Maharaashtra. In: Kiran Singh and Schiere, J.B. (eds), *Fibrous Crop Residues as Animal Feed.* Indian Council of Agricultural Research (ICAR), New Delhi, pp. 207-212.

van Raay, J.G.T. and de Leeuw, P.N. (1971) *Rural Planning in a Savanna Region.* The University Press, Rotterdam, The Netherlands.

Wallace, R.J. (1994) Rumen microbiology, biotechnology and ruminant nutrition: Progress and problems. *Journal of Animal Science* 72: 2992-3003.

Ward, J.K. (1978) Utilization of corn and grain sorghum residues in beef cow forage systems. *Journal of Animal Science* 46: 831-840.

Williams, T.O. (1989) *Livestock Development in Nigeria: A Survey of the Policy Issues and Options.* ALPAN Paper No. 21, International Livestock Centre for Africa (ILCA), Addis Ababa, Ethiopia.

World Bank (1992) *World Development Report 1992.* Oxford University Press, New York, USA.

Acknowledgements

The authors would like to thank Adrian Egan, Parthasarathy Rao and Peter de Leeuw for their valuable comments on an earlier version of this paper.

3. Crop Residues in Tropical Africa: Trends in Supply, Demand and Use

P.N. de Leeuw

International Livestock Research Institute, PO Box 30709, Nairobi, Kenya

Abstract

Crop residues (CRs) are the plant materials that remain after food crops have been harvested. They may be left in the field as grazing for livestock and/or as mulch, or transported to the homestead for stall feeding or use as fencing, building and roofing materials or as fuel. The importance of CRs as potential livestock feed varies with the type of crops grown—cereals, grain legumes, roots/tubers—and also with the proportion of land under food crops and with the yields of the relevant plant parts. The output of CRs tends to rise with rural population density. The proportion of total CRs allocated as feed depends on such factors as livestock density and rules of access. These in turn are influenced by land tenure and the relative importance of livestock in the farming system.

Supply of and demand for CRs have been analysed at different spatial levels: country, region, village and individual farmer. At the country level, statistics on the area of cropland per tropical livestock unit (TLU, 250 kg liveweight) are available. The output of crops per TLU is a good indicator of CR supply, provided adequate ratios of crop yield to CR yield can be developed. To illustrate this approach several examples are given, arranged along a continuum of rising population density, with farming systems in the semiarid and subhumid zones of West Africa and Zimbabwe at the low end and systems in the Ethiopian and Rwandan highlands at the high end. Although these profiles highlight potential supplies per hectare and when paired with livestock density potential demand, accurate data require the monitoring of grazing livestock throughout the year to record their sources of feed. Being time-consuming to collect and analyse, few such data sets exist.

To facilitate calculations using ratios, relationships between grain and stover yields have been established for the principal cereals (pearl millet, sorghum, maize and wheat). It has been found that CR yields rise more slowly than grain yields, causing a decline in grain:stover ratios. This decline is particularly steep in coarse-grained cereals. Nonetheless, stover and grain outputs do show parallel trends.

Apart from the impact of higher crop production on CR output, this paper also discusses the implications of intensifying land use, including issues such as access to CRs in different land tenure systems, the association between livestock holdings and cropping patterns and areas, and trade in feedstuffs. Finally, some speculations and prognoses are volunteered as to the future of the livestock and food crop sectors within the political and economic context of tropical Africa.

Introduction

Crop residues (CRs) are roughages that become available as livestock feeds after crops have been harvested. They are distinct from agricultural by-products (such as brans, oil cakes, etc), which are generated when crops are processed. Residues can usually be grouped by crop type—cereals, grain legumes, roots and tubers, and so on (World Bank, 1989; Nordblom and Shomo, 1995).

Apart from being a source of animal feed, residues are used as building, roofing and fencing materials, as fuel, and as fertilizer or surface mulch in cropland (van Raay and de Leeuw, 1970, 1974). Their value when used as feed depends on the demand from livestock owners, which varies with the overall supply and demand situation for feeds. This in turn depends on the livestock density, usually expressed in tropical livestock units per square kilometre (TLU km^{-2}) and the supply of other feed resources, in particular forage and browse from natural vegetation (de Leeuw and Rey, 1995). The supply of CRs is a function of the proportion of land used for cropping and of the edible feed yields per unit of land. Where consumable feed from CRs exceeds grazing from natural pastures (expressed in tonnes of dry matter per hectare, t DM ha^{-1}), expansion of cropland has a positive effect on overall feed supplies.

Estimates of consumable yields of natural pastures differ according to ecological zone: from 0.2 and 0.5 t DM ha^{-1} in the arid and semiarid zones, rising to 0.72 and 0.76 t DM ha^{-1} in the subhumid/humid zones and the highlands respectively (Winrock, 1992). Models developed by the Food and Agriculture Organization of the United Nations (FAO) for consumable pasture biomass in Kenya and West Africa use length of growing period (LGP) rather than ecozone as the basis for estimates. In the Kenya model, biomass increases progressively from 0.3 t DM year^{-1} at an LGP of < 75 days to 3.6 t DM year^{-1} at 270 days (Kassam *et al.*, 1991a). In West Africa, biomass for an LGP of up to 120 days is estimated as extremely low (0.1 t DM ha^{-1}), rises to 3 t DM ha^{-1} by 210 days but declines again to 0.7 t DM ha^{-1} by 300 days because of disease constraints (Kassam *et al.*, 1991b).

To illustrate the importance of CRs within a zonal and farming systems context, CR feed budget profiles can be described for a number of tropical ecoregions, along a gradient of increasing rural population density. At one end is the arid zone, where cropping is rare or intermittent, with pastoral livestock systems predominating and common property natural pastures being the sole source of feed. It is estimated that this zone harbours 30% of all TLUs in tropical Africa, including 90% of all camels, 30% of cattle and 40% of the small ruminant population (de Leeuw *et al.*, 1995a). At the other extreme are the smallholder farmers of the Ethiopian highlands, who keep cattle and equines for traction, manure, fuel and milk and rely on CRs as the major source of feed. Profiles from West Africa, Zimbabwe, Kenya and Rwanda occupy intermediate positions along the continuum, each illustrating how cropping patterns, crop/livestock interactions and population density affect CR supply and use.

Crop selection, cropping practices (e.g. double- and intercropping) and level of inputs influence CR yields. More importantly, changes in cropping patterns may affect access and use, while intensification of production (through improved varieties, increased fertilizer use, better tillage and weeding) should increase yields of both crops and residues. This paper reviews the major crops in tropical Africa in terms of responses to changes in management and input levels and the resultant impact on CR feed supplies. In addition, broad trends in land use and tenure and their likely impact on CR access and use are discussed.

Country Profiles

Nordblom and Shomo (1995) calculated feed supplies for 17 countries in the West Asia-North Africa (WANA) region, including Ethiopia and Sudan. Three major sources of feed—crop residues, forage from natural pastures and feed grain/other concentrates—were combined to calculate national feed budgets. Conversion factors from grain to CR were derived from Kossila (1988), while edible herbage from natural pastures was taken as 0.5 t DM ha^{-1} year^{-1}. Despite a substantial increase in crop residue output arising from the expansion of cropped land, total output from natural pastures was held constant across time. The amount of feed available per TLU was calculated from livestock statistics produced by the FAO. As Ethiopia and Sudan are at the extreme ends of a continuum from tropical lowlands to cool highlands, annual feed budgets for these two countries show very great contrasts (Fig. 3.1).

No similar country-by-country analyses of feed supplies have been published for tropical Africa, but ruminant stocking rates by regions have been matched with estimated herbage outputs from natural pastures to assess levels of herbage utilization (e.g. Breman and de Ridder, 1991; de Leeuw and Reid, 1995; de Leeuw and Rey, 1995) and of nutrient removal by livestock (de Leeuw *et al.*, 1995b). Rough estimates of the CR supply have been made for individual countries, including Tanzania (Lwoga and Urio, 1987) and Malawi (Munthali and Dzowela, 1987). The usual procedure consisted of estimating the hectares of land devoted to each crop and using a CR:grain yield ratio to calculate national CR output. These ratios vary greatly—an issue that will be addressed below.

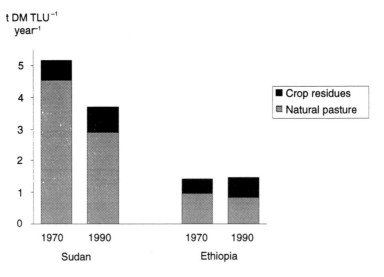

t DM TLU^{-1} year^{-1}

Fig. 3.1. Biomass from natural pastures and crop residues in Ethiopia and Sudan (Nordblom and Shomo, 1995).

Two kinds of ratio can be used to link grain and CR yield. The first is a simple one in which grain yield is divided by an agreed factor expressing the harvest index, or proportion of grain to total above-ground biomass. Kossila (1988) and Nordblom and Shomo (1995) used ratios of this kind. CR:grain ratios usually decline with increasing grain yield and are therefore higher in Africa than in Asia (for wheat: 2.0 compared with 1.3) or South America (for maize: 3.0 compared with 2.0) (Kossila, 1988). However, to evaluate CRs as a feed resource a second ratio is needed related to "edibility", since it makes little sense to include inedible plant parts in potential feed budgets. To estimate the consumable fraction of a CR, data are required on such parameters as the likely removal rates by grazing animals or the refusal rates of stall-fed livestock.

Potential supplies of CRs in Africa can be approximated from country statistics on the proportion of land cultivated (e.g. World Bank, 1989; WRI, 1990), combined with yield estimates for the grains, tubers and so on of the major crops (World Bank, 1989). In the manner of Nordblom and Shomo (1995), these can then be expressed as a percentage of the total supply (assuming a given output for natural pastures and/ or uncultivated land) and, by using data on livestock populations (Winrock, 1992), as the amount of feed available per TLU and year. These estimates can only be approximations. They would come closer to reality if appropriate ratios of grain to CR were available. Further caution is needed because data obtained in this way tend to ignore within-country heterogeneity in terms of ecozones. As a result, estimates for small countries may be more realistic than for large ones like Nigeria, Ethiopia, Sudan or Tanzania. For such countries, estimates need to be made on the basis of a zonal stratification (de Leeuw and Rey, 1995).

Stocking densities on cropland vary widely, from 21 TLU ha^{-1} in arid/semiarid Somalia to 1.5 in Gambia and 0.9-1.1 TLU ha^{-1} in Sahelian countries such as Niger and Burkina Faso. In countries where smallholder cropping is relatively important, densities are lower, mainly in the range of 0.35 to 0.55 TLU ha^{-1} (Table 3.1). Statistics on the production of major crops by country indicate a dominance of cereal crops in the drier ecoregions and highlands, and the increasing importance of roots and tubers (cassava, sweet potatoes, yams) with increasing rainfall. In countries such as Togo, Benin, Rwanda and Uganda, tubers produce up to twice as much tonnage as cereals. Cash crops, in particular groundnuts and cotton, and pulses need to be included, since although smaller in amount their residues are important for their relatively high quality (Table 3.1).

Table 3.1. Crop production and cropland per TLU for four tropical African countries.

	Crop production (t TLU^{-1}year^{-1})					TLU ha^{-1} of cropland
Country	Sorghum/ millet	Maize	Total cereals	Grain legumes	Roots and tubers	
Senegal	0.69	0.08	0.84	0.57	0.12	2.97
Benin	0.14	0.34	0.50	0.12	1.39	0.43
Malawi	0.20	1.60	1.84	0.33	0.30	0.33
Rwanda	0.32	0.20	0.54	0.34	1.29	0.53

Source: World Bank (1989).

As regards their use by livestock, the three major sources of CRs—cereals, legumes and roots/tubers—differ in one important aspect: while most cereal stover is grazed in situ, leguminous residues are often collected as hay for sale or stall-feeding. Similarly, the foliage of roots and tubers is harvested and only becomes available, together with other by-products, when the crops are processed as human food. There is thus a crop-dependent continuum from communal to owner-controlled access, which is also related to the quality and market value of the residue or by-product.

Although the country approach followed by Nordblom and Shomo (1995) appears suitable for the WANA region, zonal stratification seems preferable for tropical Africa, with its greater contrasts in ecozones and farming systems and their associated crop/livestock subsystems (de Leeuw and Rey, 1995; de Leeuw *et al.*, 1995b). To capture these zonal differences, I will now present five case studies from different parts of sub-Saharan Africa.

Subregional Profiles

West Africa

Partitioning daily grazing time was a method commonly used in Nigeria and Mali to assess the importance of CRs in cattle diets. Herds were followed and their sources of feed (browse, grass, CRs by crop) and grazing locations (fallows, uplands, valleys) were recorded, usually at monthly intervals for an entire year. In northern Nigeria, this was done in settled and semi-settled agropastoral herds in the semiarid zone and in sedentary herds in the northern and southern subhumid zone (Table 3.2).

The CR grazing period usually extended over 3-4 months from late November to March. During this period CR grazing averaged 50-80% of total grazing time and 13-24% of the entire annual total. The highest levels of daily CR grazing were reached

Table 3.2. Composition of crop residue grazing (as a percentage of grazing time) of agropastoral cattle herds in Nigeria.

	% of grazing time by period in each zone			
Residues	Semiarid[1]	Semiarid[2]	Subhumid (north)	Subhumid (south)
Sorghum	49	61	28	9
Millet	4	3	-	16
Maize	-	-	-	1
Rice	-	-	-	17
Groundnut	5	7	4	-
Cowpea	1	-	6	-
Soyabean	-	-	-	8
Cotton	13	10	-	-
Total	72	81	38	51
Period (months)	3.0	3.5	4.0	4.0
% of annual grazing time	18.0	23.6	12.7	17.0
References	van Raay and de Leeuw (1970)	van Raay and de Leeuw (1974)	van Raay and de Leeuw (1974)	Powell and Bayer (1985); Powell (1986)

[1] Settled farmers.
[2] Agropastoralists.

in December, soon after crop harvest, and gradually declined thereafter. The availability of each CR depended on harvest date, fields of cotton and rice being freed for grazing some 2 months later than those of sorghum and millet.

In the semiarid north (LGP 120-180 days), millet contributed little to livestock feed, as the grain was harvested early (late August) and the residue was inaccessible because of intercropping with later-maturing sorghum or cotton (van Raay and de Leeuw, 1974). However, in the southern subhumid zone, millet is sown in seedbeds before being transplanted, and is harvested from mid-November onwards (Powell and Bayer, 1985). In this zone millet residues were harvested or grazed by livestock from harvested fields. Two months later, rice residues became available in harvested fields and at communal threshing sites. Legume haulms contributed much less than grain crops, but because of their high feeding and/or cash value they were harvested separately and often taken to the homestead for storage (Powell, 1986).

Reliance on rice straw and fields after harvest was very high in the areas surrounding the large irrigation projects along the Niger River in central semiarid Mali (Wilson *et al.*, 1983). Here herds grazed upland savanna in the rainy season, but from November to May relied on rice straw (57%), millet residues (12%) and harvested rice fields (20%). In total, crop-related grazing accounted for 41% of the total annual feed supply (Dicko-Toure, 1980).

In most of the dry semiarid Sahel (annual rainfall, 400-500 mm; LGP, 75-120 days), millet stover is the dominant residue; it is usually sole-cropped and grazed from early November onwards (Williams *et al.*, 1995). However, when intercropped with spreading varieties of cowpea (usually at low density), plants are uprooted and transported to the homestead (Stoop, 1986).

During the late 1980s, the CR situation in the cotton-growing region of Mali was analysed along a northwest-southeast transect covering over 3 degrees of latitude (13°20-10°30), with annual rainfall ranging from 700 to 1100 mm (Leloup and Traore, 1989, 1991). The population density was fairly uniform (20-30 people km^{-2}). Livestock density decreased from north to south (from 20 to 14 TLU km^{-2}), but reached 35 TLU km^{-2} where cotton production is most intensive, generating high demand for work oxen (Bosman *et al.*, 1995). Cropping patterns varied along the rainfall gradient. Cotton covered 17% of the land, becoming less important where rainfall was <800 mm. Millet and sorghum predominated throughout the transect, but were partially replaced by maize and upland rice in the south, occupying only 21% and 16% of the cropland respectively at the southern end. Similarly, groundnut and cowpea increased from 6 to 15% with increasing rainfall. Cereals dominated feed output in terms of both yield and percentage of total cropland (68% in the south, 93% in the north). There was a fairly stable ratio between feed from CRs as a fraction of total feed and the percentage of cropped land, the fraction increasing by 1.6% (SD±0.15) with each 1% rise in area cultivated (Leloup and Traore, 1989, 1991).

In the extreme northeast of the Malian transect, the herbaceous layer in the communally used natural pastures consists mostly of annual species. Hence herbage yields are lower than in the south and, once the rains are over, disappearance rates (due to leaf and seed fall, termites, etc) are high (Hiernaux, 1989). Consequently, reliance on CRs was much higher than further south: the CR grazing period extended from November to May, with the cattle diet consisting of 90-100% CRs in November-December and falling to 45% in April-May; CRs accounted for 35% of annual grazing time (Leloup and Traore, 1991).

A similar transect (560-975 mm of rain) was described in Burkina Faso during 1975-1982 (Vierich and Stoop, 1990). Over this period, population density increased from 34 to 45 people km^{-2}, cotton became more important and pastoralists moved further south, reducing manure exchanges and prompting greater use of inorganic fertilizers. Decreasing rainfall encouraged the adoption of short-duration sorghum and cultivation of lower slopes and valleys. These changes were most pronounced at the southern end, as in the north pearl millet remained the major crop.

In southern Nigeria, small-scale farmers keep small flocks (5-8 head) of goats (and some sheep) as a secondary enterprise, the cropping of cassava, maize and cocoa being their major occupations (Platteeuw and Oludimu, 1993). Smallstock may be free-ranging, partially tethered or entirely confined, leading in the latter two cases to increasing levels of supplementary feeding. Cassava and maize by-products are the major supplements, although many other feedstuffs are fed, especially in the humid forest zone (Table 3.3). The average level of supplementation covered about half the total feed requirements of an adult goat weighing 15 kg (Bosman, 1995).

Southern Africa

In the late 1970s, the proportion of land under cultivation in the communal areas of Zimbabwe was 15-20% in the subhumid and 4-6% in the semiarid zone (Clatworthy, 1987). However, much higher levels were reported in the late 1980s (Cousins, 1992; see Table 3.4). By this time, over 60% of the land was planted with maize, which was replaced by sorghum and millet at the dry end of the continuum (Clatworthy, 1987). Groundnut and cotton were the major cash crops, occupying 15 and 2% of cropland respectively. Likewise, in Malawi, the major rainfed crops included maize (81%), sorghum/millet (4%), groundnut (10%) and cotton (4%) (Munthali and Dzowela, 1987). Smallholder mixed production systems were predominant in Zimbabwe's communal lands (Clatworthy, 1987). Cattle herds were small (average 5-7 head) and oxen comprised 15-30% of the herd (Steinfeld, 1988; Cousins, 1992).

Steinfeld (1988) recorded livestock wealth and farm size in two contrasting environments in Zimbabwe, one in the subhumid zone (700 mm annual rainfall, 130 m altitude) and the other in the semiarid zone (520 mm, 900 m). Cattle wealth, which ranged from 0 to > 25 head per farm, was closely linked with farm size, area cropped with maize and resulting grain yields, which more than doubled across the wealth gradient in both locations, as the richer farmers had more manure and could afford additional inputs (Fig. 3.2). Nonetheless, maize stover appeared unimportant as a feed source, since the amount of maize grain produced per head of cattle declined with increasing cattle wealth from 390 to 160 kg head^{-1} year^{-1} at the subhumid location and from 200 to 60 kg head^{-1} year^{-1} at the semiarid location.

Table 3.3. Composition (%) of supplementary feed offered to goats kept by smallholders in subhumid and humid southwest Nigeria.

Zone	Subhumid	Humid
Ingredient:		
Cassava tuber	29	19
Cassava peel	44	21
Maize bran	23	29
Other[1]	4	31
Total supplement intake (g day $^{-1}$ head $^{-1}$)	320	370

[1] Other feeds consisted of browse and herbage (31%), by-products from banana and fruits (25%) and food processing (14%), yam peels (18%) and others (11%).
Source: Bosman and Ayeni (1993).

Concerns about the dangers of overgrazing on communal lands prompted research in feed resource management and herding practices. In five villages, grazing time was recorded to assess how different habitats were used (Table 3.4). Herding or enclosing cattle in fenced paddocks was obligatory during the 6-month growing season, so as to avoid crop damage, but during the dry months animals were free-ranging or subject to intermittent low-intensity herding. Although cropland with CRs was accessible, use by grazing cattle was low compared with other habitats. Preference rates (percentage of grazing time as a ratio of percentage land) averaged 0.60 for cropland, whereas valleys and home compounds showed rates of 4.4 and 2.2 respectively. These valleys (and other low-lying areas) covered only 4-16% of the village land. Time spent in home compounds was relatively high, as these were often fenced and used as holding grounds when herding labour was scarce. Stored crop residues were sometimes fed there. Consequently, cattle spent around half their total grazing time in these two habitats. It seems likely that, due to active herding during the cropping season, cattle become used to trekking between home compounds and water points in the valleys, avoiding cropland. They continue this habit when left unherded during the dry season. In addition, cropland may have become less attractive or less accessible due to dry-season ploughing (Cousins, 1992). Consequently, CRs provided only 12% of the annual feed supply from 22% of the land.

Eastern Kenya

Long-term changes in land use, cropping patterns and livestock populations are well documented for Kenya's Machakos District (Tiffen *et al.*, 1994). This area has bimodal rainfall, which varies from 500 to 1000 mm annually along steep altitudinal

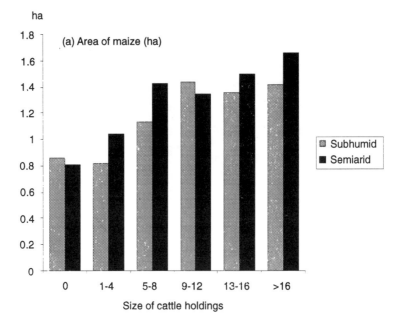

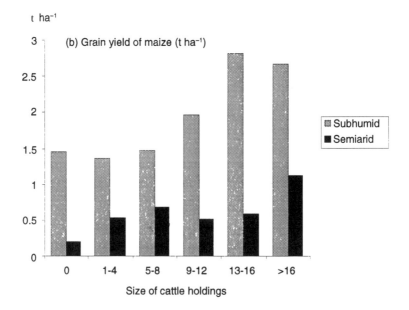

Fig. 3.2. Effect of size of cattle holdings on land in maize and maize grain yields in two communal areas of Zimbabwe (adapted from Steinfeld, 1988).

Table 3.4. Annual utilization rates (% of grazing time) and habitat preferences of cattle in five communal villages in Zimbabwe.

Location (natural region)[1]	Cropland		Homesteads		Valleys		Cattle (head km⁻¹)
	% area	% grazing time	% area	% grazing time	% area	% grazing time	
Chamatamba (II)	14	5	3	21	3	20	32
Mutakwa (III/IV)	28	8	8	12	14	45	65
Mangenzi (IV)	38	19	4	12	4	21	26
Mazvihwa (IV)	20	9	9	8	16	68	23
Mazuihwa (IV)	12	17	1	3	6	48	23
Mean	22	12	5	11	9	40	34

[1] Natural regions: II = subhumid zone; III = intermediate; IV = semiarid.
Sources: Locations 1, 2 and 3 from Cousins (1992); 4 and 5 from Scoones (1989); stocking rates of locations 4 and 5 from Scoones (1993).

gradients (1000-2000 m), giving rise to considerable variability in the potential of the land. As a result, population densities range from 300-400 inhabitants km⁻² in high-altitude areas to 20-50 km⁻² in low-altitude areas. People have been moving down into the relatively underpopulated drier areas, resulting in population growth rates of 11% per year compared to 2.5% in the high- and 3.2% in the medium-potential zones.

Two major farming systems have developed: a high-intensity system relying on coffee, vegetables, fruits trees and dairying with crossbred cows, and a low-input mixed crop/livestock system in which maize and pulses are the main crops. Due to the rapid colonization of semiarid land, food production has kept pace with rural population growth, the cultivated area having remained at 0.27 ha capita⁻¹ since 1970 (Mbogoh, 1991). In contrast, livestock wealth has steadily declined, falling from 0.43 TLU capita⁻¹ in 1974 to 0.26 TLU capita⁻¹ in 1989. Both cattle and small ruminant populations have remained constant over the long term, though falling and rising during relatively dry and wet periods (Ackello-Ogutu, 1991). Since the area of land under food crops nearly doubled during the period 1970-1988, the ratio of cropland to TLUs rose from 0.31 ha TLU⁻¹ in 1970-1974 to 0.55 ha TLU⁻¹ in

1985-1988. Over 95% of the land used for food crops was devoted to maize and beans, with sorghum, millet, cassava and banana occupying the remainder. However, the maize-beans mixture became more important over time, with land planted to maize increasing by 10% and to beans by 15% annually. Average edible CRs for two seasons were estimated at 2.5 t DM ha^{-1} for maize (de Leeuw *et al.*, 1990) and 1 t DM ha^{-1} for beans and other pulses (de Leeuw and Nyambaka, 1995). Potential feed availability therefore increased from 0.6 to 1.0 t DM TLU^{-1} year^{-1}. Due to the higher proportion of pulse residue, feed quality has improved substantially.

When farm size and ecozone are taken into account, a different picture emerges. With increasing farm size, the proportion of cropped land diminishes as there is usually a labour constraint for weeding. Consequently, farmers rarely cultivate more than 4 ha of land in each season. Hence a farmer with 2 ha on average cropped 65% of his or her land, compared with only 40% for a farmer with 10 ha. Reliance on crop residues therefore increases as farm size falls, the more so as stocking rates show a similar trend, rising from 0.8 TLU ha^{-1} for a 10-ha farm to 2.0 TLU ha^{-1} for a 2-ha farm (Fig. 3.3). As maize-pulse intercrops and rangelands each produce about 3.5 t DM ha^{-1} year^{-1} (de Leeuw *et al.*, 1990), the potential feed available decreased from 4.0 to 1.7 t TLU^{-1} year^{-1}, suggesting that the farmer with only 2 ha of land barely covered two-thirds of the feed needs of his or her livestock (Fernández-Rivera *et al.*, 1995). Such farmers are forced to exploit their CRs to the full, to overstock their 0.4 ha of rangeland, to herd their cattle along roadsides and on wasteland, to rent grazing land from other farmers or, as a last resort, to purchase feed. At village level, livestock:cropland ratios increased with annual rainfall and human population density. Aggregate stocking rates in two locations with 160 people km^{-2} averaged 3.2 TLU ha^{-1}. This fell to around 1.4-1.6 TLU ha^{-1} in drier locations with 50-60 people km^{-2} (de Leeuw, 1988).

Ethiopian Highlands

The conversion of grazing land into cropland is likely to boost the importance of CRs as livestock feed. This is especially true in the highlands of Ethiopia, where oxen are commonly used for ploughing and equines for transport. The highlands cover some 425 000 km^2 and have a human population of 25 million. This population owns around 20 million TLU, composed of 19 million cattle, 28 million small ruminants and 6 million donkeys and horses (Jahnke, 1982; Gryseels, 1988). While stocking rates average about 50 TLU km^{-2}, there are many areas with densities of more than 100 TLU km^{-2}. As a result, fallows have all but disappeared and private and communal pastures have become a small fraction of the total land (Table 3.5). Residues from cereal crops and pulses combined with post-harvest stubble grazing account for over 90% of all feed.

At higher altitudes (e.g. Debre Berhan in Table 3.5), stocking rates remain high even though less than half the land is cropped and livestock derive over half their feed from pasture and fallow grazing and from hay (Gryseels, 1988). In high-density

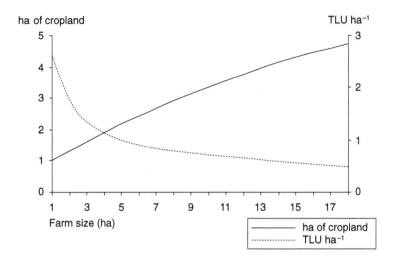

Fig. 3.3. Effect of farm size on land cropped and stocking rate (TLU ha⁻¹) in Eastern Province, Kenya (adapted from Jaetzold and Schmidt, 1983).

Table 3.5. Characteristics of the feed supply in crop/livestock systems in the highlands of Ethiopia.

Locations	ILCA sites [1]		Ada District [2]		Debre Berhan [3]	
Farm size (ha)	2.3		3.5		6.2	
Stocking rate (TLU ha⁻¹)	1.5		1.0		1.3	
Feed (t DM TLU⁻¹)	0.97		1.57		0.93	
Feed (t DM ha⁻¹)	1.50		1.62		1.16	
Feed sources	Land (%)	Yield (%)	Land (%)	Yield (%)	Land (%)	Yield (%)
Crop residues	87	42	90	76	45	37
Stubble and weeds	87	42	90	18	-	-
Fallows and pastures	13	16	10	6	55	63

[1] Averaged data for three sites (Dogollo, Ginchi, Inewari). For details see Asamenew (1991).
[2] Average data for Ada District, adapted from Shiferaw (1991).
[3] Adapted from Gryseels (1988).

areas, keeping sufficient oxen for soil tillage claims first priority. As herds are small, 70% of farmers possess fewer than five head of cattle, of which 50% may be oxen. Where, as at Debre Berhan, the proportion of cropped land is relatively low, oxen are less numerous (20% of the herd), allowing intensive milk production from cross-bred cows (Gryseels, 1988). Similarly, in Selale District, where only 40% of the land is cultivated, 40% of the feed consists of stored hay but only 12% is straw from crops. Grazing, mostly during the rains, accounts for 35% of total feed. Although some cut-and-carry green oats and grass are used, increased milk output from cross-bred cows is generated mainly by feeding concentrates (Varvikko, 1991).

Rwanda and Western Kenya

Very high densities of rural people are also found in the "Great Lakes" region of Rwanda, Western Kenya and Burundi. In Rwanda, there were 290 people km^{-2} in 1991. Since only 50% of the land can be cultivated and the rate of urbanization is low (8%), cropland has become increasingly scarce, amounting to only 0.18 ha per rural inhabitant (WRI, 1990; Rwanda, 1991). According to FAO (1993), the live-stock population in 1992 included 610 000 cattle, 395 000 sheep and 1 100 000 goats (577 000 TLU). Over the past decade, livestock mass has declined by 17% and cattle numbers have fallen by 26%, but sheep and goat populations have risen by 30 and 27% respectively. The current stocking rate is 0.23 TLU ha^{-1} for all land, but only 0.52 TLU ha^{-1} of cropland.

Excellent statistics on land use and cropping patterns (Rwanda, 1991), combined with analyses of total and edible residue yields of the major crops, allow estimates to be made of the feed budget of the entire Rwandan farming sector. About 65% of farmland is under crops, of which 83% is devoted to crops that produce residues for feeding livestock (Table 3.6). Bananas are the dominant crop, followed by cereals. Together these provide 70% of the roughage and occupy 42% of the cropped land (Table 3.7). High-quality feed is produced from beans, sweet potato and cassava (leaves, vines and peelings, the latter as leftovers from processing). Despite the high population pressure, 35% of the land within farms remains un-cultivated, consisting of pastures, fallows and woodlots. These provide 45% of the total feed supply. Total feed output averages about 1 t DM ha^{-1}, totalling 1.2 million t DM $year^{-1}$. This is equivalent to 2.1 t TLU^{-1} $year^{-1}$ or 5.7 kg DM day^{-1} (Table 3.8).

Considering that 50% of the country remains outside the farming cycle (WRI, 1990), it would appear that Rwanda is not overstocked. However, further fragmen-tation of farms will diminish feed output, as pastures become cropland and fallows disappear. Already, the upper quartile of farmers defined in terms of farm size (>1.8 ha) crop 60% of their land, whereas the bottom quartile (< 0.4 ha) cultivate 78%. As a result, only 27% of farmers own cattle, whereas 53% still keep goats (Rwanda, 1991). This skewed distribution of livestock wealth may imply that the feed theo-retically available on stockless farms may not be available in the "real world", be-cause markets and exchange mechanisms are poorly developed.

Table 3.6. Average cropping pattern and residue yields in Rwanda, 1989.

Crop	Area cultivated (%)	Edible portion	Yield (t DM ha⁻¹ year⁻¹) Total	Edible	%[1]
Banana	26.2	Stems, leaves, skins	3.70	1.30	41
Cassava	8.5	Leaves, peel	1.10	0.55	5
Sweet potato	10.4	Vines, peel	2.10	1.05	13
Maize	7.8	Stover, sweepings	1.50	0.60	11
Sorghum	8.5	Stover, dregs	4.10	0.94	19
Phaseolus bean	22.0	Leaves, pods	0.40	0.20	11
Coffee	6.1	-	-	-	-
Other	10.5	-	-	-	-

[1]As a percentage of a total edible feed yield of 651 500 t DM (= 100).
Sources: Rwanda (1987, 1991).

In western Kenya, feed budgeting was used as a tool in the development of on-farm feed resources in a project to introduce dual-purpose dairy goats into the very small, resource-poor farms of this densely populated region (Semenye and Hutchcroft, 1992). Emphasis was on fully utilizing the potential of maize, pulses and cassava as both food and fodder crops. The output of feed per hectare was strongly influenced by altitude and soil fertility, maize grain and residue yields being highest at Maseno,

Table 3.7. Major sources of crop residues in Rwanda.

	Banana	Roots and tubers	Cereals	Grain legumes	Total (%)
Land (%)	31	23	20	26	100
CR yield (%)	41	19	30	10	100
Amount (t DM ha⁻¹year⁻¹)	1.30	0.82	1.55	0.40	-

Table 3.8. Total livestock feed production from farmland in Rwanda, 1989.

Land use	Area ('000 ha)	Feed (t DM ha^{-1}year^{-1})	Total feed ('000 t DM)
Cropped land	776	0.84	651.5
Pastures or fallows	233	2.00	466.8
Woodlots	115	0.50	30.9
Other (compounds)	62	0.50	30.9
Total	1186	-	1206.8

Sources: Table 3.7 and Rwanda (1991).

which has deep fertile nitrosols in contrast to other sites where soils are sandier (Table 3.9). Among the pulses, pigeonpea is grown for its high yield, ability to produce a ratoon crop and production of green leaf or pea through opportunistic harvesting when other foods are scarce. Onim *et al.* (1986) confirmed the high grain and residue yields obtained from tall (up to 3.5 m) local double-cobbing maize varieties at Maseno, which produce 3.9 t ha^{-1} of grain in the short and 6.3 t ha^{-1} in the long rains.

In addition to the high residue yield obtained after grain harvest (Table 3.9), green feed can also be produced during the growing season: two thinnings of excess plants early in the season can realize 1 t DM ha^{-1}, while herbage collected during weeding may yield a further 0.5 t DM ha^{-1}. Later in the season, leaf stripping (removal of one leaf per week) and cutting off the top part of each plant (above the cob) during the hardening grain stage can generate 0.5 and 2.0 t ha^{-1} of dry feed respectively (Onim *et al.*, 1992). Although these practices reduce the feed value of the post-harvest stover, they help maintain a year-round feed supply, especially since feed from natural pastures and roadsides is very limited.

Grain, Stover and Livestock Feed

Millet

Pearl millet is the principal grain crop in the dry semiarid zone (LGP 75-120 days), comprising 65% of total output of seven West African Sahelian countries and exceeding 75% in Mali, Chad and Niger (World Bank, 1989). Grain yields are low, ranging from 200 kg ha^{-1} (McIntire *et al.*, 1992) to 800 kg ha^{-1} (Williams *et al.*, 1995), and variable across years due to erratic rainfall. On-farm residue yields are mostly in the range of 1 to 2 t DM ha^{-1}.

The potential for increases in grain and residue yields appear quite promising, as shown by the effects of incorporating crop residues, manuring and applying inorganic fertilizers, especially when these inputs are combined and in years when rainfall is adequate (Bationo *et al.*, 1995). While control yields averaged 189 kg ha^{-1}, the use of inorganic fertilizers combined with CR mulching produced 1108 kg of grain. The CR yield rose from 1.3 to 3.8 t DM ha^{-1}, giving CR:grain ratios of 6.9 and 3.4 respectively. Responses to manure were variable, averaging 54 kg of grain (range 22-169) and 0.22 t of CR t^{-1} of manure applied (Williams *et al.*, 1995). These responses indicated that modest increments in inputs can have relatively large effects on millet CR production (Fig. 3.4).

Powell and Fussell (1993) partitioned the stover of a harvested millet crop yielding 500 kg ha^{-1} of grain into upper parts (including threshing residues, upper stems and grainless tillers) and lower stems. Of the upper parts, 1.1 t (40% of the total biomass) were destined as livestock feed, containing 7.8% crude protein and having a digestibility coefficient of 50.4%. The lower stems (a further 1.1. t ha^{-1}) were left in the field to improve soil conservation and organic matter. Together with

Table 3.9. Grain and fodder yields (t ha^{-1}) at three sites in high- and medium-potential areas of western Kenya.

	Maseno		Kaimosi		Masumbi	
Altitude (m)	1680		1815		1200	
Rainfall (mm)	980		1015		1110	
	Grain	Fodder	Grain	Fodder	Grain	Fodder
Pure crops:						
Maize[1]	9.3	10.4	4.8	6.3	2.4	3.2
Phaseolus bean	1.4	0.6	-	-	0.4	0.5
Pigeonpea	-	5.4	-	4.8	-	5.6
Intercrops:						
Maize+	9.3	14.0	5.0	5.0	1.6	3.1
Phaseolus bean	0.8	0.3	-	-	0.2	0.3
Maize+	10.5	13.8	5.7	4.6	1.6	2.9
Pigeonpea	-	3.6	-	2.8	-	7.6

[1] Plants population of maize: 37 000 ha^{-1}; fertilizer: 40 kg P$_2$O$_5$ ha^{-1} as triple superphosphate.
Sources: Mathuva *et al.* (1986); Semenye and Hutchcroft (1992).

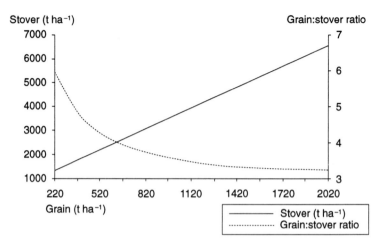

Fig. 3.4. Relationships between grain and stover yields and grain:stover ratios of millet in the semiarid zone of Niger (Bationo *et al.*, 1995; Williams *et al.*, 1995).

115 kg ha⁻¹ of bran produced in milling, total livestock feed amounted to 1.23 t DM ha⁻¹ or 45% of the total biomass of 2.75 t DM ha⁻¹.

Sorghum

In subhumid Nigeria, sorghum CR yields increased from 4 to 9.5 t DM ha⁻¹ when grain yield rose from 1 to 3 t ha⁻¹, with corresponding leaf fractions of 1 and 2 t DM ha⁻¹ respectively (Powell, 1984, 1986). This is in line with the edible yields of sorghum CRs recorded in the semiarid zone of northern Nigeria by van Raay and de Leeuw (1970) (see Table 3.2). When CR yields rose from 2.5 to 8.7 t DM ha⁻¹, grazeable forage rose much less, from 1.5 to only 2.1 t ha⁻¹.

Although CR:grain ratios decline with rising grain yields, CR yields usually do increase, albeit at a much lower rate than grain. For example, McIntire *et al.* (1988) determined grain and stover yields in 78 samples of brown, red and white sorghum varieties harvested from different fields in several highland locations in Ethiopia. The grain yields ranged from 1 to 7 t ha⁻¹, although most (82%) of the samples yielded between 2 and 6 t grain ha⁻¹. While CR:grain ratios fell from 2.3 to 1.2, actual stover yield increased from 5.7 to 7.0 t DM ha⁻¹. According to Osafo *et al.* (1993), the proportion of leaf and sheath increased at the same rate, averaging 45% of the total stover (Fig. 3.5). These ratios were much higher in tall West African sorghums: 4.2 at 1 t ha⁻¹ of grain and 4.7 at 3 t ha⁻¹ (Powell, 1984). Figure 3.6 shows the results of on-farm experimentation in subhumid Nigeria, in which the effect of stylosanthes intercropping and stylo-enriched fallow on subsequent local sorghum yields was tested (Tarawali and Mohammed-Saleem, 1994). The relationship between grain and stover yields is shown over a range of grain yields from 0.3 to 2.2 t ha⁻¹. Accompanying stover yields varied from 1.6 to 8.8 t DM ha⁻¹.

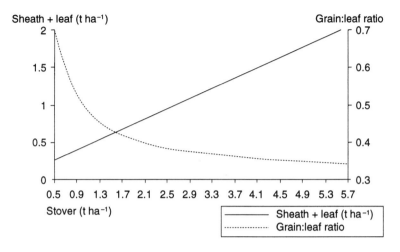

Fig. 3.5. Relationship between total stover and sheath plus leaf yields and grain:leaf ratios in Ethiopian sorghum varieties (Osafo *et al.*, 1993).

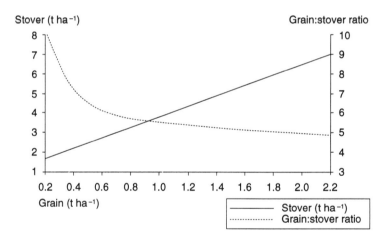

Fig. 3.6. Relationships between grain and stover yield and grain:stover ratios in sorghum in the subhumid zone of Nigeria (Tarawali and Mohammed-Saleem, 1994).

Diet selectivity of grazing cattle has been assessed in grazing behaviour studies (see Table 3.2) and in feeding trials. Powell (1986) counted bites of zebu cattle grazing CRs of sorghum and millet in subhumid Nigeria and found that at the start of the CR grazing season, 90% of the selected forage was leaves, declining to 60-70% in late March. Stems were initially avoided, but gradually increased to up to 20% in the diet for sorghum and 30% for millet; over the entire 4 months, 60 and 80% respectively of the CR diet consisted of leaves (Powell and Bayer, 1985).

In feeding trials with Ethiopian sorghum varieties containing 38% leaves, steers offered 10 kg of forage daily ingested only 4.7 kg, most of which was leaf (Reed *et al.*, 1988). Sheep reacted similarly when offered sorghum stover containing 32% leaves and sheaths at 25, 50 and 75% above their recorded intake level. At all levels, around 97% of the offered leaves were consumed. At low and high levels of offer, however, 98 and 69% of the sheaths and 86 and 26% of the stems were selected respectively (Aboud *et al.*, 1994). This clearly demonstrated the sheep's preference for leaves.

Maize

As shown above, maize residues are of much greater importance in eastern and southern than in West Africa. In the latter region, it is only since the late 1970s that maize has begun to replace sorghum, mainly in the subhumid zone. Because short-duration varieties are preferred and these are usually intercropped with later-maturing crops, residues may be inaccessible to livestock and hence lost as feed (Powell, 1986; Jabbar, 1993; Jabbar *et al.*, 1995).

Grain yields in eastern and southern Africa vary from below 500 kg ha^{-1} to as high as 10 t ha^{-1} (see Table 3.9), depending on ecozone, variety, planting density, fertilizer application and overall management. Low-input maize yielded from 0.2 to 1.1 t ha^{-1} in the semiarid zone and from 1.4 to 2.9 t ha^{-1} at higher rainfall levels in Zimbabwe (Fig. 3.3). In bimodal rainfall areas in semiarid Kenya, average yields approached 1 t ha^{-1} per season, yielding on average 2.7 t of CR, half of which was consumable by cattle. At grain yields between 1 and 2 t ha^{-1}, the inedible portion changed little, consisting of 40% stems and 20% empty cobs (de Leeuw and Nyambaka, 1995). In the subhumid coastal plain of Kenya, on-station grain yields ranged from 0.2 t ha^{-1} (when short rains were poor) to 4.9 t ha^{-1} during adequate long rains, with a corresponding decline in CR:grain ratios from 2.6 to 0.8 (Fig. 3.7). Powell (1984), in Nigeria, showed that, from 1 to 3 t ha^{-1} of grain, leaf yield increased from 0.7 to 1.3 t DM ha^{-1}, representing 31-32% of the total CR.

For grain yields of 3.2 to 5.4 t ha^{-1} in the Kenyan highlands, CR:grain ratios fell slightly from 0.99 to 0.85. Crop residues averaged 3.8 t ha^{-1}, 70% or 2.7 t ha^{-1} of which were removed by the farmer for feed and bedding for his or her dairy cows (Ransom *et al.*, 1995). A similar range of ratios is shown in Table 3.6 where, for grain yields of 4 t ha^{-1} or more, the CR:grain ratio is close to unity, rising to 2.0 when grain yields fall to 1.6 t ha^{-1} with decreasing rainfall and soil fertility.

The Cool Tropics

Of the 13.7 million t of CRs produced as DM in Ethiopia, wheat, barley and teff, the typical highland crops, accounted for 6, 10 and 17% of the total respectively. The residues from maize and from sorghum/pearl millet growing in the mid- to low-

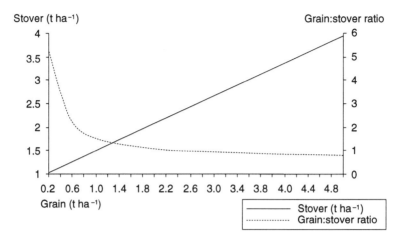

Fig. 3.7. Relationships between grain and stover yields and grain:stover ratios of maize in subhumid coastal Kenya (Mureithi *et al.*, 1994).

altitude zones appeared more important (39 and 36% of the total), probably because high CR:grain ratios were assumed (3.0 for maize, 5.0 for sorghum; Kossila, 1988; Nordblom and Shomo, 1995). Residues from faba bean, other pulses and cotton contributed the remaining 2%.

At medium altitudes in the highlands (1500-2000 m), teff and wheat are the major cereals, covering up to 85% of the cultivated areas. Various pulses make up the balance. The average cereal CR yield is about 2 t DM ha⁻¹, 70% of which is edible. Pulses produce less (0.8-1.5 t ha⁻¹) (McIntire *et al.*, 1988). When fed to livestock, these residues are mixed with cereal straw. However, they are often used as fuel or as brooms for sweeping.

At higher altitudes, barley becomes the major cereal, yielding 1 t ha⁻¹ of straw, while reported yields of wheat and pulses are lower (Asamanew, 1991). However, since wheat is considered both a subsistence and a cash crop, research and extension to increase yield have been intensified since the late 1970s. Whereas grain yields of local varieties are about 1 t ha⁻¹, yields of up to 4 t ha⁻¹ can be achieved with improved varieties, earlier planting, better weeding and higher fertilizer applications. Demonstration trials with a recommended package achieved yields of 2.6 t ha⁻¹ for two new varieties, compared to 1.5 t ha⁻¹ for the local control (Adugna *et al.*, 1991). Yields of straw increased proportionally (Fig. 3.8), but its quality as animal feed differed between varieties, according to farmers (Hailu and Chilot, 1992). Feed quality is likely to be related to the proportion of leaf, sheaths and stems in the straw. Apparently, stem content in exotic varieties in Ethiopia is higher than in local ones (63% compared with 53%, according to Jutzi and Mohamed-Saleem, 1992). This is also higher than in varieties in the Near East (38-48% stem according to Capper *et al.*, 1988), and in Australia (58% stem, according to Pearce *et al.*, 1988). However, these tall stemmy varieties are appreciated for roofing and sometimes as fuel.

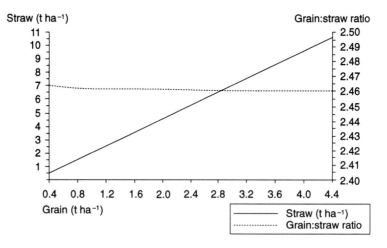

Fig. 3.8. Relationships between wheat grain and stover yields and grain:stover ratios in the Ethiopian highlands (adapted from Asamenew, 1991).

Other Crops

Three other major sources of residues have been briefly mentioned above: (i) haulms and hays from grain legumes, in particular groundnut and cowpea; (ii) foliage, peels and other by-products of root and tuber crops such as cassava, sweet potato and yams; and (iii) cotton residues.

Relatively high yields of grain for groundnuts and pulses (mostly cowpea) are reported in the Sahelian countries (Table 3.1). These can be used as proxies for haulm yield and its importance as a livestock feed. In Nigeria, Powell (1986) recorded kernel yields of 300-1220 kg ha^{-1} producing 0.7-1.5 t DM of haulms. In Mali, kernel yields were similar (0.7-1.0 t ha^{-1}), with haulm yields ranging from 0.4 to 0.7 t DM ha^{-1}. Cowpea produces less than groundnut, especially when spreading varieties are used. Forage yields of 0.4-1.2 t ha^{-1} were reported in Mali (Leloup and Traore, 1989, 1991) and of 0.2-0.5 t ha^{-1} in Burkina Faso, for plant densities of 5000-20 000 plants ha^{-1} when intercropped with sorghum (Stoop, 1986).

Much higher yields of legume forage were recorded in the densely populated area around Kano, northern Nigeria, where crop residues supply up to 80% of livestock feed (Mortimore, 1991). Haulms from cowpea and groundnut produced 1.5 t DM ha^{-1}, accounting for 30% of the total roughage supply and 75% of the total crude protein output (Hendy, 1977). Cropland is heavily manured, which may account for the high hay yields (de Leeuw *et al.*, 1995b).

The potential of cassava-based diets for ruminants, pigs, poultry and rabbits has been reviewed by several authors. In most experimental diets, cassava flower was used as an energy source and a substitute for feed grain, while cassava peels were tried in goats' diets (Ifut, 1992). Smith (1992) studied the inclusion of cassava foliage in cattle diets, while sweet potato vines have been intensively tested in western Kenya as ingredients in the diets of dual-purpose goats (Semenye and Hutchcroft, 1992).

During the past 25 years, cassava production in sub-Saharan Africa has doubled to 25 million t, of which half is produced in Nigeria and Zaire. Over the same period sweet potato production rose by 90% to 6.4 million t, most of which is grown in eastern and Central Africa (Scott and Suarez, 1992), with explosive expansion in Uganda (218%) and Kenya (287%) (Scott and Ewell, 1992). In view of the existing tradition of supplementary feeding of root crops to small ruminants in humid West Africa (Table 3.3), there can be little doubt that the utilization of root crops will expand still further in the future, given the rapidly rising urban demand for animal products, especially in the coastal region.

In 1987, cotton was an important cash crop in at least 20 African countries and seed production in tropical Africa amounted to 2.7 million t. Half of this total was produced in nine West African countries, with Mali and Côte d'Ivoire as the major producers. The balance was grown in eastern and southern Africa, with Sudan (0.52 million t) and Zimbabwe (0.3 million t) the biggest suppliers (World Bank, 1989).

Little information is available on post-harvest cotton residues. Usually they remain in the field, at least until the second half of the cropping season. Harvesting is a gradual process, often taking several weeks to complete. Hence livestock access is delayed. Residues recorded in northern Nigeria in February averaged 280 kg DM ha^{-1}, 360 kg for high- and 200 kg for low-yielding fields (van Raay and de Leeuw, 1970). Cotton residues supplied 22 and 17% of the total available residues in January and February respectively (Table 3.2).

In Mali, edible cotton residues fell from 0.7 t DM ha^{-1} when rainfall was high to 0.3 t DM ha^{-1} when it was low (Leloup and Traore, 1989, 1991). These yields also reflected the higher input levels (fertilizer, pesticides, ox ploughing) used in Mali during the late 1980s, compared with those used in northern Nigeria in 1970. In on-station trials in Nigeria, much higher residue yields were obtained in research plots than in farmers' fields, due to the higher level of inputs used. Medium-input plots produced 0.8-1.6 t DM ha^{-1}, whereas high-input plots yielded up to 3.7 t DM ha^{-1}, 50% of which was leaf litter and 36% husks left after the picking of cotton bolls. Feeding trials showed that most of this material was acceptable to sheep.

Access, Conflicts and Trade

Scientists making feed budgets, whether for countries, districts or villages, tend to assume that feeds are readily accessible to livestock irrespective of who crops or owns the land. Free dry-season access to all unfenced land, be it natural pastures, fallows or cropped land after harvest, prevailed in most of West Africa during the 1970s (e.g. van Raay and de Leeuw, 1970, 1974; Toulmin, 1992). Access has become more limited in recent years, being subject to transactions in which crop residue grazing is exchanged for manure (Powell, 1986) and/or milk (van Raay and de Leeuw,

1970). In Mali, wealthy settled farmers have dug wells to attract transhumant Fulani herdsmen, whose cattle manure fields close to their homesteads. These fields are cropped with millet varieties selected for their high response to heavy manure inputs (Toulmin, 1992).

In Nigeria, at the wetter end of the subhumid zone, problems of access are common, giving rise to conflicts between crop farmers and livestock owners that may be exacerbated by ethnic differences. As in northern Côte d'Ivoire and southern Mali, two groups of people share the same habitat: semi-settled Fulani pastoralists, with herds averaging 50 cattle, farm on average 3.5 ha of land, while smallholder Yoruba farmers, owning few animals, each cultivate about 5 ha of land (Jabbar *et al.,* 1995). The CRs of the latter group are in high demand. However, with an almost continuous growing season of 9-10 months, fields are rarely without crops. Maize is usually double-cropped, both crops being mixed with long-duration yams, cassava or sorghum. Eyoh (1992) points out that the market value of these mixtures is at least 50% higher than that of sole-cropped grains or yams. A high market value increases the likelihood of conflicts.

In Côte d'Ivoire, the scene is set for similar conflicts, again arising between transhumant Fulani herdsmen and settled farmers, this time the Senoufo. Fulani herds, which are often large (60-80 head), move deeply into the country's humid south during the dry season. They are herded by a single hired herder who is underpaid and overworked—typically a migrant from the north engaged on a short-term insecure contract (Bassett, 1994). Most water sources for cattle are located in valleys and are difficult to access as they are surrounded by fields of wet rice belonging to Senoufo farmers (Landais, 1983). These fields are not harvested until late in the dry season. Herds trespassing on rice fields accounted for a quarter of all damages recorded through interviews with farmers. Similarly, cotton, having been interplanted with earlier-maturing crops, is picked as late as February and accounted for a further 40% of recorded damage claims (Bassett, 1994).

Conflicts may also arise in purely settled communities where there are large differences in livestock wealth. In central Mali, relatively wealthy farmers (with several oxen and donkey carts) may possess 25 TLU and cultivate up to 17 ha of cropland. They live amidst poor farmers with few livestock, who crop only 4-10 ha (Bonnet, 1988). Disparity in livestock wealth was even more extreme in the cotton belt of northern Côte d'Ivoire. Here overall livestock populations were as high as 0.7 TLU ha^{-1} of cropped land, but 20% of the farmers owned an average of 14 TLU, maintained at 2.1 TLU ha^{-1}, while the other 80% of farmers had few animals but surplus CRs, since their stocking rate was only 0.12 TLU ha^{-1} of cropped land (Schuetterle and Coulibaly, 1988). A similar situation was found in The Gambia where, in a sample of 82 farmers, 16% owned on average 48 TLU, 42% owned 6 and 42% only 2. The CRs available per TLU showed a reverse trend, increasing from 0.55 to 3.65 t TLU^{-1} with decreasing livestock wealth (Tanner *et al.,* 1993). In such situations conflicts may be inevitable, especially when herders are withdrawn and animals are allowed to roam freely across all unfenced land as soon as most crops have been harvested.

In eastern Africa, conflicts arising from crop damage appear to be infrequent. However, stock raiding, sometimes combined with murder and pillage, are more common. In northeast Kenya, raiding occurs between primarily pastoral tribes (e.g. the Pokot versus the Turkana; see Dietz, 1987). Elsewhere, it occurs between pastoral and agropastoral groups (e.g. the Datoga/Barbaig versus sedentary Sukuma crop/ livestock farmers and other groups in west Tanzania). During 1981-1985, an estimated 151 000 cattle were stolen in such raids, many of them being subsequently marketed illegally in neighbouring countries (Ndagala, 1991).

The restriction of access to CRs is a common component of intensification. It takes several forms. Active herding may continue during the dry season, so as to prevent crop damage and allow farmers to use their feed resources as they see fit. At the same time, manure contracts are more easily enforced and controlled when stock is herded. For instance, smallholder farmers in subhumid Nigeria invited specific Fulani herdsmen to allow their animals to manure only fields selected for growing ginger as a cash crop during the next growing season (Powell, 1986). Monocropping is becoming more common, partly to facilitate CR grazing at the optimum time. This is advantageous where there are long rainy seasons allowing two maize crops, as in the humid zone of Nigeria (Jabbar, 1993).

In the Ethiopian highlands, hay and straws are stored in large stacks in farm compounds, facilitating exchanges between livestock-rich and -poor farmers. Farmers renting or borrowing draught oxen are obliged to feed them. Demand for livestock feed varies between seasons and years. Asamanew (1991) showed that the market prices of straw and hay in 1988 rose steeply between the time of harvest and the next growing season, from US\$ 40 t^{-1} to US\$ 130 t^{-1} for straws and from US\$ 48 t^{-1} to US\$ 105 t^{-1} for grass hay. The market price for teff grain at the time was US\$ 500 t^{-1}, while for sorghum and wheat it was US\$ 300 t^{-1} (Webb *et al.*, 1992). (These prices reflect an overvalued exchange rate of EB 2.10 = US\$ 1.00.) This shows that storing straw until the next growing season can be a substantial source of income for stockless farmers. McIntire *et al.* (1988) showed a similar trend for sorghum during the drought period of 1984/85. Ratios between CR and grain values were 0.25 at harvest but doubled to 0.50 during the following growing season. In more normal years, the ratio was 0.35. Thus, if the harvest were 2.2 t ha^{-1} of grain and 3.3 t ha^{-1} of CRs, the value of the CRs would be 30% of that of the total crop. A similar calculation for millet in Niger indicated that, because of low grain yield (0.22 t ha^{-1}), residues accounted for more than 50% of the total market value, even though the CR:grain price ratio was only 0.24 (McIntire *et al.*, 1992).

Transporting feed to large urban centres is common in Niger (Niamey) and in Mali (Bamako, Niono). In northern Nigeria, bundles of groundnut and cowpea hay are offered for sale along major highways. As mentioned earlier, in the densely populated Machakos District of Kenya, grazing land is often rented (Tiffen *et al.*, 1994). In western Kenya, mixed bundles of fresh feed (2-4 kg DM) are offered for sale consisting of Napier grass, roadside grasses, green maize stover (residues from green cob consumption) and sugarcane tops. Prices ranged from US\$ 0.06 to 0.14 kg^{-1} DM (Onim *et al.*, 1987).

Discussion

The importance of CRs as animal feeds is mostly related to local circumstances. It depends greatly on the supply and demand situation in time and space of individual farmers within their communities. The proportion of cropped land determines the extent of reliance on natural pastures as an alternative source of feed, their availability being subject to local rules of access and user rights (IFAD, 1995). As the edible yields of animal feed from cropped and uncropped land are broadly similar, total feed supplies are not greatly affected. However, increased cropping may encroach on the better natural pastures, such as those on lower slopes and in valleys (Vierich and Stoop, 1990), reducing the common feed pool.

The privatization of CRs will intensify as the links between privately owned livestock and cropland are strengthened, or in other words as farmers tend to feed their CRs to their own livestock. This fits in with a broader trend: livestock holdings in Africa over the past 25 years have shifted from mobile transhumant pastoralists to sedentary farmers, absentee civil servants and traders (Jabbar *et al.*, 1995).

Total feed demand depends on the overall local stocking rate, but the ratio of supply to demand varies across seasons and years as well as between individual farms. Variability in ratios between farms is greatest where communal grazing land is scarce, as for example in the Ethiopian highlands, Rwanda and Kenya. Where access and use of feed is entirely farmer-controlled, benefits from intensification of the crop subsystem can translate directly into higher livestock output. As I have shown, modest increments in input levels can double the output of cereal CRs, which, having full control, the farmer can manipulate and manage so as to increase feed supplies and effective use by livestock. When farmers perceive the true value of their CRs they often reassess them as a marketable commodity and start to engage in trade in feeds. This allows stock-poor producers to extract added value when intensifying their own cropping enterprises. Orderly market transactions will help reduce the risk of conflict in mixed farming systems as damage assessment becomes less arbitrary and compensation costs rise. The wealthier transhumant pastoralists of West Africa may find it more economic to pay for temporary grazing rights on post-harvest CRs than to trespass and raid—activities that risk prohibitive fines from rent-seeking police and court officials.

The marketing of feedstuffs may also have a profound impact on the largely untapped potential of the feed processing industry in cassava, sweet potatoes and yams. Women are at the centre of this activity, functioning as the producers, marketers and processors of these crops in addition to owning and managing livestock. Both they and their animals stand to benefit greatly from the growth in this subsector that is likely to occur as marketable surpluses are created by new technology.

The use of CRs as livestock feed has an impact on the environment, contributing to the depletion of nutrients and organic matter and the risks of erosion in farmers' fields (through the removal of plant cover). However, if cereal CRs can be partitioned into edible and non-edible portions and values can be attached to the feed and mulch components, management techniques could evolve that would encourage the dual-purpose use of CRs for both livestock feed production and soil fertility maintenance.

If the profitability of local livestock production rises, then high-quality livestock feeds such as groundnut and cottonseed cake could be sold to local producers instead of being exported, as they often are at present. Much of this export trade merely provides foreign exchange for the elite to spend on unnecessary luxuries. The benefits of retaining feeds for use in developing the domestic economy could be immense. Already, a large local feed industry has evolved in Kenya, because of strong demand from local producers.

Here as elsewhere, strong urban demand for milk and meat has encouraged specialized small- and large-scale dairy, poultry and pig production. In addition, declining urban incomes among certain groups have prompted a search for supplementary income from such sources as backyard livestock enterprises. These enterprises, operating in a (peri)-urban context, often cannot produce their own feed and depend on supplies brought in from the countryside. This gives rise to new transport and marketing channels, providing added value to crop producers within the peri-urban orbit. It also encourages the development of flexible small-scale feed mills.

Such small-scale livestock enterprises abound in the sprawling urban agglomerations of tropical Africa. In West Africa, several million sheep and goats are reared and sold to satisfy prescribed Muslim sacrifice rites (Kolff and Wilson, 1985). In greater Addis Ababa and Nairobi, thousands of cows in backyards provide milk directly to urban consumers. A daily stream of donkeys (to Addis Ababa), camels (to Niamey), horse- or donkey-drawn carts (to Bamako and Niono), and of bicycles, pick-ups, taxis and buses everywhere brings grass and legume hay, straw and browse into the cities.

The rising demand for milk and meat may encourage many more producers to seek a foothold in (peri-)urban markets. Depending on transport costs, producers within 50 km of urban centres may be able to deliver fresh milk to consumers, often using cooperative arrangements to market their produce. For instance, in Ethiopia the main road from Moyale, on the Kenya border, to Addis Ababa has encouraged trading centres to evolve, increasing local demand for milk. In addition, butter and draught oxen are supplied to the densely populated highlands and associated urban populations. While liquid milk is supplied by producers close to the bigger towns, butter is provided from more distant sources.

Developments in the production, utilization and marketing of CRs are and will be influenced by many—often interacting and conflicting—factors. Some of these are linked with overall trends in the world economy, such as fluctuations in grain prices, in particular of wheat and rice, of which tropical African countries are large net importers. The rising world market prices of wheat and rice might well produce a shift in demand towards locally produced grains (maize, sorghum and millet), which in turn would increase and intensify their production, thereby boosting the production of CRs. This trend could be reinforced by reduced donor-financed food imports for famine relief and resettlement programmes, and by further devaluations of local currencies brought about by structural adjustment.

Assuming that prices of imported milk and meat continue their upward trend, demand for local products will strengthen, increasing the need for locally produced feed resources, including CRs. Reduced subsidies on beef imports from the European Union, combined with the devaluation of the CFA franc, have already revitalized the livestock trade between the Sahelian producers of francophone West Africa and the subregion's coastal consumers. Continuing erosion of the Ghanaian and Nigerian currencies is encouraging similar trends there.

Given the elasticities of demand for livestock products, consumer demand in Africa is difficult to predict. If there is a upturn in the economy, demand will probably rise. However, this depends largely on who will benefit from such an upturn. It is widely believed that structural adjustment has widened the already severe gap between the region's rich and poor. Overall demand for livestock products could well decline if the proportions of urban and rural poor in the total population rise. Quite apart from the effects of structural adjustment, Africa's corrupt ruling elites and the poor quality of government in so many of the region's countries could well see to it that the long-promised recovery of African economies fails to materialize. There are signs of hope—but we cannot be certain that the future of Africa's beleagured economies will be brighter than their past.

Conclusions

The following major conclusions may be drawn:

1. The potential of CRs as a livestock feed increases with rising population density, while the demand for them depends on the livestock population density and the alternative functions of CRs in the farming system.

2. At the village level, stocking rates differ greatly between individual farmers. Stock-poor farmers may have excess feed, while stock-rich farmers, despite having more cropland and higher crop yields, may be short of feed. Where land tenure is individualized and oxen are used for tillage, stocking rates and the demand for feed rise with diminishing farm size.

3. In mixed cropping systems with long growing seasons, intercropping may restrict the access of livestock to CRs, such that the residues of early-maturing crops may decompose in situ without being grazed. Increased livestock holdings will stimulate monocropping or the intercropping of crops of similar cycle length.

4. In the future, rising demand for locally grown crops and for livestock products may lead to higher use of inputs, resulting in higher crop and CR yields, feed budgeting and the allocation of feeds to different classes of stock ranked according to their revenue-earning capacity. New cropping patterns may evolve that allocate larger shares of land to grain legumes and roots/tubers. This will diversify and enhance CR quality, a process that can be further promoted through the inclusion of by-products in feeds.

References

Aboud, A.A.O, Owen, E., Reed, J.D. and Said, A.N. (1994) Influence of amount of feed offered on growth, intake and selectivity: Observations on sheep and goats. In: Lebbie, S.H.B., Rey, R. and Irungu, E.K. (eds), *Small Ruminant Research and Development.* Proceedings of a Conference, 7-11 December 1992, Arusha, Tanzania. Technical Centre for Agricultural and Rural Cooperation (CTA) and International Livestock Centre for Africa (ILCA), Addis Ababa, Ethiopia, pp 157-162.

Ackello-Ogutu, C. (1991) Livestock production. In: Tiffen, M. (ed.), *Environmental Change and Dry Land Management in Machakos District, Kenya, 1930-90: Production Profile.* ODI Working Paper No. 55, Overseas Development Institute, London, UK, pp. 45-89.

Adugna, H., Workneh, N. and Bisrat, R. (1991) Technology transfer for wheat production in Ethiopia. In: Hailu Gebre-Mariam, Tanner, D.G. and Mengistu Hulluka (eds), *Wheat Research in Ethiopia: A Historical Perspective.* Institute of Agricultural Research (IAR) and Centro Internacional de Mejoramiento de Maiz y Trigo (CIMMYT), Addis Ababa, Ethiopia, pp. 277-299.

Asamenew, G. (1991) A Study of the Farming Systems of some Ethiopian Highland Vertisol Locations. Working Document. International Livestock Centre for Africa (ILCA), Addis Ababa, Ethiopia.

Bassett, T.J. (1994) Hired herders and herd management in Fulani pastoralism (northern Cote d'Ivoire). *Cahier d'Etudes Africaines* 34: 147-173.

Bationo, A., Buerkert, A., Sedego, M.P., Christianson, B.C. and Mokwunye, A.U. (1995) A critical review of crop-residue use as soil amendment in the West African semiarid tropics. In: Powell, J.M, Fernández-Rivera, S., Williams, T.O. and Renard, C. (eds), *Livestock and Sustainable Nutrient Cycling in Mixed Farming Systems in sub-Saharan Africa.* Proceedings of an International Conference, 22-26 November 1993, International Livestock Centre for Africa (ILCA), Addis Ababa, Ethiopia, vol. 2, pp. 305-322.

Bonnet, B. (1988) Etude de l'élevage dans le développement des zones cotonnières. PhD thesis, University of Montpellier. Centre de Coopération Internationale en Recherche Agronomique pour le Développement (CIRAD) and Institut d'Elevage et de Médecine Vététerinaire Tropicale (IEMVT), Montpellier, France.

Bosman, H.G. (1995) Productivity assessments in small ruminant improvement programmes: A case study of the West African dwarf goat. PhD thesis, State Agricultural University, Wageningen, The Netherlands.

Bosman, H.G. and Ayeni, A.O. (1993) Zootechnical assessment of innovations as adapted and adopted by the goat keeper. In: Ayeni, A.O. and Bosman, H.G. (eds), *Goat Production Systems in the Humid Tropics.* Proceedings of an International Workshop, 6-9 July 1992, Ife-Ife, Nigeria. Pudoc, Wageningen, The Netherlands, pp. 45-57.

Bosman, R., Bengaly, M. and Defoer, T. (1995) Pour un système durable de production au Mali-sud: Accroître le rôle des ruminants dans le maintien de la matière organique des sols. In: Powell, J.M, Fernández-Rivera, S., Williams, T.O. and Renard, C. (eds), *Livestock and Sustainable Nutrient Cycling in Mixed Farming Systems in sub-Saharan Africa.* Proceedings of an International Conference, 22-26 November 1993, International Livestock Centre for Africa (ILCA), Addis Ababa, Ethiopia, vol. 2, pp. 171-182.

Breman, H. and de Ridder, N. (eds) (1991) *Manuel sur les Pâturages des Pays Sahéliens.* Karthala, Paris, France.

Capper, B.S., Thompson, E.F. and Herbert, F. (1988) Genetic variation in feeding value of barley and wheat straw. In: Reed, J.D., Capper, B.S. and Neate, P.J.H. (eds), *Plant Breeding and the Nutritive Value of Crop Residues.* Proceedings of a Workshop, 7-10 December 1987, International Livestock Centre for Africa (ILCA), Addis Ababa, Ethiopia, pp. 177-193.

Clatworthy, J.N. (1987) Feed resources for small-scale livestock producers in Zimbabwe. In: Kategile, J.A., Said, A.N. and Dzowela, B.H.(eds), *Animal Feed Resources for Small-scale Livestock Producers.* Proceedings of the Second PANESA Workshop, 11-15 November 1985, International Livestock Centre for Africa (ILCA), Nairobi, Kenya, pp. 44-60.

Cousins, B. (1992) *Managing Communal Rangeland in Zimbabwe: Experiences and Lessons.* Commonwealth Secretariat, London, UK.

de Leeuw, P.N. (1988) Livestock and Fodder Resources of Smallholder Farms in Semiarid Eastern Kenya. Mimeo, International Livestock Centre for Africa (ILCA), Nairobi, Kenya.

de Leeuw, P.N. and Nyambaka, R. (1995) An assessment of maize residues as a source of feed for livestock in smallholder farms in semiarid Kenya. *Feed Resources Newsletter* 5 (2): 1-7. International Livestock Research Institute (ILRI), Nairobi, Kenya.

de Leeuw, P.N. and Reid, R. (1995) Impact of human activities and livestock on the African environment: An attempt to partition the pressure. In: Wilson, R.T., Ehui, S. and Mack, S. (eds), *Livestock Development Strategies for Low-income Countries.* Proceedings of the Joint FAO/ILRI Round Table, 27 February-2 March 1995, International Livestock Research Institute (ILRI), Addis Ababa, Ethiopia, pp. 29-39.

de Leeuw, P.N. and Rey, B. (1995) An analysis of current trends in the distribution patterns of ruminant livestock in sub-Saharan Africa. *World Animal Review* 83: 47-59.

de Leeuw, P.N., Dzowela, B.H. and Nyambaka, R. (1990) Budgeting and allocation of feed resources. In: *Proceedings of the First Joint Workshop of PANESA/ARNAB,* December 1988, Lilongwe, Malawi. International Livestock Centre for Africa (ILCA), Addis Ababa, Ethiopia, pp. 222-223.

de Leeuw, P.N., McDermott, J.J and Lebbie, S.H.B. (1995a) Monitoring of livestock health and production in sub-Saharan Africa. *Preventive Veterinary Medicine* 25: 195-212.

de Leeuw, P.N., Reynolds, L. and Rey, B. (1995b) Nutrient transfer from livestock in West African Agricultural production systems. In: Powell, J.M., Fernández-

Rivera, S., Williams, T.O. and Renard, C. (eds), *Livestock and Sustainable Nutrient Cycling in Mixed Farming Systems in sub-Saharan Africa.* Proceedings of an International Conference, 22-26 November 1993, International Livestock Centre for Africa (ILCA), Addis Ababa, Ethiopia, vol. 2, pp. 371-391.

Dicko-Toure, M.S. (1980) The contribution of browse to cattle fodder in the sedentary system of the Office du Niger. In: Le Houérou, H.N. (ed.), *Browse in Africa: The Current State of Knowledge.* International Livestock Centre for Africa (ILCA), Addis Ababa, Ethiopia, pp. 313-319.

Dietz, T. (1987) *Pastoralists in Dire Straits: Survival Strategies and External Interventions in a Semiarid Region at the Kenya/Uganda Border, Western Pokot, 1900-1986.* Netherlands Geographical Studies No. 49, University of Amsterdam, The Netherlands.

Eyoh, D.L. (1992) Reforming peasant production in Africa: Power and technological change in two Nigerian villages. *Development and Change* 23: 37-66.

FAO (1993). *Production Yearbook 1992.* Rome, Italy.

Fernández-Rivera, S., Williams, T.O., Hiernaux, P. and Powell, J.M. (1995) Faecal excretion by ruminants and manure availability for crop production in semiarid West Africa. In: Powell, J.M., Fernández-Rivera, S., Williams, T.O. and Renard, C. (eds), *Livestock and Sustainable Nutrient Cycling in Mixed Farming Systems in sub-Saharan Africa.* Proceedings of an International Conference, 22-26 November 1993, International Livestock Centre for Africa (ILCA), Addis Ababa, Ethiopia, vol.2, pp. 149-169.

Gryseels, G. (1988) Role of livestock on mixed smallholder farms in the Ethiopian highlands. PhD thesis, State Agricultural University, Wageningen, The Netherlands.

Hailu, B. and Chilot, Y. (1992) Vertisol farming systems in north Shewa. In: Franzel, S. and van Houten, H. (eds), *Research with Farmers: Lessons from Ethiopia.* CAB International, Wallingford, UK, pp. 97-108.

Hendy, C.R.C. (1977) *Animal Production in Kano State, Nigeria.* Land Resource Report No. 21. Land Resource Division, Tolworth, UK.

Hiernaux, P. (1989) Note sur l'évolution de la biomasse des pailles au cours de la saison sèche. Working Document, International Livestock Centre for Africa (ILCA), Bamako, Mali.

IFAD (1995) *Common Property Resources and the Rural Poor in sub-Saharan Africa.* International Fund for Agricultural Development, Rome, Italy.

Ifut, O.J. (1992) The potential of cassava peels for feeding goats in Nigeria. In: Hahn, S.K., Reynolds, L. and Egbunike, G.N. (eds), *Cassava as Livestock Feed in Africa.* Proceedings of a Workshop, 14-18 November 1988, International Institute of Tropical Agriculture (IITA), Ibadan, Nigeria, pp. 53-72.

Jabbar, M.A. (1993) Evolving crop/livestock farming systems in the humid zone of West Africa: Potential and research needs. *Outlook on Agriculture* 22: 13-21.

Jabbar, M.A., Reynolds, L. and Francis, P.A. (1995) Sedentarization of cattle farmers in the derived savanna region of southwest Nigeria: Results of a survey. *Tropical Animal Health and Production* 27: 55-64.

Jaetzold, R. and Schmidt, H. (1983) *Farm Management Handbook of Kenya,* vol.2: East Kenya. Ministry of Agriculture, Nairobi.

Jahnke, H.E. (1982) *Livestock Production Systems and Livestock Development in Tropical Africa.* Kieler Wissenschaftsverlag Vauk, Kiel, Germany.

Jutzi, S.C. and Mohamed-Saleem, M.A. (1992) Improving productivity on highland verstisols: The oxen-drawn broadbed maker. In: Franzel, S. and van Houten, H. (eds), *Research with Farmers: Lessons from Ethiopia.* CAB International, Wallingford, UK, pp. 97-108.

Kassam, A.H., van Velthuizen, H.T., Fisher, G.W. and Shah, M.M. (1991a) Agro-ecological Land Resources Assessment for Agricultural Development Planning: A Case Study of Kenya. World Soil Resources Report No. 71, Food and Agriculture Organization of the United Nations (FAO), Rome, Italy.

Kassam, A.H., van Velthuizen, H.T. and Mohamed-Salem, M.A. (1991b) Fodder Productivity Assessment for West Africa. Report 4, International Livestock Centre for Africa (ILCA), Kaduna, Nigeria, and Food and Agriculture Organization of the United Nations (FAO), Rome, Italy.

Kolff, H.E. and Wilson, R.T. (1985) Livestock production in central Mali: The *mouton de case* system of smallholder sheep fattening. *Agricultural Systems* 16: 217-230.

Kossila, V. (1988) The availability of crop residues in developing countries in relation to livestock populations. In: Reed, J.D., Capper, B.S. and Neate, P.J.H. (eds), *Plant Breeding and the Nutritive Value of Crop Residues.* Proceedings of a Workshop, 7-10 December 1987, International Livestock Centre for Africa (ILCA), Addis Ababa, Ethiopia, pp. 29-39.

Landais, E. (1983) *Analyse des Systèmes d'Elevage Bovin Sédentaire du Nord de la Côte d'Ivoire,* vol. 2. Institut d'Elevage et de Médicine Vétérinaire des Pays Tropicaux (IEMVT), Maisons Alfort, France.

Leloup, S. and Traore, M. (1989) *La Situation Fourragère dans le Sud-est du Mali,* vol. 1: Région CMDT de Sikasso et de Koutiala. Institut d'Economie Rurale (IER), Sikasso, Mali, and Royal Tropical Institute, Amsterdam, The Netherlands.

Leloup, S. and Traore, M. (1991) *La Situation Fourragère dans le Sud-est du Mali,* vol. 2: Région CMDT de San. Institut d'Economie Rurale (IER), Sikasso, Mali, and Royal Tropical Institute, Amsterdam, The Netherlands.

Lwoga, A.B. and Urio, N.A. (1987) An inventory of livestock feed resources in Tanzania. In: Kategile, J.A., Said, A.N. and Dzowela, B.H. (eds), *Animal Feed Resources for Small-scale Livestock Producers.* Proceedings of the Second PANESA Workshop, 11-15 November 1985, International Livestock Centre for Africa (ILCA), Nairobi, Kenya, pp. 23-34.

Mathuva, M., Onim, J.F.M., Otieno, K. and Fitzhugh, H.A. (1986) The effect of environment on biomass productivity of intercropped maize with food beans, finger millet, pigeonpea and sesbania in a maize cropping system. In: *Proceedings of the Fifth Small Ruminant CRSP Workshop,* 4-6 November 1986, Small Ruminant Collaborative Research Support Program (SR-CRSP), Nairobi, Kenya, pp. 21-31.

Mbogoh, S.G. (1991) Crop production. In: Tiffen, M. (ed.), *Environmental Change* and *Dry Land Management in Machakos District, Kenya, 1930-90: Production Profile.* ODI Working Paper No. 55, Overseas Development Institute, London, UK, pp. 1-44.

McIntire, J., Reed, J.D., Fedla, A., Jutzi, S. and Kebede, Y. (1988) Evaluating sorghum cultivars for grain and straw yield. In: Reed, J.D., Capper, B.S. and Neate, P.J.H. (eds), *Plant Breeding and the Nutritive Value of Crop Residues.* Proceedings of a Workshop, 7-10 December 1987, International Livestock Centre for Africa (ILCA), Addis Ababa, Ethiopia, pp. 283-304.

McIntire, J., Bourzat, D. and Pingali, P. (1992) *Crop/Livestock Interaction in sub-Saharan Africa.* World Bank Regional and Sectorial Studies. Washington DC, USA.

Mortimore, M. (1991) A review of mixed farming systems in the semiarid zone of sub-Saharan Africa. Livestock Economics Division, Working Document 17. International Livestock Centre for Africa (ILCA), Addis Ababa, Ethiopia.

Munthali, J.T. and Dzowela, B.H. (1987) Inventory of livestock feeds in Malawi. In: Kategile, J.A., Said, A.N. and Dzowela, B.H. (eds), *Animal Feed Resources for Small-scale Livestock Producers.* Proceedings of the Second PANESA Workshop, 11-15 November 1985, International Livestock Centre for Africa (ILCA), Nairobi, Kenya, pp. 61-69.

Mureithi, J.G., Tayler, R.S. and Thorpe, W. (1994) The effects of alley farming with *Leucaena leucocephala* and of different management practices on the productivity of maize and soil chemical properties in lowland coastal Kenya. *Agricultural Systems* 27: 31-51.

Ndagala, D.K. (1991) The unmaking of the Datoga: Decreasing resources and increasing conflict in rural Tanzania. *Nomadic People* 28: 71-82.

Nordblom, J.L. and Shomo, F. (1995) *Food and Feed Prospects to 2020 in the West Asia-North Africa Region.* ICARDA Social Science Paper No. 2, International Centre for Agricultural Research in the Dry Areas, Aleppo, Syria.

Onim, J.F.M., Mathuva, M., Otieno, K. and Fitzhugh, H.A. (1986) Comparing some maize populations to standard varieties for their feed and grain yields in western Kenya. In: *Proceedings of the Fifth Small Ruminant CRSP Workshop,* 4-6 November 1986, Small Ruminant Collaborative Research Support Program (SR-CRSP), Nairobi, Kenya, pp. 9-20.

Onim, J.F.M., Semenye, P.P., Fitzhugh, H.A. and Mathuva, M. (1987) Research on feed resources for small ruminants on smallholder farms in western Kenya. In: Kategile, J.A., Said, A.N. and Dzowela, B.H. (eds), *Animal Feed Resources for Small-scale Livestock Producers.* Proceedings of the Second PANESA Workshop, 11-15 November 1985, International Livestock Centre for Africa (ILCA), Nairobi, Kenya, pp. 149-158.

Onim, J.F.M., Fitzhugh, H.A. and Getz, W.R. (1992) Development and use of feed resources. In: Semenye, P.P. and Hutchcroft, T. (eds), *On-farm Research and Technology for Dual-purpose Goats.* Small Ruminant Collaborative Research Support Program (SR-CRSP) and Kenya Agricultural Research Institute (KARI), Nairobi, Kenya.

Osafo, E.L.K., Owen, E., Said, A.N., Gill, M., McAllan, A.B. and Kebede, Y. (1993) Sorghum stover as a ruminant food in Ethiopia: Effect of cultivar, site of growth, pre-harvest leaf stripping and storage on yield and morphology. In: Gill, M., Owen, E., Pollot, G.E. and Lawrence, T.L.J. (eds), *Animal Production in Developing Countries*. Occasional Publication of the British Society of Animal Production No. 16, pp. 188-189.

Pearce, G.R., Lee, J.A., Simpson, R.J. and Doyle, P.T. (1988) Sources of variation in the nutritive value of wheat and rice straws. In: Reed, J.D., Capper, B.S. and Neate, P.J.H. (eds), *Plant Breeding and the Nutritive Value of Crop Residues*. Proceedings of a Workshop, 7-10 December 1987, International Livestock Centre for Africa (ILCA), Addis Ababa, Ethiopia, pp. 195-231.

Platteeuw, W.L. and Oludimu, O.L. (1993) Economic assessment of more intensive goat keeping systems. In: Ayeni, A.O. and Bosman, H.G. (eds), *Goat Production Systems in the Humid Tropics*. Proceedings of an International Workshop, 6-9 July 1992, Ife-Ife, Nigeria. Pudoc, Wageningen, The Netherlands, pp. 58-76.

Powell, J.M. (1984) Assessment of dry-matter yield from grain yield in the West African savannah zone. *Journal of Agricultural Science* (Cambridge) 103: 695-698.

Powell, J.M. (1986) Crop/livestock interactions in the subhumid zone of Nigeria. In: von Kaufmann, R., Chater, S. and Blench, R. (eds), *Livestock Systems Research in Nigeria's Subhumid Zone*. Proceedings of a Symposium, 20 October-2 November 1984, Kaduna, Nigeria. International Livestock Centre for Africa (ILCA), Addis Ababa, Ethiopia, pp. 268-303.

Powell, J.M. and Bayer, W. (1985) Crop residue grazing by Bunaji cattle in central Nigeria. *Tropical Agriculture* (Trinidad) 62: 302-304.

Powell, J.M. and Fussell, L.K. (1993) Nutrient and structural carbohydrate partitioning in pearl millet. *Journal of Agronomy* 85: 862-866.

Ransom, J.K., Ojiem, J. and Kanampiu, F.K. (1995) Nutrient flux between maize and livestock in a maize-coffee-livestock system in central Kenya. In: Powell, J.M., Fernández-Rivera, S., Williams, T.O. and Renard, C. (eds), *Livestock and Sustainable Nutrient Cycling in Mixed Farming Systems in sub-Saharan Africa*. Proceedings of an International Conference, 22-26 November 1993, International Livestock Centre for Africa (ILCA), Addis Ababa, Ethiopia, vol. 2, pp. 411-417.

Reed, J.D., Kebede, Y. and Fussell, L.K. (1988) Factors affecting the nutritive value of sorghum and millet crop residues. In: Reed, J.D., Capper, B.S. and Neate, P.J.H. (eds), *Plant Breeding and the Nutritive Value of Crop Residues*. Proceedings of a Workshop, 7-10 December 1987, International Livestock Centre for Africa (ILCA), Addis Ababa, Ethiopia, pp. 233-251.

Rwanda (1987) *Etude du Sous-secteur Elevage,* vol. 2. Kigali.

Rwanda (1991) *Enquête Nationale Agricole.* Kigali.

Schuetterle, A. and Coulibaly, L. (1988) The socioeconomic aspects of livestock production in villages in northern Côte d'Ivoire. In: *Livestock Production in Tsetse-infested Areas of Africa.* International Laboratory for Research on Animal Diseases

(ILRAD) and International Livestock Centre for Africa (ILCA), Addis Ababa, Ethiopia, pp. 389-398.

Scoones, I. (1989) *Patch Use by Cattle in the Dry Lands of Zimbabwe: Farmer Knowledge and Ecological Theory.* ODI Pastoral Development Network Paper No. 28b. Overseas Development Institute, London, UK.

Scoones, I. (1993) Why are there so many animals? Cattle population dynamics in the communal areas of Zimbabwe. In: Behnke, R.H., Scoones, I. and Kerven, C. (eds), *Range Ecology at Disequilibrium.* Overseas Development Institute (ODI), London, UK, pp. 62-76.

Scott, G.J. and Ewell, P. (1992) Sweet potatoes in African food systems. In: Scott, G., Ferguson, P.I. and Herrera, J.E. (eds), *Product Development for Root and Tuber Crops,* vol. 3: Africa. Proceedings of the Workshop on Processing, Marketing and Utilization of Root and Tuber Crops in Africa, 26 October-2 November 1991. International Institute of Tropical Agriculture (IITA), Ibadan, Nigeria, and Centro Internacional de la Papa (CIP), Lima, Peru, pp. 91-103.

Scott, G.J. and Suarez, V. (1992) Transforming traditional food crops: Product development for roots and tubers. In: Scott, G., Ferguson, P.I. and Herrera, J.E. (eds), *Product Development for Root and Tuber Crops,* vol. 3: Africa. Proceedings of the Workshop on Processing, Marketing and Utilization of Root and Tuber Crops in Africa, 26 October-2 November 1991. International Institute of Tropical Agriculture (IITA), Ibadan, Nigeria, and Centro Internacional de la Papa (CIP), Lima, Peru, pp. 3-23.

Semenye, P.P. and Hutchcroft, T. (1992) *On-farm Research and Technology for Dual-purpose Goats.* Small Ruminant Collaborative Research Support Program (SR-CRSP) and Kenya Agricultural Research Institute (KARI), Nairobi, Kenya.

Shiferaw, B. (1991) Crop/livestock interactions in the Ethiopian highlands and effects on sustainability of mixed farming: A case study from Ada District. MSc thesis, Agricultural University of Norway, Oslo.

Smith, O.B. (1992) A review of ruminant responses to cassava-based diets. In: Hahn, S.K., Reynolds, L. and Egbunike, G.N. (eds), *Cassava as Livestock Feed in Africa.* Proceedings of a Workshop, 14-18 November 1988, International Institute for Tropical Agriculture (IITA), Ibadan, Nigeria, and Centro Internacional de la Papa (CIP), Lima, Peru, pp. 39-53.

Steinfeld, H. (1988) *Livestock Development in Mixed Farming Systems: A Study of Livestock Production Systems in Zimbabwe.* Kieler Wissenschaftsverlag Vauk, Kiel, Germany.

Stoop, W.A. (1986) Agronomic management of cereal/cowpea cropping systems for major toposequence land types in the West African Savanna. *Field Crops Research* 14: 301-319.

Tanner, J.C., Holden, S.J., Dampha, K. and Jallow, A. (1993) Potential for livestock supplementation in smallholder farms in The Gambia. In: Gill, M., Owen, E., Pollot, G.E. and Lawrence, T.L.J. (eds), *Animal Production in Developing Countries.* Occasional Publication No. 16, British Society of Animal Production, pp.190-191.

Tarawali, G. and Mohamed-Saleem, M.A. (1994) Establishment techniques for stylo-associated cropping systems. In: de Leeuw, P.N., Mohamed-Saleem, M.A. and Nyamu, A.M. (eds), *Stylosanthes as a Forage and Fallow Crop*. Proceedings of the Regional Workshop on the Use of Stylosanthes in West Africa, 26-31 October 1992, Kaduna, Nigeria. International Livestock Centre for Africa (ILCA), Addis Ababa, Ethiopia, pp. 183-201.

Tiffen, M., Mortimore, M. and Gishuki, F. (1994) *More People, Less Erosion*. ACTS Press, Nairobi, Kenya.

Toulmin, C. (1992) *Cattle, Women and Wells: Managing Household Survival in the Sahel*. Clarendon Press, Oxford, UK.

van Raay, H.G.T. and de Leeuw, P.N. (1970) The importance of crop residues as a fodder: A resource analysis in Katsina Province, Nigeria. *Tydschrift voor Economische und Sociale Geographie* 61: 137-147.

van Raay, H.G.T. and de Leeuw, P.N. (1974) *Fodder Resources and Grazing Management in a Savanna Environment: An Ecosystems Approach*. Occasional Paper No. 5, Institute of Social Studies, The Hague, The Netherlands.

Varvikko, T. (1991) Nutrition constraint to improved smallholder milk production in the Ethiopian highlands: The Selale experience. In: *Proceedings of the Fourth National Livestock Improvement Conference*, 13-15 November 1991, Institute of Agricultural Research (IAR), Addis Ababa, Ethiopia, pp. 43-50.

Vierich, H.I.D. and Stoop, W.A. (1990) Changes in West African Savanna agriculture in response to growing population and continuing low rainfall. *Agriculture, Ecosystems and Environment* 31: 115-132.

Webb, P., von Braun, J. and Johannes, Y. (1992) *Famine in Ethiopia: Policy Implications of Coping with Failure at National and Household Levels*. Research Report No. 92, International Food Policy Research Institute (IFPRI), Washington DC, USA.

Williams, T.O., Powell, J.M. and Fernández-Rivera, S. (1995) Manure utilization, drought cycles and herd dynamics in the Sahel: Implications for cropland productivity. In: Powell, J.M., Fernández-Rivera S., Williams, T.O. and Renard, C. (eds), *Livestock and Sustainable Nutrient Cycling in Mixed Farming Systems in sub-Saharan Africa*. Proceedings of an International Conference, 22-26 November 1993, International Livestock Centre for Africa (ILCA), Addis Ababa, Ethiopia. vol. 2, pp. 392-409.

Wilson, R.T., de Leeuw, P.N. and de Haan, C. (eds) (1983) *Recherches sur les Systèmes des Zones Arides du Mali: Résultats Préliminaires*. Research Report No. 5, International Livestock Centre for Africa (ILCA), Addis Ababa, Ethiopia.

Winrock (1992) Assessment of Animal Agriculture in sub-Saharan Africa. Mimeo, Airlington, Virginia, USA.

World Bank (1989) *Sub-Saharan Africa: From Crisis to Sustainable Growth*. Washington DC, USA.

WRI (1990) *World Resources 1990-1991*. World Resources Institute/Oxford University Press, New York, USA, and Oxford, UK.

Acknowledgements

The author wishes to thank Mrs N. Marciana for processing text and tables, and Mrs L. Masharia, Mr M. Kenyanjui and Mr D. Njubi for preparing figures and other materials for presentation.

4. Cowpea and its Improvement: Key to Sustainable Mixed Crop/Livestock Farming Systems in West Africa

B.B. Singh[1] and S.A. Tarawali[2]
[1]*International Institute of Tropical Agriculture, IITA Kano Station, PMB 3112, Kano, Nigeria*
[2]*International Livestock Research Institute, PMB 5320, Ibadan, Nigeria*

Abstract

Cowpea (*Vigna unguiculata* (L.) Walp.) is the most widely cultivated food legume in semiarid West Africa, where rainfall is scanty and soils are sandy and relatively infertile. Surveys conducted in the region have shown that most households keep livestock and grow cowpea mostly as an intercrop with millet and sorghum. Cowpea grains are consumed as food and the haulms are fed to livestock as a nutritious fodder. The cowpea plant's ability to fix atmospheric nitrogen helps maintain soil fertility, its deep roots improve soil structure and its drought tolerance ensures normal growth even in relatively dry areas. Manure from livestock is returned to the field and animals provide milk, meat and traction for land preparation, weeding and transport. Cowpea is thus an integral component of crop/livestock farming systems in West Africa.

However, cowpea grain and fodder yields are very low for various reasons, including shading by cereals in intercropping systems, numerous diseases, insect pests, parasitic weeds, drought and low soil fertility. The International Institute of Tropical Agriculture (IITA) and the International Livestock Research Institute (ILRI) are working together to develop improved dual-purpose cowpea varieties with resistance to biotic and abiotic stresses and better nutritional attributes. Improved varieties are evaluated as sole crops and intercrops, both with and without insecticides, so as to select the best variety for different conditions. Considerable progress has been made and several varieties with combined resistance to important diseases, aphid, bruchid, thrips and *Striga* have been developed. These have potential yields of over 2 t ha^{-1} of grain and 3 t ha^{-1} of fodder in areas with 500 mm or more annual rainfall and over 1 t ha^{-1} of grain and 1 t ha^{-1} of fodder in very dry areas with less than 300 mm annual rainfall. Preliminary experiments have also shown significant varietal differences in fodder quality and its effect on milk yield. These differences provide opportunities to develop dual-purpose cowpea varieties with high yield potential and enhanced fodder quality.

Introduction

Cowpea (*Vigna unguiculata* (L.) Walp) is said to have originated in Africa, where it has become an integral part of traditional cropping systems, particularly in the semiarid West African savanna (Steele, 1972). Of the crop's estimated world total area of about 10 million ha, Africa alone accounts for over 7.5 million ha, of which about 70% lies in West and Central Africa (Singh *et al.*, 1996). The crop is primarily grown in mixtures with cereals, especially maize in the moist savanna and sorghum and millet in the dry savanna. Cowpea grain, which is valued for its high nutritive quality and short cooking time, serves as a major source of protein in the daily diets of the rural and urban poor. Its tender leaves are eaten as a spinach-like vegetable, while its immature pods and seeds are also consumed as vegetables. Farmers in the dry savanna use cowpea haulms as a nutritious fodder for their livestock. The plant's ability to fix atmospheric nitrogen helps maintain soil fertility, while its tolerance to drought extends its adaptation to drier areas considered marginal for most other crops (Singh *et al.*, 1995).

At least 30% of sub-Saharan Africa's significant populations of ruminant livestock are found in West Africa, where over 50% of cattle, sheep, goats and camels are raised in the semiarid and arid zones. Cattle are managed either as large, pastoral herds with some transhumance in the dry season, or smaller, sedentary herds belonging to mixed farmers, for whom traction and manure use are important components of the farming system. Most of the small ruminants are kept as small flocks around the permanent home, although some transhumance does occur (Bourn *et al.*, 1994). Throughout the Sudano-Sahelian region, the availability of nutritious fodder is limited due to low and erratic rainfall and the long dry season extending from October to May (Sivakumar, 1990). Crop residues are an essential fodder resource, and farmers deliberately opt for crop combinations and management practices that optimize residue production, especially from the more nutritious grain legumes such as cowpea and groundnut, which respectively contain about 21% and 17% crude protein in the dry haulms. These residues are harvested and conserved, either for dry-season feeding to the farmer's own animals or for sale to other farmers during the critical period of feed scarcity in the mid- to late dry season. Some farmers obtain up to 25% of their annual cash income from the sale of grain legume residues during this period (ICRISAT, 1991).

Trends and projections for the region indicate that livestock numbers have risen and are likely to continue to do so. Human populations are likewise expanding rapidly, leading to increased demand for food and livestock products. Because of the limited potential for increasing the area under cultivation, this extra demand must be met through intensification. Increasing both the grain yield and the quantity and quality of crop residues of cowpea is an important option, since it would increase human food supplies while simultaneously sustaining soil fertility and decreasing the cost of raising livestock

Against this background, the International Institute of Tropical Agriculture (IITA) and the International Livestock Research Institute (ILRI) have launched a joint project to develop improved dual-purpose cowpea varieties with higher grain and fodder yields and to extend intensive cowpea cultivation in the Sudano-Sahelian region. This paper describes the progress made so far.

Cowpea in Traditional Cropping Systems

Traditional cropping systems in West Africa are highly variable, differing not just from region to region but from farmer to farmer. Several authors have described these systems in general terms (Steele, 1972; Norman, 1974; Abalu, 1976; Renard *et al.*, 1987; N'tare, 1989), but quantitative details on component crops and their productivity are scarce.

To obtain a better understanding of the role of cowpea in crop/livestock interactions, IITA and ILRI undertook a quantitative study in the Minjibir and Gezawa local government areas of Kano State, Nigeria, in 1992-1993. The study areas are located about 30 km northeast of Kano city (12°N) and represent the transition zone between Sudan savanna and the Sahel proper. Over 90% of the fields had mixed cropping involving millet, sorghum, cowpea and groundnut with occasional fields of maize, cassava and bambara groundnut. The major crop mixtures were millet-cowpea (22%), millet-sorghum-cowpea (15%), millet-cowpea-groundnut (12%) and sorghum-cowpea-groundnut (6%). Cowpea was thus a predominant component of all crop mixtures. The planting pattern differed from farmer to farmer, but cowpea was generally planted in alternate rows or within the cereal rows, occupying 33 to 50% of the land area in each field. There was great diversity in the varieties grown, but these could be divided into two main groups: (i) early-maturing varieties, grown for grain; and (ii) late-maturing varieties, grown for fodder. Both types were planted in the same field. One of the most commonly practised rotations was millet-grain cowpea-millet-fodder cowpea (Singh, 1993). In this system, millet is planted first in widely spaced rows (1.5-3 m apart) at the onset of the rains (May-June), with 1 m hill-to-hill distance within the rows, reaching a density of about 4000 to 6000 plants ha^{-1}. Grain-type early cowpea varieties are planted between alternate millet rows at a hill-to-hill distance of 1 m once the rains are well established, towards the end of June. Fodder-type late-maturing cowpea is then planted in mid-July, in the remaining alternate rows. This pattern is depicted in Fig. 4.1. Sorghum and groundnut can either replace or supplement the millet and grain cowpea to give a more complex mixture.

The grain cowpea and millet are harvested at the end of August or the beginning of September, while the late cowpea or sorghum (which is also planted late in this

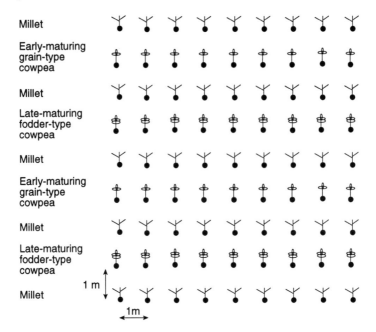

Fig. 4.1. Traditional intercropping system.

system) is left in the field until the onset of the dry season (October-November). The farmers wait until the cowpea leaves show signs of wilting before they cut cowpea plants at the base and roll the plants into bundles with all the leaves still intact. These bundles are kept on roof-tops or in tree-forks for drying, before being fed to livestock or sold to other farmers in the peak dry season (March-May), when fodder prices are high. If there are rains in October-November or if sufficient residual moisture is available, the late cowpea produces a reasonable amount of grain and fodder. Thus the existing cropping system in northern Kano State uses rainfall from May to October quite effectively to produce both human food and animal feed.

The grain and fodder yields of component crops varied from field to field and also depended on the crop mixture. In general, cowpea grain yield ranged from 0 to 132 kg ha^{-1} and fodder yield from 120 to 1820 kg ha^{-1}. Grain and fodder yields of groundnut ranged from 0 to 197 kg ha^{-1} and from 144 to 1976 kg ha^{-1} respectively. The grain yield of millet ranged from 131 to 2600 kg ha^{-1} and the stover yield from 259 to 6995 kg ha^{-1}. The grain and stover yields for sorghum ranged from 0 to 4903 kg ha^{-1} and from 538 to 13 015 kg ha^{-1} respectively. The zero yields of cowpea and groundnut were due to insects and diseases respectively. Those of late sorghum were due to terminal drought.

The highest yields came from fields located near the home compound, which had received some manure and/or inorganic fertilizers. These fields had grain yields of 3513 kg ha^{-1} and biomass yields of 13 241 kg ha^{-1}, compared to 729 kg ha^{-1} and

3870 kg ha⁻¹ for fields which were further away from the compound and did not receive manure or fertilizer. The average quantity of manure applied ranged from 3 to 6 t ha⁻¹, but not all fields received manure each year. At the end of the rains, all the biomass was taken out of the field; cowpea and groundnut haulms were stored for fodder, while millet and sorghum stover were used for fencing, thatching or as fuel. Occasionally, the dry leaves from millet and sorghum stovers were stripped and used as fodder (Powell, 1986).

Thus, cowpea and groundnut haulms are a valuable commodity. They may either be fed to the farmer's own livestock or sold in the market for cash. During the peak dry season, the average price of cowpea haulms can be as high as 30 to 50% of the cowpea grain price on a dry weight basis, bringing substantial additional income to farmers.

A concurrent study of 100 farming households in the same areas, conducted by Agyemang *et al.* (1993), revealed that all households had livestock, with 58% owning cattle, 92% goats, 96% sheep and 24% donkeys. The average herd size was 6 for cattle, 11 for goats and 10 for sheep. None of the households reported growing crops other than fodder-type cowpea specifically for animal feed, because of the shortage of land. All households collected and stored crop residues for feed, with cowpea and groundnut haulms playing important roles. The majority of households planted one to five fields, but some had up to 11 fields. Each field was less than 0.55 ha. The predominant crops were millet, cowpea, sorghum and groundnut, in different mixtures. Again, it was found that all households applied manure in some fields each year, but not in all because of insufficient manure. The average yields of all crops were 20-55% higher in fields with manure than in those without. Among the cattle owners, 69% owned one to three pairs of oxen, which were used for ploughing, ridging, weeding and transport of farm produce. Even though all households cultivated and stored crop residues, 86% of them still had to purchase feed to supplement their own supplies during the dry season.

The current farming systems in these areas have reached a relatively high level of integration between crops and livestock. This has undoubtedly helped to sustain food production, albeit at a relatively low level. However, human and livestock populations are increasing rapidly and there is an urgent need to raise yields. This is especially the case for drought-tolerant and soil-improving legumes such as cowpea, which grows in low-rainfall areas where cereal yields are low. Increasing the productivity of cowpea would provide food and fodder for more people and livestock, without depleting the resource base. The development of improved cowpea varieties will therefore contribute greatly towards the well-being of small-scale, resource-poor farmers in the marginal drylands of West Africa.

Key Constraints

Cowpea grain and fodder yields are very low in West Africa. The reasons are several, but the most important ones are: (i) low density of cowpea and shading by cereals in intercropping systems; (ii) diseases, insect pests and parasitic weeds such as *Striga gesnerioides* and *Alectra vogelii;* (iii) drought stress and low soil fertility; (iv) lack of inputs and infrastructure.

Low Density and Shading

The major reason for low yields is low plant density and the shading of cowpea by cereals in intercropping systems. The cowpea population typically ranges from 1000 to 6000 plants ha^{-1}, intermixed with fast-growing cereals such as millet and sorghum. Even with good management and full insecticide protection, the potential yield of 2 t ha^{-1} achieved by a good variety grown as a sole crop is reduced to less than 400 kg ha^{-1} in a 1 millet:1 cowpea plant intercrop (Singh, 1993). The effect of shading is pronounced when the cowpea is planted 3-4 weeks after the millet, as is the common practice in the West African dry savanna. Terao *et al.* (1995b) noticed drastic reduction in branching and growth of cowpeas when planted 3 weeks after millet compared to simultaneous planting. They therefore recommended simultaneous planting or strip cropping to minimize shading in the early stages of cowpea growth. However, they also observed that simultaneous planting in areas with less than 400 mm rainfall resulted in lower millet yields, due to competition with cowpea for moisture. Since the water requirement of cowpea is less than that of millet and since grain cowpea matures earlier, the intercropping system in areas with less than 400 mm rainfall would be more efficient if it had more cowpea and less millet. Changing the current 1 millet:1 cowpea ratio to a 2 millet:4 cowpea or a 1 millet:4 cowpea ratio (strip cropping) would ensure higher grain and fodder yields of cowpea with only a small reduction in millet yields. It could even be advantageous to plant cowpea as a sole crop in the first year followed by sole-crop or intercropped millet in the second year, so as to maximize the productivity of both crops in terms of time and space.

Diseases, Insect Pests and Parasitic Weeds

The major diseases affecting cowpea are viruses, web blight, *Cercospora,* brown blotch, *Septoria* and scab in the moist savanna and bacterial blight, false smut and ashy stem blight in the dry savanna (Emechebe and Shoyinka, 1985). These diseases not only reduce the plant's growth and biomass production but also affect the quality of both grain and fodder.

Cowpea is also attacked by several insect pests, of which aphid, thrips, *Maruca* pod borer, pod bugs and bruchid are the most widespread, causing substantial yield losses and reduced quality (Singh and Jackai, 1985). *Maruca* pod borer is not important in areas with less than 300 mm annual rainfall, but is very damaging in areas with more than 500 mm, where it may cause up to 100% yield losses. It has been observed that *Maruca* adults move from coastal West Africa towards the northern savannas with the rains (Bottenberg *et al.,* 1996). They appear in light traps at Kano and Nguru in northern Nigeria from June to October, whereas they are observed in the field at Ibadan throughout the year.

Cowpea is parasitized by *Striga gesnerioides* (Willd) Vatke and *Alectra vogelii* (Benth.), which cause considerable yield reduction. *A. vogelii* is more prevalent in the moist savanna, whereas *Striga* is more widespread in the dry, particularly in the Sahelian zone where soils are sandy and infertile (Singh and Emechebe, 1991). Several different strains of *S. gesnerioides* have been observed in West Africa, causing different levels of parasitization in different varieties (Lane *et al.,* 1995).

Through collaborative research by IITA and national and regional programmes, considerable progress has been made in breeding cowpea varieties for multiple disease and pest resistance. Sources of resistance to major viruses, *Cercospora,* brown blotch, bacterial blight, *Septoria* and scab have been identified and are being used in breeding programmes (Singh *et al.,* 1984, 1987; Abadassi *et al.,* 1987; Singh, 1993). In addition, lines with good resistance to aphid and bruchid and moderate resistance to thrips have been selected and these too are being used in breeding programmes (Adjadi *et al.,* 1985; Singh *et al.,* 1985; Bata *et al.,* 1987; Singh and Singh, 1990; Singh, 1993). Similarly, good progress has been made in breeding for resistance to *S. gesnerioides* and *A. vogelii* (Aggarwal, 1991; Singh and Emechebe, 1990, 1991; Singh, 1993; Atokple *et al.,* 1995; Berner *et al.,* 1995). Efforts are now under way to identify sources of resistance to *Maruca* pod borer and pod bugs, which are a major problem in the moist savannas and more humid regions.

Drought Stress and Low Soil Fertility

Cowpea is inherently more tolerant to drought than other crops, which is why it is cultivated in low-rainfall areas. However, the late-maturing fodder-type varieties planted as a sole crop or a relay intercrop with millet may suffer severe drought stress due to the early cessation of rains. Early-planted cowpeas may also be affected by intermittent drought during the early cropping season. Improved varieties with deep rooting systems and drought tolerance will give more stable yields over a wider range of environments. They will also enable cowpea cultivation to be extended further into the arid zone. Several lines with improved drought tolerance have now been identified and efforts are under way to incorporate this trait into existing improved varieties (Singh, 1987, 1993; Hall *et al.,* 1992; Singh *et al.,* 1995; Terao *et al.,* 1995a; Watanabe *et al.,* 1995).

Throughout the Sahel, cowpea growth is retarded due to the poor fertility of the region's sandy soils, which are especially low in phosphorus and micronutrients (Bationo *et al.,* 1991). Although cowpea can fix sufficient atmospheric nitrogen to meet most of its requirements, the limited availability of manure and inorganic fertilizers, coupled with drought stress and high temperatures, frequently result in very low yields. Increasing the population of cowpea and groundnut in the cropping system could partially alleviate this problem by furthering the integration of crop and livestock production, thereby increasing the quantity and quality of manure available, as well as by increasing the below-ground biomass. Efforts are being made to develop cowpea varieties with better nitrogen-fixing ability (Miller Jr. and Fernandez, 1985; Walker and Miller, 1986). Significant variability in the number and weight of nodules among cowpea varieties has been observed at IITA's Kano Station and a number of lines with better nodulation are being further studied. In addition, newly identified germplasm lines which show more growth in sandy soils than other varieties are being investigated.

Lack of Inputs and Infrastructure

Lack of improved seeds, fertilizers and farm chemicals is a major bottleneck in African agriculture in general and in the West African dry savanna in particular. Cowpea and millet yields can be increased several-fold just by applying a little fertilizer and two sprays of a suitable insecticide (Bationo and Mokwunye, 1991; Singh, 1993). However, even when farmers have cash and are eager to purchase these inputs, they often cannot do so because the inputs themselves are not available locally. Lack of transport and credit facilities further add to the problems of obtaining inputs and bringing farm produce to the market. A few non-governmental organizations (NGOs) have initiated projects to help farmers obtain inputs, but there is a great need to persuade governments to allocate more resources to agricultural development.

Research Strategy

In view of the constraints outlined above, the main thrust of IITA and ILRI's cowpea research is to develop improved varieties and cropping systems that can provide higher yields with limited or little use of purchased inputs. The major objectives are to:

1. Incorporate into existing improved cowpea varieties resistance to diseases, insect pests and parasitic weeds, as well as adaptive traits such as drought tolerance, better nitrogen fixation, enhanced and rapid growth under low soil fertility conditions, and shade tolerance.
2. Develop a range of cowpea varieties combining higher grain and fodder yield potential with diverse plant types and times to maturity to fit into different cropping systems and ecological niches.
3. Evaluate new breeding lines as sole or intercrops, with and without the use of insecticides, to select lines on the basis of both grain and fodder yields as well as superior nutritional quality.

4. Develop improved cropping systems and planting patterns with higher cowpea densities, so as to maximize cowpea grain and fodder production and increase overall economic output per unit area and per unit time.

5. Develop early-maturing, fast-growing, drought-tolerant cowpea varieties with resistance to aphid, thrips, bruchid and *Striga* for sole cropping and strip cropping with millet or sorghum in areas of the Sahel with less than 300 mm rainfall.

Varietal Improvement

Early efforts in cowpea breeding focused mainly on developing grain varieties for sole cropping (Singh and N'tare, 1985), but during the past decade greater emphasis has been placed on intercropping situations and on the leaves and fodder (Paul *et al.*, 1988; Akundabweni *et al.*, 1989; N'tare, 1989; Singh, 1991, 1993, 1994; Blade *et al.*, 1992; N'tare and Williams, 1992; Shetty, 1993; Ehlers, 1994; Tarawali *et al.*, 1996). IITA has established a station at Kano, Nigeria, with the major goal of developing cowpea varieties for intercropping with millet and sorghum. The scientists at this station work closely with Kano-based scientists of the International Crops Research Institute for the Semi-Arid Tropics (ICRISAT) as well as with national scientists from Nigeria, Niger, Cameroon, Benin, Ghana, Mali and Burkina Faso.

Considerable progress has been made. The performance of promising varieties from different maturity groups at Minjibir (Kano) Nigeria in the 1995 cropping season is presented in Tables 4.1 to 4.4. The total rainfall during the season was 559 mm, from June to September. The crop was planted in the first week of July and harvested in September-October, depending on maturity group. The varieties were evaluated in three systems: (i) sole crop, with insecticide applied at flowering and podding; (ii) sole crop, without insecticide; and (iii) intercrop with millet, approximating farmers' practice of alternate rows of millet and cowpea, about 1 m apart and without insecticide protection. The results indicate that the most promising varieties yielded between 2 and 2.5 t ha^{-1} of grain and 2 and 3.5 t ha^{-1} of dry fodder if planted as sole crops and provided with minimal insecticide protection. When unprotected, the cowpea tends to be more vegetative because the flowers and pods are damaged during translocation of carbohydrates from leaves to pods, such that grain yields decrease and fodder increases. The grain and fodder yields of all varieties were drastically reduced when they were intercropped with millet, ranging from 157 to 233 kg ha^{-1} for grain and 251 to 477 kg ha^{-1} for fodder. Millet yields ranged from 1654 to 2198 kg ha^{-1} of grain and from 3441 to 4421 kg ha^{-1} of stover. Since cowpea fodder is six to eight times more valuable (in cash terms) than millet stover, sole-cropped cowpea with two sprays would be the most profitable option. However, in the absence of fertilizers and chemicals, farmers normally grow mixtures.

The data in the tables reveal significant differences among cowpea varieties for both grain and fodder yields, as sole crops and as an intercrop. Some new varieties, including IT93K1140, IT93K2045-29, IT90K277-2, IT93K385-3,

Table 4.1. Grain and fodder yields (kg ha^{-1}) of promising early-maturing cowpea varieties at Minjibir, Nigeria, 1995[1].

Variety	Sole crop (2-spray)		Sole crop (no spray)		Intercrop (no spray)	
	Grain	Fodder	Grain	Fodder	Grain	Fodder
IT93K1140	2408	2515	363	1389	85	218
IT86D719	2220	1680	79	2306	238	406
IT90K284-2	2079	2661	135	2542	70	140
IT93K2045-29	1983	1398	451	1986	246	437
IT93K165-3	1746	1166	425	3722	190	159
Mean	1594	1520	229	1625	157	251
LSD (5%)	557	717	293	1038	97	204

[1] Early-maturing: 60-70 days.

Table 4.2. Grain and fodder yields (kg ha^{-1}) of promising medium-maturing cowpea varieties at Minjibir, Nigeria, 1995[1].

Variety	Sole crop (2-spray)		Sole crop (no spray)		Intercrop (no spray)	
	Grain	Fodder	Grain	Fodder	Grain	Fodder
IT90K277-2	2523	1837	147	4042	263	500
IT93K385-3	2303	1973	462	1676	194	390
IT93K573-1	2168	2098	175	4028	235	1328
IT93K596-12	1932	1714	254	2028	174	203
IAR48	2050	1450	62	2528	115	406
Mean	1887	1751	149	2061	176	477
LSD (5%)	518	1126	289	1438	144	354

[1] Medium-maturing: 70-80 days.

IT93K573-1, IT93K23, IT86D716, IT93K2271-4 and IT93K608-13-1, were superior to the local varieties, namely IAR48, Danila and Kanannado. Of these, IT90K277-2, IT90K391, IT93K573-1 and IT93K608-13-1 were also among the highest yielders when intercropped. These data, together with the results from previous years of on-station and on-farm trials, indicate that improved cowpea

Table 4.3. Grain and fodder yields (kg ha⁻¹) of promising medium-maturing semi-determinate cowpea varieties at Minjibir, Nigeria, 1995[1].

Variety	Sole crop (2-spray)		Sole crop (no spray)		Intercrop (no spray)	
	Grain	Fodder	Grain	Fodder	Grain	Fodder
IT93K23	2739	3277	21	4416	144	406
IT90K277-2	2571	1492	163	4250	293	437
IT92KD37-1	2316	3590	36	4139	42	171
IT90K391	2278	3423	8	4000	423	703
IT90K365	2026	2588	28	3861	237	437
IT93K621-7	1944	2818	347	1833	117	406
Danila	1835	1429	157	1222	151	265
Mean	1869	2241	87	2887	172	265
LSD (5%)	458	521	180	2414	129	535

[1] Medium-maturing: 75-85 days.

Table 4.4. Grain and fodder yields (kg ha⁻¹) of promising dual-purpose late-maturing cowpea varieties at Minjibir, Nigeria, 1995[1].

Variety	Sole crop (2-spray)		Sole crop (no spray)		Intercrop (no spray)	
	Grain	Fodder	Grain	Fodder	Grain	Fodder
IT86D716	1983	2640	1	5167	353	609
IT93K2271-4	1808	3383	1	5167	353	609
IT93K398-2	1800	3298	4	5250	331	593
IT93K608-13-1	1720	3736	2	3378	752	718
Kanannado	1037	2745	1	2972	87	421
Mean	1541	2577	20	2919	233	335
LSD (%)	434	1213	NS	1710	204	341

[1] Late-maturing: 80-100 days.

varieties can produce between 50 and 200% more grain and fodder than what farmers are achieving at present from traditional varieties. Most of the improved varieties have combined resistance to thrips, aphid, bruchid and *Striga* and are therefore not only higher-yielding but also more stable in their performance over different years and at different locations. The locations used for testing have included Samaru, Minjibir and Mallammadori in Nigeria, Olelewa, Maradi and Niamey in Niger, Maroua in Cameroon and Bamako in Mali.

The grain and fodder yields of a few selected varieties at Olelewa, a very dry location 100 km north of Zinder, in Niger, are presented in Tables 4.5 to 4.7. In 1995, several varieties yielded between 700 and 1468 kg ha^{-1} of grain and 833 and 2916 kg ha^{-1} of fodder within just 70 days and with only 257 mm rainfall and no insecticide protection. The only insects observed were aphid and thrips, and most of the new lines are resistant to these, as also to *Striga*. The most promising varieties for this area were IT89KD349, IT88D867-11, IT93K699-1 and IT93K398-2, which yielded close to 1000 kg ha^{-1} of both grain and fodder. The adjacent plot with a traditional millet-cowpea intercrop yielded 190 kg ha^{-1} of cowpea grain, with the millet drying prematurly due to drought. These data suggest that there is good scope for increasing cowpea grain and fodder production without the use of purchased inputs in the really marginal areas of the Sahel.

Cropping Systems Improvement

To overcome the problems of low plant density and shading, several studies are in progress to develop improved cropping systems using higher cowpea densities and different crop combinations. Table 4.8 summarizes the results of an experiment involving ten cowpea varieties planted as sole crops and intercropped with millet in three different arrangements with and without insecticide protection. The data

Table 4.5. Performance of improved medium-maturing semi-determinate cowpea varieties without insecticide at Olelewa, Niger, 1995.

Variety	Grain (kg ha^{-1})	Fodder (kg ha^{-1})	Days to maturity
IT89KD349	1022	1916	66
IT88D867-11	856	1249	67
IT90K365	736	1333	65
IT90K277-2	665	1499	67
IT92KD371-1	622	2416	67
Danila	726	1246	63
Mean	613	1276	66
LSD (5%)	462	755	3

Table 4.6. Performance of improved medium-maturing dual-purpose cowpea varieties without insecticide at Olelewa, Niger, 1995.

Variety	Grain (kg ha^{-1})	Fodder (kg ha^{-1})	Days to maturity
IT93K699-1	1468	1249	64
IT93K398-2	1086	833	68
IT93K608-13-1	731	1083	70
IT93K621-2	702	1333	65
Danila	764	1333	67
Mean	638	1012	69
LSD (5%)	312	699	2

Table 4.7. Performance of improved medium-maturing cowpea varieties without insecticide at Olelewa, Niger, 1995.

Variety	Grain (kg ha^{-1})	Fodder (kg ha^{-1})	Days to maturity
IT93K876-30	652	1416	67
IT93K876-12	622	1416	67
IT93K573-1	618	1249	67
IT93K621-7	604	2916	71
IT93K614-4	516	1083	70
IAR48	245	749	70
Mean	407	1162	69
LSD (5%)	312	699	2

on grain and fodder yields indicate significant differences between treatments and increased yields with increased cowpea populations. As expected, yields were lowest in the unsprayed intercrop and highest in the sprayed sole crop. Spraying increased grain yields by 70%, but had no effect on fodder. Grain and fodder yields differed significantly in different cropping systems and the cropping system x variety interaction was also highly significant. The spray x cropping system interaction was significant. The best-performing cowpea varieties across all systems were IT90K277-2 and IT89KD349, both of which outyielded the local check Danila. Both these varieties are resistant to aphid, thrips and bruchid, have multiple disease

Table 4.8. Mean grain and fodder yields (kg ha⁻¹) of cowpea and millet in different cropping systems with and without insecticide spray at Minjibir, Nigeria, 1995.

Component	1 millet:1 cowpea		2 millet:4 cowpea		1 millet:4 cowpea		Sole crop		LSD (5%)
	Spray	No spray	Spray	No spray	Spray	No spray	Spray	No spray	
Cowpea grain	312	170	847	503	1042	608	1459	882	142
Cowpea fodder	802	675	1094	959	1275	1377	1914	2138	361
Millet grain	2507	2421	1650	1532	1179	1189	-	-	NS
Millet stover	5971	5529	3771	3542	2906	3165	-	-	NS
IT90K277-2 grain	243	139	1079	745	1332	682	1816	1155	370
IT90K277-2 fodder	967	434	1066	950	1476	1343	2651	2084	1030
IT89K-D349 grain	743	267	1032	423	1329	825	2094	991	370
IT89K-D349 fodder	550	267	1932	1749	2060	1559	2735	4102	1030
Danila grain	262	205	602	291	727	553	714	504	370
Danila fodder	800	400	1577	811	1499	1573	2914	2801	1030
LSD (5%):									
Cowpea grain	455	316	455	316	455	316	455	316	
Cowpea fodder	1109	1076	1109	1076	1109	1076	1109	1076	

resistance and are moderately resistant to *Striga*. Considering the high prices fetched by cowpea grain and fodder compared to those of millet grain and stover, total income would be highest for sole-cropped sprayed cowpea. This would be followed by a strip cropping system combining 1 millet:4 cowpea with 2 millet:4 cowpea, provided the best cowpea variety for each system is used.

In another experiment, several cowpea varieties were evaluated using a ratio of 1 millet or sorghum:4 cowpea to select the best varieties for strip cropping and to study the effect of different cereals on adjacent cowpea rows. The results indicated that cowpea varieties can (but do not always) perform differently in sorghum and millet strip cropping systems (Table 4.9). For example, the varieties IT93K734 and IT93K596-3 yielded higher in the sorghum than in the millet system, whereas Danila and Aloka local did worse. IT86D719 and IT90K277-2, in contrast, performed well with both the cereals. Overall, the sorghum-cowpea strip cropping system gave higher grain and fodder yields for both crops than did millet-cowpea strip cropping. The yield differences between millet and sorghum may be primarily due to time to maturity, since the millet took only 90 days whereas the sorghum required over 128 days. Different degrees of competition for moisture, nutrients and light may have been responsible for the cowpea variety x cereal crop interaction.

These results suggest that it is possible to increase cowpea grain and fodder production by developing a more efficient cropping system. However, more experiments need to be conducted at different locations to identify appropriate crop and variety combinations, which are likely to be highly site-specific.

Double Cropping in the Sudan Savanna

In areas with 600 to 1000 mm annual rainfall, it is possible to plant a second crop of cowpea on residual soil moisture, provided the first crop can be harvested before 10 September.

Farmers around Kano normally plant short-duration millet in May-June, when the early rains set in, and intercrop cowpea some 3 to 4 weeks later. The millet and early cowpea are harvested at the end of August or the beginning of September, while the late cowpea remains in the field until October/November and is finally harvested at the onset of drought. This system makes use of rains from May to November, but its overall productivity is low.

To ascertain whether a second crop of cowpea can be successfully grown using the new cowpea varieties, a millet-early cowpea intercropping trial was planted at IITA's Kano Station research farm, Minjibir, Nigeria, at the onset of rains in 1995. The millet was planted in rows 2 m apart on 15 June, while the cowpea varieties were planted between the millet rows on 1 July. After the millet and cowpea varieties had been harvested in the first week of September, the same cowpea varieties were planted as sole crops in the same plots on 12 September. The last rain, some 13 mm, was received on 25 September, after which the crop developed successfully on residual soil moisture. Overall, this system yielded 1934 kg ha^{-1} of millet grain,

Table 4.9. Grain and fodder yields (kg ha⁻¹) of cowpea, sorghum and millet in strip cropping, Minjibir, 1995¹.

Cowpea variety	Cowpea in sorghum				Cowpea in millet				Sorghum in cowpea		Millet in cowpea	
	Grain		Fodder		Grain		Fodder					
	Adjacent²	Middle	Adjacent	Middle	Adjacent	Middle	Adjacent	Middle	Grain	Stover	Grain	Stover
IT93K734	1615	2258	690	1157	835	761	320	450	2108	4791	843	2818
IT93K596-3	1316	1486	1023	1669	1237	1225	850	1691	2029	4818	863	2818
IT93K621-7	811	915	1157	1624	588	660	846	1580	2154	4783	960	2933
IT86D719	688	1303	610	1402	529	803	890	1491	2524	5582	635	2622
Aloka local	231	280	334	458	783	966	214	231	2187	5014	671	2534
IT86D715	1326	2223	1201	2047	1311	1479	1380	1825	2607	7396	493	2454
Danila	869	1160	601	1313	1517	1560	801	1179	2446	4040	849	2747
IT90K277-2	1005	2161	690	1157	1706	1937	779	1046	2462	5403	363	2303
Mean	983	1473	788	1353	1063	1174	760	1187	2315	5228	710	2666
LSD (5%)	114	206	302	613	114	126	177	302				

¹ Planting pattern: 1 cereal row:4 cowpea rows. Sorghum and millet were planted on 1 July, cowpea on 17 July. Millet was harvested end September, sorghum end November.
² Rows adjacent to millet or sorghum.

3818 kg ha^{-1} of millet stover, 176 kg ha^{-1} of cowpea grain and 477 kg ha^{-1} of cowpea fodder from the first cowpea crop, followed by 360 kg ha^{-1} of cowpea grain and 684 kg ha^{-1} of cowpea fodder from the second. These results are well above the normal yields achieved by farmers in this area. The best cowpea variety for the second crop was IT90K277-2, which yielded 918 kg ha^{-1} of grain and 1194 kg ha^{-1} of fodder when sprayed twice and 266 kg ha^{-1} of grain and 1900 kg ha^{-1} fodder in no-spray plots within 76 days of planting. In another field, where eight varieties were planted on 2 September 1995, the average grain yield was over 500 kg ha^{-1} of grain, without insecticide sprays. The yields would have been even higher if there had been more rain in September. The total rainfall during the year at Minjibir was only 559 mm, compared to the norm of over 800 mm. Similar experiments need to be conducted at more locations, since the results of this one have indicated what looks like a promising option for intensifying agriculture in the Sudan and moist savannas.

Dry-season Cowpea

Several West African countries have developed their irrigation sectors and/or have *fadamas*—low-lying riverbeds where cowpea can be grown on residual soil moisture during the dry season (Blade and Singh, 1994). Normally rice is grown in both the *fadamas* and the irrigated areas during the rainy season, while a range of crops, including wheat and vegetables, is grown during the dry season. Cowpea is a suitable alternative dry-season crop for such situations, as it requires only moderate amounts of water and matures in 80-90 days. The major constraints to dry-season cowpea cropping are viruses, leaf thrips, nematodes and aphids, for all of which resistant varieties are available. Several such varieties, developed at IITA, were evaluated under irrigation at Wudil (30 km east of Kano) and at Kadawa (40 km south of Kano) and in a wetland area at Nguru, Nigeria, in on-farm dry-season trials from 1991 to 1995. The grain yields of several cowpea varieties were between 1 and 2 t ha^{-1}, while fodder yields were between 4 and 10 t ha^{-1}. One of the most promising varieties is IT89KD288, which has combined resistance to aphids, thrips, bruchids, viruses and nematodes. Starting with 200 g of seeds of this variety in 1992, one farmer from Bunkure village (near Kano) was able to multiply sufficient seeds to distribute to 47 farmers, who planted about 0.5 ha each in 1995. The grain yield ranged from 800 to 1700 kg ha^{-1} and the fodder yield from 1100 to 2900 kg ha^{-1}. The crop matured at the end of April or the beginning of May, when the prices of cowpea grain and fodder were very high. Cattle herders purchased the standing fodder after the pods had been harvested and grazed their cattle in situ. Farmers made so much profit that this technology is taking off by itself. At the time of writing, we have already visited over 100 farms with dry-season cowpea for the 1996 season in Bunkure village alone. With a little support in seed production and input supply, cowpea cultivation during the dry season should spread fast, boosting much-needed food and fodder supplies in the region.

Nutritional Quality of Cowpea Fodder

A range of values has been reported, but the average protein content of cowpea fodder is 21%, with 60% dry-matter degradability. This compares with 4 to 7.5% protein in cereal stovers with less than 50% degradability (Powell, 1986; Tarawali *et al.*, 1996). IITA and ILRI are conducting expriments to compare the quantity and quality of fodder from cowpea with that from other forage legumes being promoted by ILRI as fodder resources in the dry subhumid zone. Results show that cowpea fodder is usually better than forage legumes in terms of both quantity and quality in semiarid areas (Tarawali *et al.*, 1996). Preliminary investigations into the effects of time of harvest and storage on fodder quality have revealed that quality at harvest is maintained under local storage conditions (roof-tops or tree-forks), so it is worthwhile harvesting fodder immediately the rains cease, before quality starts to deteriorate (Tarawali *et al.*, 1996).

Studies are under way to assess the genetic variability of quality attributes of cowpea fodder in different varieties. In a trial conducted at Ibadan in 1995, four cowpea varieties were evaluated for grain and fodder yields as well as their effect on milk yield. Whereas IT81D994 had the highest fodder yield (4 t ha^{-1} of dry matter), IT86D719 and IT86D716 were able to give about 500 kg ha^{-1} of grain as well as 1.5 to 2.5 t ha^{-1} of fodder (Fig. 4.2). A preliminary feeding trial to assess the effects of this cowpea hay on milk yield has indicated that there are significant differences in the milk yield obtained after feeding fodder of different varieties. On average, 1 kg of fodder of IT86D719 gave 1.11 l of milk, whereas IT89KD391 gave only 0.76 l. Studies are in progress to confirm these findings. Establishing

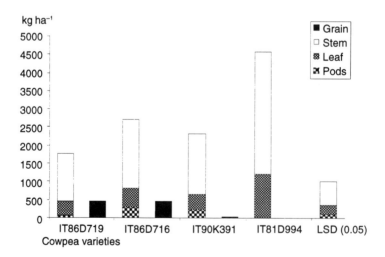

Fig. 4.2. Mean grain and fodder yields (dry matter) of different cowpea varieties harvested 85 days after planting at Ibadan, Nigeria, 1995.

genotypic differences in fodder quality offers exciting opportunities for breeding improved dual-purpose cowpeas.

Conclusions

Cowpea will continue to be a major source of food and fodder in West Africa, where it is also important in ensuring the stability and sustainability of mixed crop/livestock systems. Successful breeding efforts to develop drought tolerance and resistance to diseases, insect pests and parasitic weeds will enable cowpea cultivation to extend into drier areas, resulting in increased food and fodder production without harming the natural resource base. The current collaborative efforts between IITA and ILRI will further enable plant breeders to incorporate improved fodder quality traits into the background of existing productive varieties, allowing more efficient conversion of fodder into milk and meat and improved maintenance of animals required for the production of traction and manure.

References

Abadassi, J.A., Singh, B.B., Ladeinde, T.A.O., Shoyinka, S.A. and Emechebe, A.M. (1987) Inheritance of resistance to brown blotch, *Septoria* leaf spot and scab in cowpea *(Vigna unguiculata* (L.) Walp). *Indian Journal of Genetics* 47: 299-303.

Abalu, G.O.I. (1976) A note on crop mixtures under indigenous conditions in northern Nigeria. *Journal of Development Studies* 12: 212-220.

Adjadi, O., Singh, B.B. and Singh, S.R. (1985) Inheritance of bruchid resistance in cowpea. *Crop Science* 25: 740-742.

Aggarwal, V.D. (1991) Research on cowpea *Striga* resistance at IITA. In: Kim, S.K. (ed.), *Combating Striga in Africa.* Proceedings of the International Workshop, 22-24 August 1988, International Institute of Tropical Agriculture (IITA), Ibadan, Nigeria, pp. 90-95.

Agyemang, K., Little, D.A. and Singh, B.B. (1993) Emerging evidence of highly integrated crop/livestock farming systems in northern Nigeria: A case study from Kano State. *Cattle Research Network Newsletter* 12: 7.

Akundabweni, L.S., Peter-Paul, C. and Singh, B.B. (1989) Evaluation of elite lines of cowpea *(Vigna unguiculata* (L.) Walp.) for leaf/fodder plus grain (i.e. dual purpose). *Tropical Agriculture* (Trinidad) 67: 133-136.

Atokple, I.D.K., Singh, B.B. and Emechebe, A.M. (1995) Genetics of resistance to *Striga* and *Alectra* in cowpea. *Journal of Heredity* 86: 45-49.

Bata, H.D., Singh, B.B., Singh, S.R. and Ladeinde, T.A.O. (1987) Inheritance of resistance to aphid in cowpea. *Crop Science* 27: 892-894.

Bationo, A. and Mokwunye, A.V. (1991) Alleviating soil fertility constraints to increased crop production in West Africa. *Fertilizer Research* 29: 195-217.

Bationo, A., Ndunguru, B.J., N'tare, B.R., Christianson, B.C. and Mokwunye, A.V. (1991) Fertilizer management strategies for legume-based cropping systems in the West African semiarid tropics. In: Johansen, C., Lee, K.K. and Sahrawat, K.L. (eds), *Phosphorous Nutrition of Grain Legumes in the Semiarid Tropics.* Proceedings of an International Workshop, 8-11 January 1990, International Crops Research Institute for the Semi-Arid Tropics (ICRISAT), Patancheru, India, pp. 213-225.

Berner, D.K., Kling, J.G. and Singh, B.B. (1995) *Striga* research and control: A perspective from Africa. *Plant Disease* 79: 652-660.

Blade, S.F. and Singh, B.B. (1994) Providing options: Improving dry-season cowpea production in northern Nigeria. *Agronomy Abstract* 1994. American Society of Agronomy, Madison, Wisconsin, USA, p. 75.

Blade, S.F., Mather, D.E., Singh, B.B. and Smith, D.L. (1992) Evaluation of yield stability of cowpea under sole and intercrop management in Nigeria. *Euphytica* 61: 193-201.

Bottenberg, H., Tamo, M., Aroudokoun, M.D., Jackai, L.E.N., Singh, B.B. and Youm, O. (1996) Population dynamics and migration of cowpea pests in northern Nigeria. Implications for integrated pest management. *Proceedings of the Second World Cowpea Conference,* 3-8 September 1995, Accra, Ghana. International Institute of Tropical Agriculture (IITA), Ibadan, Nigeria.

Bourn, D., Wint, W., Blench, R. and Woolley, E. (1994) Nigerian livestock resources survey. *World Animal Review* 78: 49-58.

Ehlers, J.D. (1994) Correlation of performance of sole-crop and intercrop cowpeas with and without protection from insect pests. *Field Crops Research* 36: 133-143.

Emechebe, A.M. and Shoyinka, S.A. (1985) Fungal and bacterial diseases of cowpea in Africa. In: Singh, S.R. and Rachie, K.O. (eds), *Cowpea Research, Production and Utilization.* John Wiley & Sons, Chichester, UK, pp. 173-192.

Hall, A.E., Mutters, R.G. and Farquhar, G.D. (1992) Genotypic and drought-induced differences in carbon isotope discrimination and gas exchange of cowpea. *Crop Science* 32: 1-6.

ICRISAT (1991) *ICRISAT West African Programs Annual Report 1990.* International Crops Research Institute for the Semi-Arid Tropics, Niamey, Niger.

Lane, J.A., Moore, I. H.M., Child, D.V., Cardwell, K.F., Singh, B.B. and Baily, J.A. (1995) Virulence characteristics of a new race of the parasitic angiosperm *Striga gesnerioides* from southern Benin on cowpea. *Euphytica* 72: 183-188.

Miller, J.C. Jr and Fernandez, G.C.J. (1985) Selection for enhanced nitrogen fixation in cowpea. In: Singh, S.R. and Rachie, K.O. (eds), *Cowpea Research, Production and Utilization.* John Wiley & Sons, Chichester, UK, pp. 137-325.

Norman, D.W. (1974) Rationalizing mixed cropping under indigenous conditions: The example of northern Nigeria. *Journal of Development Studies* 11: 3-21.

N'tare, B.R. (1989) Evaluation of cowpea cultivars for intercropping with pearl millet in the Sahelian Zone of West Africa. *Field Crops Research* 20: 31-40.

N'tare, B.R. and Williams, J.H. (1992) Response of cowpea cultivars to planting pattern and date of sowing in intercrop with pearl millet in Niger. *Experimental Agriculture* 28: 41-48.

Paul, C.P., Singh, B.B. and Fatokun, C.A. (1988) Performance of dual-purpose cowpea varieties. *Tropical Grain Legume Bulletin* 35: 28-31.

Powell, S.M. (1986) Yields of sorghum and millet and stover consumption by livestock in the subhumid zone of Nigeria. *Tropical Agriculture* 62: 77-81.

Renard, C., N'tare, B.R. and Fussell, L.K. (1987) Progrès de la recherche sur les systèmes culturaux. In: *Les Cultures Associées au Mali.* Proceedings of a Seminar, 15-16 September 1987, Institut d'Economie Rural (IER) and International Crops Research Institute for the Semi-Arid Tropics (ICRISAT), Bamako, Mali, pp. 91-104.

Shetty, S.V.R. (1993) *Cropping Systems Research for the West African Semiarid Tropics: Current Status and Future Priorities at ICRISAT Sahelian Center.* International Crops Research Institute for the Semi-Arid Tropics, Niamey, Niger.

Singh, B.B. (1987) Breeding cowpea varieties for drought escape. In: Menyonga, J.M., Bezuneh, T. and Youndeowei, A. (eds), *Food Grain Production in Semiarid Africa.* Organization of African Unity (OAU) and Semi-Arid Food Grains Research and Development Network (SAFGRAD), Ouagadougou, Burkina Faso, pp. 299-306.

Singh, B.B. (1991) Improving cowpea varieties for traditional intercropping systems of the West African savanna. *Agronomy Abstract.* American Society of Agronomy, Madison, Wisconsin, USA.

Singh, B.B. (1993) *Cowpea Breeding: Archival Report (1988-1992) of Grain Legume Improvement Program.* International Institute of Tropical Agriculture (IITA), Ibadan, Nigeria, pp. 10-53.

Singh, B.B. (1994) Breeding suitable cowpea varieties for the West and Central African Savanna. In: Menyonga, J.M., Bezuneh, T.B., Yayock, J.Y. and Soumana, I. (eds), *Progress in Food Grain Research and Production in Semiarid Africa.* Organization of African Unity (OAU) and Semi-Arid Food Grains Research and Development Network (SAFGRAD), Ouagadougou, Burkina Faso, pp. 77-85.

Singh, B.B. and Emechebe, A.M. (1990) Genetics of *Striga* resistance in cowpea genotype B 301. *Crop Science* 30: 879-881.

Singh, B.B. and Emechebe, A.M. (1991) Breeding for resistance to *Striga* and *Alectra* in cowpea. In: Ranson, J.K., Musselman, L.J., Worsham, A.D. and Parker, C. (eds), *Proceedings of the Fifth International Symposium on Parasitic Weeds,* 24-30 June 1991, Centro Internacional de Mejoramiento de Maíz y Trigo (CIMMYT), Nairobi, Kenya, pp. 303-305.

Singh, B.B. and N'tare, B.R. (1985) Development of improved cowpea varieties in Africa. In: Singh, S.R. and Rachie, K.O. (eds), *Cowpea Research, Production and Utilization.* John Wiley & Sons, Chichester, UK, pp. 105-115.

Singh, B.B. and Singh, S.R. (1990) Breeding for bruchid resistance in cowpea. In: Fujii, K., Gatehouse, A.M.R., Johnson, C.D., Mitchel, R. and Yoshida, T. (eds),

Bruchids and Legumes: Economics, Ecology and Co-evolution. Kluwer Academic Publishers, The Netherlands, pp. 219-228.

Singh, B.B., Singh, S.R. and Jackai, L.E.N. (1984) Cowpea breeding for disease and insect resistance. In: *Breeding for Durable Disease and Pest Resistance.* FAO Plant Production and Protection Paper No. 55, Food and Agriculture Organization of the United Nations, Rome, Italy, pp. 139-152.

Singh, B.B., Singh, S.R. and Adjadi, O. (1985) Bruchid resistance in cowpea. *Crop Science* 25: 736-739.

Singh, B.B., Thottappilly, G. and Rossel, H.W. (1987) Breeding for multiple virus resistance in cowpea. *Agronomy Abstract.* American Society of Agronomy, Madison, Wisconsin, USA.

Singh, B.B., Mai-Kodomi, Y. and Terao, T. (1995) A simple screening method for drought tolerance in cowpea. *Agronomy Abstracts 1995.* American Society of Agronomy, Madison, Wisconsin, USA, p. 71.

Singh, B.B., Sharma, B.M. and Chambliss, O.L. (1996). Recent advances in cowpea breeding. In: *Proceedings of the Second World Cowpea Research Conference, 5-8 September 1995, Accra, Ghana.* International Institute of Tropical Agriculture (IITA), Ibadan, Nigeria.

Singh, S.R. and Jackai, L.E.N. (1985) Insect pests of cowpea in Africa: Their life cycle, economic importance and potential for control. In: Singh, S.R. and Rachie, K.O. (eds), *Cowpea Research, Production and Utilization.* John Wiley & Sons, Chichester, UK, pp. 217-231.

Sivakumar, M.V.K. (1990) Exploiting rainy season potential from the onset of rains in the Sahelian zone of West Africa. *Agriculture and Forest Meteorology* 51: 321-332.

Steele, W.M. (1972) Cowpea in Nigeria. PhD thesis, University of Reading, UK.

Tarawali, S.A., Singh, B.B., Peters, M. and Blade, S.F. (1996) Cowpea haulms as fodder. In: *Proceedings of the Second World Cowpea Research Conference, 5-8 September 1995, Accra, Ghana.* International Institute of Tropical Agriculture (IITA), Ibadan, Nigeria.

Terao, T., Matsunaga, R. and Singh, B.B. (1995a) Important root features of cowpea for adaptation to drought. *Agronomy Abstracts 1995.* American Society of Agronomy, Madison, Wisconsin, USA, p. 100.

Terao, T., Watanabe, I. and Singh, B.B. (1995b) Branching: A significant factor for cowpea performance under intercropping. *Agronomy Abstracts 1995.* American Society of Agronomy, Madison, Wisconsin, USA, p. 101.

Walker, D.W. and Miller, J.C. Jr (1986) Influence of water stress on nitrogen fixation in cowpea. *Journal of American Society of Horticultural Science* 111: 451-458.

Watanabe, I., Terao, T. and Singh, B.B. (1995) Screening cowpea lines for drought tolerance. *Agronomy Abstracts 1995.* American Society of Agronomy, Madison, Wisconsin, USA, p. 101.

5. Dynamics of Feed Resources in Mixed Farming Systems in Southeast Asia

Domingo B. Roxas[1], Metha Wanapat[2] and Md. Winugroho[3]
[1]Institute of Animal Science, University of The Philippines, Los Baños College, Laguna 4031, The Philippines
[2]Department of Animal Science, Faculty of Agriculture, Khon Kaen University, Khon Kaen 4002, Thailand
[3]Research Institute for Animal Production, PO Box 221, Bogor 16002, Indonesia

Abstract

Mixed crop/livestock farming systems are predominant in Southeast Asian countries. Animal production contributes a variable but significant proportion to farmers' incomes. This paper discusses the role of crop residues as feed for ruminants in relation to other feed resources available on the farm. In general, crop residues are not fully utilized as animal feed for a number of reasons, such as seasonal availability, low quality and collection and storage problems. Better utilization is essential if the prospects for the efficient use of feed resources in the next century are to improve. This depends greatly on whether scientists will be able to generate new technologies appropriate for small farms.

Introduction

Animal production is a very important component of agriculture in Southeast Asia. Farming systems in most countries are characterized mainly by integrated crop/livestock production, and the farmer is concerned with the day-to-day feeding of his or her animals in addition to raising crops. Even in mixed systems that are largely geared towards crop production, there are ample opportunities for achieving moderate levels of livestock production (van Bruchem and Zemmelink, 1995). However, the most appropriate strategy for doing so may differ considerably according to local conditions. Thus it becomes necessary to take a critical look at the feed resources available in different agroecologies.

The stability of feed supplies is a very important consideration in livestock production. Stability increases the farmer's chances of profiting from livestock raising, whether this is for meat, milk or other products. Feed supply varies greatly according to season; hence availability may differ from one location to another. In a farming systems perspective, the first step in any project should be to identify the locally available supply of both energy and protein sources.

Ruminant Production Systems in Southeast Asia

Because of the humid conditions prevailing in Southeast Asia, the predominant ruminant production systems are all sedentary, including some with solely live-stock and others that are mixed crop/livestock systems (Mahadevan and Devendra, 1986). The present production systems have evolved as a result of various factors, such as land availability, type of crop production practised, the frequency of crop-ping, the area of uncultivable wasteland and the options for growing forages. They may be classified generally as: (i) extensive systems; (ii) systems combining arable cropping; and (iii) systems integrated with tree cropping.

In Indonesia, the animal subsector contributed 2.11% to the country's gross domestic product (GDP) in 1992—a slight decline from its previous level of 2.25% (Sabrani and Saepudin, 1995). However, as a proportion of agriculture it grew from 6.0 to 11.3%, contributing up to 60% of farmers' incomes, especially in the arid and semiarid zones.

Table 5.1 shows the animal populations of Indonesia and other countries in 1994. Among the Southeast Asian countries, Indonesia has the largest numbers of buffalo, cattle, sheep and goats. Its two main animal production sectors, traditional and modern, differ in their development modes, objectives, decision making, investment capability and technology adoption. Agroclimate is another major factor influencing livestock production and its contribution to farmers' incomes, which is greater in upland and semiarid areas (Table 5.2).

In The Philippines, livestock production contributes 2.5% to GDP (Roxas, 1995a). The ruminant sector, composed of beef cattle, carabaos and goats, is mainly in the hands of small-scale farmers, who normally own or till 1 to 5 ha and possess about the same number of large ruminants. The dairy industry is practically non-existent. The cattle population, which is currently 1.8 million head, continues to decline at the rate of 1.1% a year, due mainly to the uncontrolled slaughter of breeding animals to meet the demand for beef. Carabaos number about 2.6 million head. Goats are the predominant small ruminant, numbering 2.8 million head. About 74 000 head of feeder cattle have been imported from Australia in an effort to meet the demand for beef and to save the breeding animals. But these are only short-term measures.

In Thailand, the Department of Livestock Development reported a total of about 6.9 million cattle in 1992 (Tongthainain, 1995). The figure for 1994, estimated by the Food and Agriculture Organization of the United Nations (FAO), stood at

Table 5.1. Ruminant population ('000s) in Southeast Asia, 1994.

Country	Buffalo	Cattle	Sheep	Goats
Brunei	10	1	-	-
Cambodia	829	2 589	-	-
Indonesia	3 512	11 595	6 411	12 281
Laos	1 308	1 137	-	153
Malaysia	186	686	336	356
Philippines	2 630	1 825	30	2 800
Singapore	-	-	-	1
Thailand	4 257	7 593	98	136
Vietnam	3 009	3 438	-	300

Source: FAO (1994).

Table 5.2. Major agroclimatic zones and livestock production systems of Indonesia.

Zone/area	Production system	Contribution to farmers' incomes (%)
Wet:		
West Java Sumatra	Lowland	5-13
Lalimantan Irian Jaya	Upland tree/crop/animal	10-20
Dry:		
Central Java East Java	Lowland	10-16
Bali Sulawesi	Upland	25-41
Semiarid:		
Nusatengara West Nusatengara East East Timor	Crop/pasture/animal	60

Source: Adapted from Sabrani and Saepudin (1995).

7.6 million, an increase of about 10%. Cattle are found mainly in smallholder crop/livestock production systems in the rainfed areas. Some 15 000 head of cattle are fattened and finished for market each year. Dairy farming is increasing, with some 203 753 head of dairy cattle in 1992 producing 245 322 t year^{-1} of raw milk. The high gross margins encourage farmers to take up dairy farming.

Despite a small population base, cattle, goat and sheep numbers in Malaysia have risen, but the number of buffalo has fallen (Tajuddin and Wan Zahari, 1992). Since the widespread introduction of rice double-cropping, many ruminants have been transferred to plantations. The integration of ruminant production with tree cropping is a logical option, given the high cost of developing land for pasture. More than 3 million ha of land are under the major plantation crops, including rubber, oil palm, coconut and fruit. Agroindustrial by-products, such as palm kernel cake, palm press fibre and rice straw, have been used for feeding ruminants.

Livestock production in Indo-China (Vietnam, Laos and Cambodia) plays an important part in the predominantly agricultural economy of these countries. The main ruminant species are water buffalo and cattle (EIU, 1993). The livestock component of agriculture has been neglected until recently (Remenyi and McWilliam, 1986). Breeding farms and artifical insemination centres, as well as feed processing factories, are now being established to modernize this subsector. However, per capita meat consumption remains low.

Throughout the region, trends in large ruminant numbers show substantial growth following periods of recovery in the crops subsector (Remenyi and McWilliam, 1986). Livestock represent a good investment opportunity for most farmers, for whom they are an important source of cash, especially in emergencies (Roder *et al.*, 1995).

Feed Resources

Table 5.3 shows the total land area and its usage in Southeast Asian countries. The percentage of arable land varies from 34% in Thailand, through 12% in Malaysia and The Philippines, to 7% in Indonesia. However, in terms of actual area, Indonesia and Thailand are probably the largest and therefore produce the greatest amount of crop residues in the region.

Crop/livestock farming systems are generally practised by smallholder farmers, who constitute the vast majority of the farming population in the developing countries of Asia (Carangal, 1995). Farmers' production systems are highly complex and variable, with interactions both between crop and animal and between on-farm and off-farm activities. Production systems are normally classified and described according to agroecological zones.

The majority of small farms are found in the rainfed lowlands, which form the backbone of Asian agriculture. The key characteristics of the systems in this zone are subsistence and low income (Devendra, 1983), but participation in the market economy is growing. It is estimated that this agroecology represents about 34% of

Table 5.3. Land use ('000 ha) in different countries of Southeast Asia.

Country	Land area	Arable land	Permanent crops	Permanent pasture	Forest	Other
Brunei	527	3	4	6	450	64
Cambodia	17 652	2 350	50	2 000	11 600	1 652
Indonesia	181 157	18 900	12 087	11 800	111 774	26 596
Laos	23 080	780	25	800	12 500	8 975
Malaysia	32 855	1 040	3 840	27	22 304	5 644
Philippines	29 817	5 520	3 670	280	13 600	5 747
Singapore	61	1	-	-	3	57
Thailand	51 089	17 600	3 200	800	13 500	15 989
Vietnam	32 549	5 500	1 200	330	9 650	15 869

Source: FAO (1994).

the total riceland in South and Southeast Asia. Many of the farmers are landless people. Rice growth and productivity depend on amount and duration of rainfall, together with the depth and duration of standing water, flooding frequency, soil type and topography. Both rice and other crops, such as cereals and oilseeds, are raised in intensive cropping systems. Crop and livestock production are usually integrated, but in ways that are highly variable, depending on specific environmental characteristics and resources, including labour availability. Areas with plentiful rainfall tend to have many swamp buffalo, some cattle, and relatively few sheep and goats. In drier areas, small ruminants flourish. A few large ruminants provide draught power. Monogastric animals, such as swine, chicken and ducks, are commonly raised, and these are the main recipients of rice by-products, including bran and middlings.

Crop and livestock production are more closely integrated in rainfed lowland and in upland systems than in the lowland irrigated areas (Devendra, 1995). Use of feed resources is more diverse in these systems, in which a broader array of crops and animals is found, leading to the development of different mixed crop/livestock enterprises. In The Philippines, the rotation of legumes with rice has increased the supply of leguminous by-products for ruminant feeding. Among these legume crops are mungbean and cowpea.

Feed resources in the irrigated lowlands include large amounts of rice straw, weeds and native grasses, the latter growing on bunds and in marginal areas as well as in the fields after harvest (Roxas, 1995b). By-products such as rice bran, rice middlings and broken rice are used by farmers to feed both monogastric and ruminant stock. The proportion of total rice area devoted to irrigated rice in Asian countries is over 50%.

In Indonesia, there is a large surplus of rice straw, only some of which is used to support ruminant production. On the island of Java, for instance, only a small proportion of the available straw is used in this way (van Bruchem and Zemmelink, 1995). It is fed mainly during the dry season, in areas where no green feed is available at this time. Rice straw is primarily used as a survival feed, barely maintaining the animal.

In areas closer to the mountains, green feeds are normally available even during the dry season, because of some precipitation and the presence of trees in nearby forests. The farmers work hard to collect these green feeds instead of feeding cereal straws.

In intensive milk-producing areas, rice straw is fed to some extent, but the production system is based largely on green feeds and concentrates. The rice straw that is fed is selected carefully in terms of quality, with newly harvested materials being preferred. In effect, farmers base their feeding system on the potential productivity of their animals.

In urban dairy enterprises some rice straw may be fed, but here again rice straw is not the major source of energy but is rather used to help maintain rumen functions. Due to scarcity of green feed, concentrates are the main source of nutrients. These enterprises must purchase nearly all their feed, so there is trade in rice straw and other forages to meet the needs of this subsector.

The situation in Thailand is similar to Indonesia and The Philippines, with crop residues still far from being fully utilized (Chantalakhana, 1984). The seasonal availability of crop residues such as rice straw seems to affect their utilization. Rice straw is generated at harvest, when rice land is also freed for grazing. The many small amounts of crop residues are not normally collected but are left in the field instead. The problems of collection and storage are major factors explaining the underutilization of residues. Chantalakhana (1984) further pointed out that another factor is the lack of training and motivation on the part of farmers, who are frequently unwilling to put in extra efforts to improve the feeding of their animals. Farmers look for immediate benefits from an intervention such as urea treatment of straw, but the effect of such technology may be felt only after months of feeding an animal.

In a few countries, notably India and Bangladesh, urea treatment may appeal to farmers who raise dairy animals, since milk output is monitored daily and the effect of improved feeding may be observed more easily.

Traditionally, farmers feed their animals grasses, legume straws and tree leaves, together with rice straw and other crop residues. Practical feeding systems in Thailand include supplementation with low-cost concentrates, urea, molasses and minerals (as blocks) (Chantalakhana, 1984). Such systems are now common in larger dairy farms or beef fattening operations.

In Indo-China, rice production is maintained, primarily because the returns to rice tend to be higher than those to annual industrial crops. About half of all riceland is double-cropped. Buffalo are the main species used for draught and are fed largely on cereal straws. Roder *et al.* (1995) point out that, in Laos at least, there is a long-term potential for increasing livestock production by replacing fallow vegetation with fodder crops. Improved fallow systems, especially those in which leguminous plants and the animals themselves are used to increase soil fertility, can lead to increased fodder availability, the suppression of weeds, accelerated nutrient cycling and improved soil moisture.

Prospects for Improving Feed Resources

The economies of most countries in Southeast Asia are expanding, leading to rapid urbanization. The World Bank projects that the region's urban population will be twice its rural population by 2030 (Pingali, 1995). Urbanization will be accompanied by rising incomes, with the result that the demand for livestock products will increase, at the expense of the traditional cereal-based diet. In fact, some countries in the region are already experiencing a decline in per capita rice consumption.

The escalating demand for livestock products will push animals to the limits of their potential for growth and milk yield (Ørskov, 1995). The consequence could well be less utilization of low-quality roughages, which cannot support satisfactory levels of productivity. But this tendency will be counteracted by the shrinking amount of land available for growing forages and other feed resources. Industrialization will continue to put pressure on land use for agriculture. Thus, the best option would be to put to use as much as possible of the biomass resulting from crop production. However, to do so it is necessary to find ways of improving the feed value of crop residues, so that these provide maximum benefit. As shown in Table 5.4, Southeast Asia has ample amounts of crop residues which can be tapped for feed supplies.

As Sabrani and Saepudin (1995) emphasized, if the livestock development objectives of the Indonesian Government are to be realized, technologies that can utilize resources with low opportunity costs need to be developed. Marginal lands, animal/tree/crop integration, native pasture and crop residues are among the options discussed. While the area of permanent pastures and forest lands that could contribute to forage biomass will shrink as human population rises, it is nevertheless thought that with proper management these areas could still become significant feed sources (Table 5.5). The introduction of higher-yielding forage species is an option for such areas.

An important approach to the increased production of feed resources in Indonesia has been the development of three-strata forage systems, especially in rainfed and upland areas (Nitis *et al.,* 1992). In these systems, grasses and ground legumes form the first and lowest stratum, shrub legumes the second and fodder trees the third. The increased forage available has enabled farmers to achieve higher stocking rates (3.2 animal units ha^{-1}, compared with 2.1 under the traditional system) and liveweights (375 kg ha^{-1}, compared with 122). This led to a 31% increase in farmers' incomes. In addition, the introduction of forage legumes reduced soil erosion by as much as 75% and brought greatly improved soil fertility.

Table 5.4. Availability of some crop residues in Southeast Asia ('000 t)[1].

Feed component	Indonesia	Malaysia	Philippines	Thailand
Rice hulls[2]	5 145	300	1 225	2 780
Rice straw[3]	38 306	2 009	11 535	22 841
Maize stover[4]	12 000	27	10 155	10 656
Sorghum stover[4]	9	-	-	981
Cassava leaves[5]	286	22	138	1 020
Sugarcane bagasse[6]	2 944	120	2 576	2 929
Sugarcane tops[7]	1 962	80	1 717	1 952

[1] At 90% DM level.
[2] Extraction rate = 15% (85% regarded as unpalatable or very poor).
[3] Ratio of grain to straw = 1:1.
[4] Ratio of grain to straw = 1:3.
[5] Estimated at 6% of cassava production.
[6] Estimated at 12% of cane production.
[7] Estimated as 20% of cane production or 8% of dry yield.
Source: Adapted from Ranjihan (1986).

Table 5.5. Availability of roughage for ruminants from the grazing area of Southeast Asia, 1993.

Classification	Area (million ha)	Yield (t ha^{-1} year^{-1})	Annual DM (million t)
Permanent pasture	13.913	0.8	11.13
Forest land	161.631	0.2	32.33
Unclassified land	54.097	0.4	21.64
Total	229.641		65.10

Source: FAO (1994).

In The Philippines, the introduction of forage grasses into the cropping system has also increased available feed resources. In addition, legume species such as *Sesbania*, *Desmanthus* and *Siratro* have demonstrated their potential to increase rice yields as well as produce forage.

Pasture development forms part of the Government of Thailand's efforts to increase feed supplies (Tongthainan, 1995). It includes the improvement of grazing lands and roadside areas, the establishment of backyard pastures and the integration of crop and livestock production. The communal grazing lands and roadsides are developed by oversowing hamata stylo (*Stylosanthes hamata*) seeds, either manually or by helicopter. For backyard pastures (forage gardens), grass species such as ruzi (*Brachiaria ruziziensis*), Hamil (*Panicum maximum* cv Hamil), Guinea (*Panicum maximum*) and Napier (*Pennisetum purpureum*) are used. To ensure sufficient seed supplies, the Department of Livestock Development launched two seed production projects, one with government research stations and the other using contract farmers. The contract farmers, who were smallholders, produced ruzi and hamata seeds at guaranteed prices from the government. In 1992, around 250 t of seeds were produced by 2485 contract farmers.

Napier grass is currently the most common species planted for cut fodder, while some other grasses such as setaria (*Setaria splendida*) and Guatemala grass (*Tripsacum laxum*) have been used for erosion control, particularly in rubber and oil palm estates in Malaysia (Halim and Suhaizi, 1994) and along hedgerows in Indonesia. Hybrid Napier (*P. purpureum* x *P. glaucum*) is gaining popularity because of its higher yield compared to common Napier.

The use of nutritious tree leaves to supplement crop residues is being promoted in The Philippines, Indonesia and Thailand. Leucaena (*Leucaena leucocephala*) and gliricidia (*Gliricidia sepium*) have long been used for fodder by farmers in these countries, where other fodder tree species are also being explored for their potential feed value.

The pre-treatment of crop residues has been introduced to farmers as another option for improving their feed value. Researchers have indicated urea treatment as promising. However, a number of constraints have prevented small-scale farmers from adopting this technology. These constraints need to be addressed if pre-treatment is to appeal to them.

Another approach to improving the feed value of crop residues is varietal selection. Several researchers have conducted studies on rice straws and other residues (Roxas *et al.*, 1984, 1986; Hart and Wanapat, 1986; Winugroho, 1986). While varietal differences in rice straw quality were shown to exist, the effects were greatly influenced by environmental factors, such as soil fertility, climate, and so on. Khush *et al.* (1998) suggested that screening on the basis of in vitro digestibility may only prove useful once the effect of environment is fully understood. Selection for straw quality should be an objective in breeding programmes alongside grain quantity and quality, the traditional concerns of plant breeders.

For several reasons, the conservation of fodder through silage making has not become widespread in Southeast Asian countries (Ranjhan, 1986). In the case of rice straw, which is harvested during the rainy season, ensiling would seem to be the appropriate technology, since it can be carried out even when the straw is wet. However, few farmers have ever actually tried it. Poor silage development, due to the low availability of sugars to promote fermentation, causes low quality and substantial losses due to decay. The high initial investment costs are a further factor which has discouraged farmers from trying this technology.

Conclusions

Crop residues will continue to play an important role as a basic feed resource for ruminant production in Southeast Asia. The increasing pressure on land and growing demand for livestock products will increase the need to make more effective use of all feed resources, including crop residues. The technologies now available will be considered more seriously by development workers and policy makers. Development efforts need to be strengthened so that these technologies will reach farmers. In addition, scientists will be called upon to explore other opportunities for improving the feed value of crop residues. However, new technologies must be designed with the conditions prevailing in smallholder farming systems firmly in mind. As Chantalakhana (1984) states, "Small farmers and farming systems ought to be kept as the central point, while their potential and limitations should be understood by crop residue scientists". Only if this approach prevails will technology be adopted, validating the investments made by governments in research and development.

References

Carangal, V.R. (1995) Methodology for extending crop/animal systems findings. In: Devendra, C. and Sevilla, C. (eds), *Crop/Animal Interaction.* IRRI Discussion Paper No. 6. International Rice Research Institute, Manila, The Philippines, pp. 533-549.

Chantalakhana, C. (1984) Relevance of crop residue utilization as animal feeds on small farms. In: Wanapat, M. and Devendra, C. (eds), *Relevance of Crop Residues as Animal Feeds in Developing Countries.* Funny Press, Bangkok, Thailand, pp. 1-12.

Devendra, C. (1983) Small farm systems combining crops and animals. In: *Proceedings of the World Conference on Animal Production,* 14-19 August 1983, Japanese Society of Zootechnical Science, Tokyo, pp. 173-191.

Devendra, C. (1995) Environmental characterization of crop/animal systems in rainfed areas. In: Devendra, C. and Sevilla, C. (eds), *Crop/Animal Interaction.* IRRI Discussion Paper No. 6. International Rice Research Institute, Manila, The Philippines, pp. 43-64.

EIU (1993) *Country Profile: Indo-China (Vietnam, Laos, Cambodia).* The Economist Intelligence Unit, London, UK.

FAO (1994) *Production Yearbook.* Food and Agriculture Organization of the United Nations, Rome, Italy.

Halim, R.A. and Suhaizi, M. (1994) Comparison of yield and quality changes with maturity in three forage grasses. In: *Sustainable Animal Production and the Environment.* Proceedings of the Seventh AAAP Animal Science Congress, 11-16 July 1994, Asian-Australian Association of Animal Production Societies, Ikatan Sarjana Ilmu-Ilmu, Peternakan, Indonesia, pp. 501-502.

Hart, F.J. and Wanapat, M. (1986) Comparison of the nutritive value of straw and stubble from rice grown in the northeast of Thailand. In: Dixon, R.M. (ed.), *Ruminant Feeding Systems Utilizing Fibrous Agricultural Residues.* International Development Program of Australian Universities and Colleges (IDP), Canberra, pp. 115-122.

Khush, G.S., Juliano, B.O. and Roxas, D.B. (1988) Genetic selection for improved nutritional quality of rice straw: A plant breeder's viewpoint. In: Reed, J.D., Capper, B.S. and Neate, P.J.H. (eds), *Plant Breeding and the Nutritive Value of Crop Residues.* Proceedings of a Workshop, 7-10 December 1987, International Livestock Centre for Africa (ILCA), Addis Ababa, Ethiopia, pp. 261-282.

Mahadevan, P. and Devendra, C. (1986) Present and projected ruminant production systems of Southeast Asia and the South Pacific. In: Blair, G.J., Ivory, D.A. and Evans, T.R. (eds), *Forages in Southeast Asian and South Pacific Agriculture.* ACIAR Proceedings No. 12, Australian Centre for International Agricultural Research, Canberra, pp. 7-11.

Nitis, I.M., Lana, K., Sukanten, W., Suarna, M. and Putra, S. (1992) The concept and development of the three-strata forage system. In: Devendra, C. (ed.), *Shrubs and Tree Fodders for Farm Animals.* International Development Research Centre (IDRC), Ottawa, Canada, pp. 92-102.

Ørskov, E.R. (1995) Livestock production in industrialized and less industrialized countries. *Bulletin of Animal Science* (special edition), Gadjah Mada University, Indonesia, pp. 15-20.

Pingali, P.L. (1995) Crop/livestock systems for tomorrow's Asia: From integration to specialization. In: Devendra, C. and Sevilla, C. (eds), *Crop/Animal Interaction.* IRRI Discussion Paper No. 6. International Rice Research Institute, Manila, The Philippines, pp. 481-499.

Ranjhan, S.K. (1986) Sources of feed for ruminant production in Southeast Asia. In: Blair, G.J., Ivory, D.A. and Evans, T.R. (eds), *Forages in Southeast Asian and South Pacific Agriculture.* ACIAR Proceedings No. 12, Australian Centre for International Agricultural Research, Canberra, Australia, pp. 24-28.

Remenyi, J.V. and McWilliam, J.R. (1986) Ruminant production trends in Southeast Asia and the South Pacific, and the need for forages. In: Blair, G.J., Ivory, D.A. and Evans, T.R. (eds), *Forages in Southeast Asian and South Pacific Agriculture.* ACIAR Proceedings No. 12, Australian Centre for International Agricultural Research, Canberra, pp. 1-6.

Roder, W., Phenchanh, S., Keoboulapha, B. and Maniphone, S. (1995) Research methodology for crop/animal systems in the hilly regions of Laos. In: Devendra, C. and Sevilla, C. (eds), *Crop/Animal Interaction.* IRRI Discussion Paper No. 6, International Rice Research Institute, Manila, The Philippines, pp. 131-150.

Roxas, D.B. (1995a) Livestock and poultry industry development in The Philippines: Potentials, problems and policies. *Bulletin of Animal Science* (special edition), Gadjah Mada University, Indonesia, pp. 27-32.

Roxas, D.B. (1995b) Environmental characterization of crop/animal systems in lowland irrigated areas. In: Devendra, C. and Sevilla, C. (eds), *Crop/Animal Interaction.* IRRI Discussion Paper No. 6. International Rice Research Institute, Manila, The Philippines, pp. 65-78.

Roxas, D.B., Castillo, L.S., Obsioma, A.R., Lapitan, R.M., Momongan, V.G. and Juliano, B.O. (1984) Chemical composition and in vitro digestibility of straw from different varieties of rice. In: Doyle, P.T. (ed.), *The Utilization of Fibrous Agricultural Residues as Animal Feeds.* Third Annual Workshop of Australian-Asian Fibrous Agricultural Residues Research Network. University of Melbourne, Australia, pp. 130-135.

Roxas, D.B., Karim, S.M.R., Lapitan, R.M., Pascual, F.S., Castillo, L.S. and Carangal, V.R. (1986) Composition and in vitro digestibility of residues from selected cowpea (*Vigna unguiculata*) varieties. In: Dixon, R.M. (ed.), *Ruminant Feeding Systems Utilizing Fibrous Agricultural Residues.* Proceedings of the Fifth Annual Workshop of the Australian-Asian Fibrous Agricultural Residues Research Network. International Development Program of Australian Colleges and Universities (IDP), Canberra, pp. 127-130.

Sabrani, M. and Saepudin, Y. (1995) Livestock and poultry industry in Indonesia: Policy, potential and problems. *Bulletin of Animal Science* (special edition), Gadjah Mada University, Indonesia, pp. 53-61.

Tajuddin, Z.A. and Wan Zahari, M. (1992) Research on nutrition and feed resources to enhance livestock production in Malaysia. In: *Utilization of Feed Resources in Relation to Nutrition and Physiology of Ruminants in the Tropics.* Proceedings of the Twenty-fifth International Symposium on Tropical Agricultural Research, 24-25 September 1991, Tropical Agriculture Research Centre, Tsukuba, Japan, pp. 9-25.

Tongthainan, Y. (1995) Extension strategies for increasing dairy and beef cattle production in Thailand. In: Devendra, C. and Sevilla, C. (eds), *Crop/Animal Interaction.* IRRI Discussion Paper No. 6. International Rice Research Institute, Manila, The Philippines, pp. 551-557.

van Bruchem, J. and Zemmelink, G. (1995) Towards sustainable ruminant livestock production in the tropics: Opportunities and limitations of rice straw-based systems. *Bulletin of Animal Science* (special edition), Gadjah Mada University, Indonesia, pp. 39-51.

Winugroho, M. (1986) Intake and digestibility of the upper and lower fractions of rice straw by sheep and goats. In: Dixon, R.M. (ed.), *Ruminant Feeding Systems Utilizing Fibrous Agricultural Residues.* Proceedings of the Fifth Annual Workshop of the Australian-Asian Fibrous Agricultural Residues Research Network. International Development Program of Australian Colleges and Universities (IDP), Canberra, pp. 123-126.

6. Dynamics of Feed Resources in Mixed Farming Systems of South Asia

Kiran Singh[1], G. Habib[2], M.M. Siddiqui[2] and M.N.M. Ibrahim[3]
[1] *Indian Veterinary Research Institute, Izatnagar, Bareilly 243 122, Uttar Pradesh, India*
[2] *North West Frontier Province Agricultural University, Peshawar, Pakistan*
[3] *Department of Animal Science, Faculty of Agriculture, University of Peradeniya, Sri Lanka*

Abstract

This paper reviews the feed resources of South Asia, providing, in the case of India, a detailed assessment of the supply and demand situation for different feed types. As in other developing regions, crop residues emerge as an increasingly important resource for small-scale producers as other feed resources, notably rangeland grazing, decline. Efforts to improve the feed value of crop residues have met with success on the research station, but have been little adopted by farmers. The research required to improve the future availability of high-quality feed rations should continue to span a wide range of topics. Innovative policy interventions should not be ruled out.

Introduction

Livestock production in South Asia is mainly a small-scale rural activity that forms an integral part of an age-old system of mixed agriculture. The animal species raised include the usual cattle, buffalo, sheep, goats, horses, pigs and poultry, besides some less common species such as yak, camel and mithun. Each species subsists on a different subset of the overall feeds available. Fish production can be a particularly interesting option in certain areas such as West Bengal and Bangladesh, but it is beyond the scope of this paper.

The various types of livestock provide different products and services, such as food, draught, fuel, manure, fibre, security, and so on. Only poultry farming and specialized peri-urban dairy farming have been developed as industries, with large units involving high-producing genotypes provided with adequate housing, feed and veterinary inputs. Animals in most traditional farming systems are simply allowed to graze and/or scavenge on wasteland, roadsides, forest edges or fallow fields. In the somewhat more intensive systems they are kept in sheds and fed with grasses, leaves, straws and household leftovers.

The low genetic potential, imbalanced feeding and poor health management of livestock in South Asia are the main factors responsible for the animals' generally low productivity. Crop and animal husbandry are interdependent: large ruminants are used to plough the land needed to produce food and cash crops and the residual biomass from these crops is used to feed them; in turn, the manure produced by the animals may be returned to the land. Most ruminants still receive at least some of their feed requirements through grazing natural grasslands, but these are deteriorating due to overgrazing (Jodha, 1986). As a result, crop residues and agroindustrial by-products are slowly assuming a more important place in ruminant production systems.

This paper will first describe the importance and diversity of small-scale livestock production systems in South Asia. It will then discuss the availability of different feed resources, before outlining the main trends in research and development. The main emphasis is on India, but we will also pay some attention to the situation in Pakistan and Sri Lanka. For want of data, we have referred only in passing to conditions elsewhere in the region.

Importance and Diversity

Livestock production forms an integral part of the South Asian rural economy. The region has large numbers of livestock, including poultry. India is a special case within the region owing to its size, the diversity of its production systems and the traditional ban on the slaughter of cows. According to the 1987 census, India had a total of 196 million cattle, 77 million buffalo, 99 million goats, 45 million sheep, 1.83 million horses, mules and donkeys, 1 million camels, 10.8 million pigs and 258 million poultry—numbers that dwarf those of the region's other countries. Pakistan, which has the region's second largest livestock population, has 36.5 million large ruminants (cattle and buffalo), 67.9 million small ruminants (sheep and goats), 182.6 million poultry and 5.30 million other animals.

The region's livestock populations harbour considerable genetic diversity. In India, this is reflected in at least 26 recognized breeds of cattle, 8 of buffalo, 40 of sheep, 20 of goats and 4 of camel, besides several breeds of horses, pigs and poultry. India has some of the world's best tropical dairy breeds and draught cattle, high-quality carpet wool sheep and prolific goats. Several of the region's breeds have been exported to other parts of the world. These include India's Yamnapari goat, now well-known in Indonesia, and the world-famous dual-purpose Sahiwal zebu cattle breed, which comes from India and Pakistan. The American zebu breeds also originated in South Asia.

Throughout the region the livestock sector generates rural employment. In addition, the bulk of the draught power required in agricultural operations is provided by livestock. The draught animals of India alone contribute some 30 000 MW of energy equivalent annually, saving electricity worth an estimated INR 100 000 million (US$ 2857 million). Other products include milk, meat and eggs and industrial raw materials such as fibre, skin and hides. Farmyard manure is an important source of fuel and fertilizer.

Almost 73% of rural households keep animals of one kind or another, and livestock play a special role in household security, particularly in smallholder farming systems. In areas with high livestock populations, income from livestock accounts for 30 to 50% of total farm income. A survey carried out by the national Council of Applied Economic Research in 1990 reported that revenue from milk sales in sample rural households accounted for 15 to 30% of total household income in different agroecological zones.

The total value of output from livestock in India is estimated at INR 436.6 billion (US$ 12.47 billion) in 1990-1991, or about 25% of the value of all agricultural output (excluding the contribution of animal draught power). The value of output from livestock grew nearly threefold between 1950 and 1990, the increase comparing well with the growth in the value of overall agricultural output. However, the national economy grew much faster—nearly fivefold over the same period—suggesting considerable room for improvement.

In the arid and semiarid areas, farming systems are highly diverse but crop residue feeding is almost everywhere crucial to the maintenance of animals. These areas constitute over 20% of India and an even larger proportion of Pakistan. Livestock production is often the principal source of livelihood in such areas, which are endowed with some of the region's best breeds of cattle, sheep, goats and camel. The animals survive on meagre resources, converting crop residues and other agricultural by-products into draught, wool, milk and meat. In the arid areas, rainfed crop production is highly erratic, a year of good rainfall being followed by two or three years of scanty rain or drought. Agricultural produce is therefore utilized almost entirely for domestic consumption, with any surpluses being stored against bad years. In such years livestock production has a stabilizing effect, enabling farmers to sustain their incomes and to make use of failed grain crops. Livestock produce that is surplus to local requirements is available for trade, providing a useful cash income to farm families.

In India as in other countries, rising human population is leading to a reduction in the area of open-access grazing land, and hence to increased dependence on crop residues and by-products—mainly straws, brans and cakes. Recently, there has been increased interest in the use of fodder trees and soil protection crops as additional sources of protein-rich animal feed.

Major constraints to the utilization of straws include their bulkiness and poor nutritive value. In addition, there is a tendency for better feeds (brans, cakes) to be removed to cities, thereby depriving rural areas. Other common problems across the region are droughts or floods resulting in seasonal or local shortages of feeds.

Bulky crop residues have high transport costs, limiting their use in overcoming these shortages.

Availability of Feed Resources

The major feed resources are grass and grazing, crop residues and by-products, cultivated fodders, weeds, and fodder from trees. The amount of fodder obtained by grazing varies from region to region according to the extent of forest areas and grazing lands and their proximity to human habitations. In hilly and forested areas, large numbers of animals are taken out to graze, and pastoralism or transhumance are not uncommon. In areas with more intensive cropping, the proportion contributed to diets by grazing and by cultivated fodder is relatively small. In such areas, crop residues are becoming increasingly important. In both Pakistan and India there is some cultivation of fodders such as berseem *(Trifolium alexandrinum)* (cool season) and pilipesara *(Vigna trilobata)* (as a catch crop) and of grain crops such as sorghum or maize for use as fodder. Fibrous crop residues (straws and stover) are poor-quality roughages that need supplementation with green fodder and/or concentrates to meet maintenance and production requirements. The production of crop residues depends on the yield and area cultivated to different crops. The principal residues used as fodder are the straws of cereals and pulses. Equally important are the by-products of oilseed processing, grain milling and other agroindustrial activities. However, these tend to become available in urban centres, not in the rural areas.

India

The first attempt to determine the requirements for and availability of different feeds for different categories of livestock in the Indian subcontinent was made by Sen and Ray (1941). They calculated feed requirements on the basis of the population census data of 1941 and then estimated the feeds available during the same year. This method, frequently used over the past half century, has a drawback in that it simply adds all feeds indiscriminately. It thus "forces" good animals to eat bad feed and hides seasonal and local discrepancies in protein:energy ratios. New methods therefore need to be developed (Zemmelink *et al.,* 1992; Kelley and Parthasarathy Rao, 1994; Schiere *et al.,* 1995).

Despite this drawback, Sen and Ray estimated feed availability to fall short of requirements by around 45%. Many similar studies have since been done, including those by Amble (1965), Whyte and Mathur (1965), India's Ministry of Food and Agriculture (1974), Ranjhan (1974) and ICAR (1985). All these studies found feed deficiencies.

The estimated current availability of dry and green fodders (crop residues, cultivated fodders and grazing) for feeding livestock in India is presented in Tables 6.1 and 6.2. The figures for the production of major crops are estimated from Ministry of Agriculture data for 1991-1992. The availability of crop residues was calculated using a suitable extraction ratio. The amount of grazing available was estimated by taking the total area under grazing from data of the Food and Agriculture Organization of the United Nations (FAO), with yields per hectare of green fodder calculated on the basis of grazing trials conducted from 1956 to 1968 at Mathura Veterinary College, Uttar Pradesh. The area under cultivated fodder was assumed to be 4% of the cropped area, or 6.6 million ha, with yields averaging 40 t ha^{-1}.

Table 6.1. Production and potential availability of residues of major crops in India,1991-1992.

Category	Crop	Production (million t)	By-product (million t)	Extraction ratio
Straws	Wheat	55.09	55.09	1:1
	Rice	73.99	95.76	1:1.3
	Barley	1.65	2.14	1:1.3
Stovers	Maize	7.98	23.94	1:3
	Small millets	0.95	3.80	1:4
	Sorghum	8.36	33.44	1:4
	Pearl millet	4.64	18.56	1:4
	Finger millet	2.68	10.72	1:4
Tuber crops	Potato[1]	15.50	3.10	1:0.2
	Cassava	5.67	1.12	1:0.2
Legumes	Pigeonpea	2.19	8.76	1:4
	Chickpea	4.15	16.60	1:4
	Others	5.71	22.84	1:4
Sugarcane	Cane	239.00	59.75	1:0.25
Oilseeds	Groundnut	7.06	14.14	1:2
	Sesame	0.67	1.34	1:2
	Niger seed	0.16	0.32	1:2
	Rapeseed/mustard	5.84	11.68	1:2
	Safflower	0.19	0.38	1:2
	Sunflower	1.18	2.36	1:2
	Soyabean	2.27	4.54	1:2
	Linseed	0.30	0.60	1:2

[1] Potato tops not included, as not currently fed to livestock.
Sources: Kossila (1984) and Ministry of Agriculture (1992).

Table 6.2. Availability of green fodders in India, 1991.

Green fodders	Area (million ha)	Yield (t ha⁻¹)	Availability (million t)	Remarks
Cultivated fodders				4% of total area cultivated; 50 t for irrigated and 30 t for unirrigated land
Non-legumes and legumes	6.6	40.0	264.0	
Grazing				
Permanent pasture	12.0	5.0	60.0	
Forest land	66.7	3.0	200.0	
Other	49.5	1.0	49.5	
Total green fodder			573.5	

Source: FAO (1991).

The potential availability of concentrates is given in Table 6.3. Again, appropriate extraction rates have been used. Table 6.4 shows the availability of cereal grains for livestock feeding. We assume that some 2 to 4% of the total cereal grains produced is used in this way. Most of this is fed to poultry and high-yielding diary animals, particularly around cities.

Using the information in Tables 6.1 to 6.4, the total availability of all feed resources in India is summarized in Table 6.5. This table shows that, since the last estimate of the national Commission on Agriculture, which was based on 1976 data, there has been an overall increase in the availability of feeds. This is largely due to a significant increase in crop production. Over the past 20 years the production of cereals has increased markedly, and with it that of straws and stovers. Similarly, the total production of oilseeds and hence of cakes and meals has also increased. However, there has been little increase in green fodder production, the area devoted to it having risen from 3 to just 4% of cropped land. More recent information on this point is available in Kelley and Parthasarathy Rao (1994).

Estimates of feed requirements can be based on the nutrient requirements of livestock, combined with livestock population statistics and data on animal productivity. We have used the nutrient requirements proposed by ICAR (1985) and livestock population and productivity statistics from FAO (1991). We have assumed that the entire sheep and goat population gets all its feed requirements from grazing, except for 1 million improved animals of each species which are provided with concentrate feeding. We have also assumed that about 90% of pigs feed by scavenging and only 10% are commercially raised using high-quality feeds.

On this basis we then calculated availability and requirements in terms of digestible crude protein (DCP) and total digestible nutrients (TDN) in 1990-1991.

Table 6.3. Potential availability of concentrates in India, 1991.

Concentrate	Seed yield (million t)	Extraction rate (%)	Meal production (million t)
Groundnut cake (undecorticated)	7.06	60	4.24
Sesame cake	0.67	70	0.47
Sunflower cake	1.18	70	0.83
Linseed cake	0.30	70	0.21
Rapeseed cake	5.84	70	4.09
Safflower cake	0.19	70	0.13
Soyabean cake	2.27	80	1.82
Subtotal			11.79
Ambaoli cake (*Hibiscus cannabinus*)			0.03
Babool seeds (*Prosopis spicigera*)			0.06
Cassava starch waste			0.04
Castor bean meal			0.06
Coconut pith			0.20
Karanj cake (*Pongamia glabra*)			0.13
Kusum cake (*Schleiclera trijuga*)			0.03
Mango seed kernels			1.00
Mahua seed cake (*Bassia longifolia*)			0.30
Niger seed cake			0.10
Rubber seed cake			0.15
Sal seed cake (*Shorea robusta*)			0.07
Spent brewer's grain			0.05
Tamarind seeds (decorticated)			8.70
Tea waste			0.02
Subtotal			10.97
Wheat bran		8.0	4.72
Rice bran		8.0	5.89
Pulse dehulling by-products		3.0	0.36
Subtotal			10.97
Cottonseed cake [1]			0.49
Total			34.22

[1] Calculated as follows: cotton yield (1991-1992) = 9.835 million bales (170 kg bale^{-1}); cottonseed = 59.5 kg bale^{-1}; cottonseed cake = 49.98 kg bale^{-1} or 0.04998 t x 9.835 = 0.492 million t. Source: Ministry of Agriculture (1992).

Table 6.4. Production and potential availability of cereal grains for feeding livestock in India, 1991-1992.

Grain	Yield (million t)	Availability for livestock	
		(%)	(million t)
Wheat	55.09	2	1.102
Rice	73.66	2	1.473
Barley	1.65	10	0.165
Maize	7.98	10	0.798
Millet	0.95	10	0.095
Sorghum	8.36	5	0.418
Pearl millet	4.64	5	0.2.32
Finger millet	2.68	5	0.134
Total			4.417

Sources: Kossila (1984) and Ministry of Agriculture (1992).

Table 6.5. Potential availability of all animal feeds in India, 1991-1992.

Category	Quantity (million t)	Dry matter	
		(%)	(million t)
Dry feeds			
Straws/stovers	387.86	90	349.07
Green feeds			
Green fodders	264.00	25	66.00
Grazing	309.50	25	77.40
Subtotal	573.50		143.40
Concentrate			
Meals/bran	34.27	90	30.84
Grains			
Cereals	4.42	90	3.98
Subtotal	38.69		34.82
Total dry matter			527.29

The calculations show that there is still a substantial gap between supply and demand of about 31%. Most critical is the deficiency of DCP, which amounts to about 58%, showing a severe imbalance in the ratio of poor-quality roughages to good feeds.

Pakistan

Pakistan's livestock population is supported by feed resources derived from the crop sector, from rangeland and other grazing areas, and from agroindustrial by-products. The type, availability and utilization of these feed resources varies greatly in the country's different agroecological zones. In order of importance, the major feed resources are crop residues (46%), grazing (27%), cultivated fodders (19%), cereal/legume grains and by-products (6%) and oil cakes, meals and animal protein (2%). Most farmers (about 75%) have small land holdings on which most of the livestock population is concentrated. The smallholders' priority is to grow cereal grain for human consumption, but these crops also provide straw and stover for their animals. In the case of wheat, the value of the straw is around 60% of that of the grain, a similar ratio to that reported for India (Kelley *et al.*, 1991).

An estimated 40 million t of crop residues are produced annually in Pakistan, out of which 52.5 and 22% are contributed by wheat and rice respectively. Traditionally, cereal straws are fed to cattle and buffalo year-round, but their proportion in the ration increases during periods of feed scarcity. Treatment of straw with urea is not commonly practised by farmers, partly due to weak extension services but mainly because these technologies appear unsuitable for resource-poor farming systems (Schiere and de Wit, 1993; Schiere *et al.*, 1995).

Cultivated fodders are used as cut-and-carry feeds and may include berseem, oats, rape, barley and sometimes wheat during the winter season and maize, sorghum and millet during the summer season. Most of these crops are ready for harvesting about 2 to 3 months after sowing. Periods of scarcity occur in May-July and again in November-December. Of the total cultivated area, only 13% is devoted to fodder crop production. Despite large increases in the ruminant population (62%) during the past 20 years, the land devoted to fodder crops has declined by about 17%, with a corresponding increase in land used for food grain production. This has further increased the dependence of livestock on crop residues and by-products. Fodder crop yields are very low in traditional farming systems, primarily due to the low use of inputs by farmers. With improved seed and production technology, they can be increased by up to 250%. In recent years, farmers have shown great interest in cultivating high-yielding fodder varieties such as Napier grass, sadabahar (*Andropogon gayanus*) and multi-cut oats. Fodder conservation through silage or hay making is not commonly practised.

Farmers collect weeds and grasses for use as fodder while they tend their cultivated crops. In addition, the grazing of animals on roadsides and canal banks, and on fallow or unutilized land is a common practice in most areas. Fodder trees and shrubs also contribute to the overall feed supply, particularly in hilly areas. In the plains, sugarcane is another major resource, supplying 13.3 million t of tops. Depending on the market price, whole sugarcane plant may be fed to lactating cattle and buffalo.

Some 24.83 million ha of rangeland contribute about 16% to the total availability of feed dry matter. In some places, however, rangeland provides about 30%

of the total feed supply. Due to continuous overgrazing and lack of range improvement, carrying capacity and the quality of grasses is deteriorating alarmingly. An exception is the highland grazing areas, but use of these is constrained by climatic factors and remoteness. Transhumant and pastoral herdsmen, with their flocks of goats and sheep, are the main users.

The major sources of supplementary feed in Pakistan are by-products from cereal milling and oilseed production. By-products of animal origin play a minor part. Wheat and rice bran and rice polishings are the main milling by-products. Cottonseed cake, rapeseed cake and maize oil cake account for almost two-thirds of the total protein supplement used to feed dairy animals. Meals of vegetable and animal origin are relatively costly and used only in poultry rations. A considerable quantity of molasses is produced annually (1.14 million t year^{-1}), but its use in animal rations is limited, mainly due to alternative uses in distillery, export to other countries and difficulties in transportation and storage at local level. The introduction of molasses-urea blocks was tried as a strategy for increasing ruminant production in dry areas, but has not yet been widely adopted.

Pakistan entered the world market for animal feeds in the early 1970s. By 1991-1992, the export of animal feed ingredients (oil cakes and cereal brans) to West Asian countries was fetching US$ 5.73 million annually. In 1993-1994, however, these exports fell sharply to US$ 0.486 million. Soyabean meal is the major feed imported, used for manufacturing compound poultry feed. The country has 215 poultry feed mills, but only five mills producing compound feed for ruminants. Generally, mixed concentrate feeds for ruminants are prepared at home by farmers.

It has been calculated that the nutrients available under the present pattern of feed utilization do not meet the requirements of Pakistan's existing livestock population. There appear to be deficiencies of 39.4% of the TDN and 56.7% of the DCP requirements. There is a growing trend towards the establishment of more intensive dairy cattle and buffalo production systems in peri-urban areas, while intensification in the poultry sector has been well under way since 1970.

Sri Lanka

The feed situation in Sri Lanka resembles that of parts of India, Pakistan and other countries in the region. As in India and Pakistan, most livestock production takes place outside the commercial sector. Cows, goats, chickens and pigs produce significant amounts of food for rural and urban families, although production per head of livestock remains low. The slaughter of cows in Sri Lanka is not popular—a situation similar to that of India. Intensive commercial sectors have developed in poultry production and, to a lesser extent, urban dairy production. Rising population pressure means that domestic animals in rural areas are making increasing use of fibrous crop residues (Ibrahim and Schiere, 1986), although green feeds are also available in the wetter southwest of the country. The major traditional sources of concentrate feeds in Sri Lanka are coconut cake and rice bran.

The Government of Sri Lanka has expressed renewed interest in promoting the country's dairy industry. This is to be achieved by setting up a Dairy Development Board, with the help of special funding from the Asian Development Bank. In addition, the country's dairy cooperatives are to form a joint venture with similar organizations in India. The cooperatives are involved in the collection, marketing and processing of milk. Although compound feeds such as dairy mix and calf meal are available on the market, most dairy farmers still rely on coconut meal and rice bran. In spite of the subsidy schemes set up by the government to introduce improved pasture and fodder species, the feed situation remains largely unchanged. The failures of the pastures subsidy scheme were documented by Ibrahim (1994). Farmers in the dry zones rely on rice straw as a feed during the dry season, and in the rainy season allow animals to graze on roadsides and uncropped areas.

Table 6.6 presents the types and sources of feed available to farmers in the mid-country and coconut triangle of Sri Lanka. These resources are supplemented by various concentrates, rice bran and coconut cake being the two main traditional ones (Ibrahim, 1987). As wheat consumption rises, the availability of wheat bran is increasing. The total concentrate feed requirement was 225 000 t in 1994, but is projected to increase to 747 000 t by the year 2000. This phenomenal increase is due to the predicted expansion of the poultry and dairy cattle industries. Two other ingredients that can be produced locally are maize and rice polishings, the demand for which in 1994 was 56 000 and 67 000 t respectively. These figures are projected to rise to 187 000 and 224 000 t respectively by the year 2000.

In the poultry subsector, the layer and broiler industry expanded very rapidly indeed in 1994-1995 due to various incentive schemes (such as contract rearing) introduced by local and foreign poultry/feed firms. Much of this expansion took place in the smallholder sector (units with 50 to 500 birds). The sector suffered a setback in early 1996 due to high concentrate feed prices and the outbreak of *gumboro* viral disease. The high feed price was due to the shortage of rice bran. Despite these problems, strong growth is likely to resume. The output of day-old chicks is projected to nearly double by the year 2000, resulting in an increase in egg production from 800 million to 1285 million over the same period. The production of broiler chicks is also projected to increase substantially, from 25 million in 1994 to 56 million by 2000, while broiler meat production rises from 23 400 to 60 000 t. This phenomenal growth will greatly affect feed availability in the ruminant subsector.

Trends in Research and Development

The human dietary requirements recommended by ICAR are 200 g day^{-1} of milk and 15 g day^{-1} of meat per person. (These figures may serve as a guideline to demand, although they need adjusting to allow for vegetarians.) On the basis of these figures, a human population of 990 million would require 79.5 million t of milk and 5.4 million t of meat per year. Current annual output in India stands at

Table 6.6. Fodder type, source and availability in Sri Lanka.

Fodder type/source	Feed availability
Off-farm grasses:	
Unfertilized, mainly from roadsides and railway reservations	Available throughout the year, except the dry months (February-April, July-September). Quantity is limited
Improved/fertilized, from government livestock-coconut farms	Access limited to members of families working the farm. Cost is SLR 75 per month, quantity unlimited
Unfertilized, on land belonging to temple/mosque	Access limited to those who own the plot or who are close to the priests
On-farm grasses:	
Rice field bunds (own or others)	Good grass growth because of use of fertilizer for rice cultivation. Access throughout the year, except during land preparation, when bund is repaired
Rice field (own)	Good grass growth because of use of fertilizer for rice cultivation. Animals cannot graze while rice is being cropped. Some farmers grow vegetables after rice, in which case non-grazing period is prolonged
Highland garden	Grass production limited, but other (tree) fodders may be available
Off-farm fodder/tree legumes:	
Roadsides	Throughout the year, if regularly harvested and pruned. Alternatively, leaf fall during the dry season may be used. Trees are not tall, but women farmers may sometimes be handicapped in reaching for taller branches
On-farm fodder/tree legumes:	
Perimeter fence	Usually regularly harvested, but quantity depends on plant spacing. Farmers can offer only leaves and twigs or whole cut branches. Generally within the reach of women
Live supports for peppers	Leaves and twigs are occasionally pruned. Branches retained to support peppers. Not allowed to gain over 3.5 m in height
Shade tree for coffee/cocoa	Fodder limited by need for tree to have height and spread sufficient to give shade
Off-farm creepers:	
Roadsides, railway reservations, lands belonging to temples, etc	Access is same as for grasses. Supply is more limited
On-farm creepers:	
Rice fields and bunds	Access is same as for grasses. Supply is more limited
On-farm jack trees	Fed occasionally. Heavy defoliation might affect fruit production. Tree must be climbed to pick leaves, which therefore tend to be fed by males only

60.8 million t of milk, indicating a large shortfall. Similar, though smaller, short-falls can be predicted for most other countries in the region, and for other livestock products such as eggs and wool.

To bridge these gaps, all countries in the region need an integrated research and development effort in animal breeding, animal nutrition and physiology, animal health and the processing of animal products.

The full expression of an animal's genetic potential in terms of milk production and other parameters depends on the provision of adequate nutrition. Genetic potential, however, is not the guiding principle for farm optimization seen by the smallholder, who has limited access to additional feed resources. For smallholders and development agents alike, the fundamental challenge is to develop a combination of animals and feeds that assures satisfactory levels of growth, reproduction and lactation. This must be based in large part on the optimal use of crop residues and by-products.

The importance of improving the feeding value of crop residues has long been well understood throughout the subcontinent. India's first project on sodium hydroxide treatment of straw was submitted for funding in the 1940s. Work on straw treatment continued for many decades (Jackson, 1977; Jayasuriya and Perera, 1982; Perdok *et al.*, 1982; Ibrahim, 1983; Saadullah, 1985; Doyle *et al.*, 1986). Workshops in Bangladesh focused on straw treatment with urea and supplementation strategies in the early 1980s (e.g. Jackson *et al.*, 1981; Preston *et al.*, 1982). In the 1980s the emphasis switched to the socioeconomic applicability of crop residue feeding methods (Wanapat and Devendra, 1985; Ibrahim and Schiere, 1986; Amir *et al.*, 1987). A major catalyst to the achievement of technical results was the work of the Indo-Dutch Project on the Bioconversion of Crop Residues (BIOCON), which operated in India from 1984 to 1995. Through a series of workshops project staff assembled and synthesized existing information on straw feeding methods (Kiran Singh and Schiere, 1993, 1995). They also developed a major new line of enquiry into the manipulation of genetic and environmental factors to improve the quality and quantity of crop residues (Joshi *et al.*, 1994; Seetharam *et al.*, 1995). The project's most important contribution, however, was to establish the suitability of different straw feeding technologies under different socioeconomic and biophysical circumstances. This included the consideration of gender issues (de Boer *et al.*, 1994a,b; Yazman *et al.*, 1994; Kiran Singh and Schiere, 1995; Singh *et al.*, 1995). This research showed that, in India as elsewhere in the region, treatments to improve the feed value of crop residues had so far found little adoption among resource-poor farmers.

Increased production from livestock to satisfy the needs of the rapidly increasing human population of South Asia requires continuing efforts to narrow the gap between supply of and demand for feeds throughout the subcontinent. It may also require a rethink of the direction of livestock development and the policies used to promote it.

Without pretending to be comprehensive, the following list of options for future research may be useful:

1. Adjustment of livestock types and numbers through: (i) suitable breeding strategies; (ii) biotechnology; (iii) government policies.

2. Increased production of fodder through: (i) high-input intensive fodder crop production; (ii) introduction of high-yielding forages and cultivation practices suited to different areas/regions; (iii) low-input pasture and forage production on marginal and submarginal land; (iv) production and dissemination of quality forage seed; (v) integrated systems of fodder production under dryland agriculture; (vi) coppice forage production system using forests and plantation sites; (vii) agroforestry system combining food and fodder production; (viii) forage production in areas with saline, alkaline or acid soils and in marshy sites; (ix) rangeland production from moisture-deficit areas; (x) preservation of surplus biomass for lean periods, in forms such as hay, silage and leaf meal; (xi) blending of shrub leaves with straws, to obtain mixtures with beneficial associative effects; (xii) minimizing losses; (xiii) discouraging burning of grasses (as practised by some forest departments); (xiv) dissemination of on-the-shelf technologies for improving the digestibility, nutritive value and palatability of crop residues (bearing in mind the limited applicability of many of these technologies); (xv) reducing losses through improved conservation, storage and feeding; (xvi) improvement of crop residues through treatments which can be adopted under village conditions; (xvii) using dual-purpose crop varieties, particularly in rainfed areas (Joshi *et al.*, 1994); (xviii) use of fast-growing root and leguminous crops in sandy soils of arid and semiarid zones.

3. Tapping of new and non-conventional feed resources, requiring studies on: (i) detoxification of antinutritional elements in animal feeds; (ii) biotechnologies for improving the nutritive value of feeds; (iii) development of supplementary feeds such as synthetic amino-acids, minerals and vitamins; (iv) use of mechanical harvesters and balers for voluminous crop residues and grasses, so as to economize on transport and storage costs (Thole *et al.*, 1988).

4. Strategic movement of fodder from surplus to deficit areas, possibly involving: (i) establishment of fodder reserves in suitable locations; (ii) creation of Animal Feed Boards; (iii) monitoring of surplus and deficit areas so as to facilitate interventions during scarcity periods; (iv) encouragement of commercial and non-government organizations and farmers to prepare supplementary feeds using recently developed technology; (v) re-allocation of currently exported feed ingredients to domestic use, creating higher-value animal products (which may then be exported).

Conclusions

South Asia has a highly diverse livestock sector employing many different feed resources of varying quality and availability. Only poultry and in some cases peri-urban dairy and pig production systems are commercialized. Most small-scale farmers base their livestock enterprises on the use of crop residues, resulting in a low output of milk and meat per animal. Livestock nonetheless provide essential support to crop production through the supply of manure and draught, and also serve as a buffer against illness, drought or other misfortunes.

The need to make better use of crop residues has prompted considerable research and many promising technologies are now available. Despite the general shortage of nutrients all over the region, none of these technologies can be universally applied—all are highly system-specific. New avenues for research and policy development may lie in the adjustment of livestock types and numbers, increased production of fodder, the tapping of new or non-conventional feed resources, and the strategic movement of fodder. This last implies rethinking the current policy of exporting certain feeds. More foreign exchange could be earned by exporting animal produce.

As regards extension, it is essential to disseminate existing technologies and management practices, many of which remain "on the shelf". Due attention must be paid to the systems specificity of these innovations. In other words, not all technologies can be applied everywhere. To be really successful, research and extension agents should strengthen their capacity to diagnose local needs, rather than relying on top-down technology-driven approaches. This issue was not addressed in this paper, but it is likely to be a major key to future progress.

References

Amble, V.N. (1965) Feed requirements of bovines and possibility of meeting them. *Indian Journal of Agricultural Research* 20 (1).

Amir, P., Akhtar, A.S. and Dawson, M.D. (eds) (1987) *Livestock in Pakistan Farming Systems Research.* Proceedings of a Workshop, 8-15 April 1987, Agricultural Research Council, Islamabad, Pakistan.

de Boer, A.J., Harika, A.S., Jain, D.K., Dixit, P.K., Lotan Singh, Patil, B.R. and Schiere, J.B. (1994a) *Targeting Livestock Research and Extension in India by Use of Agroecological Zoning and Transects.* Technical Publication No. 1. Division of Animal Sciences, Indian Council of Agricultural Research (ICAR), New Delhi, India, and Department of Tropical Animal Production, Wageningen Agricultural University, The Netherlands.

de Boer, A.J., Singh, C.B., Dixit, P.K., Singh, L. and Patil, B.R. (1994b) *Rapid Rural Appraisals to Assist Focusing of Research and Extension Programs for Livestock Development in India.* Technical Publication No. 2. Division of Animal Sciences, Indian Council of Agricultural Research (ICAR), New Delhi, India, and Department of Tropical Animal Production, Wageningen Agricultural University, The Netherlands.

Doyle, P.T., Devendra, C. and Pearce, G.R. (eds) (1986) *Rice Straw as a Feed for Ruminants.* International Development Program of Australian Universities and Colleges (IDP), Canberra, Australia.

FAO (1991) *Production Yearbook 1990.* Food and Agriculture Organization of the

United Nations, Rome, Italy.

Ibrahim, M.N.M. (1983) Physical, chemical, physico-chemical and biological treatments of crop residues. In: Pearce, G.R. (ed.), *The Utilization of Fibrous Agricultural Residues*. Research for Development Seminar Three, 19-23 May 1981, Australian Development Assistance Bureau (ADAB), Los Baños, The Philippines, and Australian Government Publishing Service, Canberra, pp. 53-68.

Ibrahim, M.N.M. (1987) Rice bran as a supplement for straw-based rations. In: Dixon, R.M. (ed.), *Ruminant Feeding Systems Utilizing Agricultural Residues*. Proceedings of the Sixth Annual Workshop of the AAFARR Network, 1-3 April 1986, University of The Philippines at Los Baños, and International Development Program of Australian Universities and Colleges (IDP), Canberra, pp. 139-146.

Ibrahim, M.N.M. (1994) Livestock development programmes and their impact on small-scale dairy farming in Sri Lanka. In: Zemmelink, G., Leegwater, P.H., Ibrahim, M.N.M. and van Bruchem, J. (eds), *Constraints and Opportunities for Increasing the Productivity of Cattle in Small-scale Crop/Livestock Systems*. Proceedings of an International Mid-term Workshop, 14-19 November 1994, pp.146-152.

Ibrahim, M.N.M. and Schiere, J.B. (eds) (1986) *Rice Straw and Related Feeds in Ruminant Rations*. Proceedings of an International Workshop, 24-28 March 1986, Kandy, Sri Lanka. Straw Utilization Project, Agricultural University, Wageningen, The Netherlands.

ICAR (1985) *Nutrient Requirements of Livestock and Poultry in India*. 1st edn. Indian Council of Agricultural Research, New Delhi.

Jackson, M.G. (1977) Review article: The alkali treatment of straw. *Animal Feed Science Technology* 2: 105-130.

Jackson, M.G., Dolberg, F., Davis, C.H., Haque, M. and Saadullah, M. (eds) (1981) *Maximum Livestock Production from Minimum Land*. Proceedings of an International Conference, 2-5 February 1981, Bangladesh Agricultural University, Mymensingh.

Jayasuriya, M.C.N. and Perera, H.G.D. (1982) Urea ammonia treatment of rice straw to improve its nutritive value for ruminants. *Agricultural Wastes* 4: 143-150.

Jodha, N.S. (1986) Common property resources and rural poor in dry regions of India. *Economic and Political Weekly* 21 (27): 1169-1181.

Joshi, A.L., Doyle, P.T. and Oosting, S.J. (1994) *Variation in the Quantity and Quality of Fibrous Crop Residues*. Proceedings of a National Seminar, 8-9 February 1994, Bharatya Agro-Industrial Foundation (BAIF), Pune, Maharasthra, India.

Kelley, T.G. and Parthasarathy Rao, P. (1994) Yield and quality characteristics of improved and traditional sorghum cultivars: Farmers' perceptions and preferences. In: Joshi, A.L., Doyle, P.T. and Oosting, S.J. (eds), *Variation in the Quantity and Quality of Fibrous Crop Residues*. Proceedings of a National Seminar, 8-9 February 1994, Bharatya Agro-Industrial Foundation (BAIF), Pune,

Maharasthra, India, pp. 133-145.

Kelley, T.G., Parthasarathy Rao, P. and Walker, T.S. (1991) The Relative Value of Cereal Straw Fodder in the Semiarid Tropics of India: Implications for Cereal Breeding Programs at ICRISAT. Economics Group Progress Report 105, Resource Management Program, International Crops Research Institute for the Semi-Arid Tropics (ICRISAT), Hyderabad, India.

Kiran Singh and Schiere, J.B. (eds) (1993) *Feeding of Ruminants on Fibrous Crop Residues: Aspects of Treatment, Feeding, Nutrient Evaluation, Research and Extension.* Proceedings of a Workshop, 4-8 February 1991, National Dairy Research Institute (NDRI), Karnal, Haryana, India. Indo-Dutch Project on Bioconversion and Indian Council of Agricultural Research (ICAR), New Delhi, and Department of Tropical Animal Production, Agricultural University of Wageningen, The Netherlands.

Kiran Singh and Schiere, J.B. (eds) (1995) *Handbook for Straw Feeding Systems: Principles and Applications with Emphasis on Indian Livestock Production.* Indian Council of Agricultural Research (ICAR), New Delhi, India.

Kossila, V.L. (1984) Location and potential feed use. In: Sundstøl, F. and Owen, E. (eds), *Straw and other Fibrous By-products as Feed.* Elsevier, Amsterdam, The Netherlands, p. 4.

Ministry of Agriculture (1992) *All-India Final Estimates of Principal Crops (1991-1992).* Directorate of Economics and Statistics, New Delhi.

Ministry of Food and Agriculture (1974) *Feed and Fodder Committee Report.* Government of India, Krishi Bhavan, New Delhi.

Perdok, H.B., Thamotharam, M., Blom, J.J., van den Born, H. and van Veluw, C. (1982) Practical experiences with urea-ensiled straw in Sri Lanka. In: Preston, T.R., Davis, C.H., Dolberg, F., Haque, M. and Saadullah, M. (eds), *Maximum Livestock Production from Minimum Land.* Proceedings of an International Conference, 13-18 February 1982, Bangladesh Agricultural University, Mymensingh.

Preston, T.R., Davis, C.H., Dolberg, F., Haque, M. and Saadullah, M. (eds) (1982) *Maximum Livestock Production from Minimum Land.* Proceedings of an International Conference, 13-18 February 1982, Bangladesh Agricultural University, Mymensingh.

Ranjan, S.K. (1974) Feed and fodder requirements for target milk and meat production in India. In: *National Symposium on Agricultural Research and Development since Independence*, p.34.

Saadullah, M. (1985) Supplementing urea-treated rice straw for native cattle in Bangladesh. In: Wanapat, M. and Devendra, C. (eds), *Relevance of Crop Residues in Developing Countries.* Proceedings of an International Workshop, 21-23 November 1985, International Foundation of Science, Khon Kaen, Thailand, pp. 315-329.

Schiere, J.B. and de Wit, J. (1993) Feeding standards and feeding systems. *Animal Feed Science and Technology* 43: 121-134.

Schiere, J.B., de Wit, J. and Steenstra, F. A. (1995) Matching animals and feeds for maximum farm system output in low-input agriculture: Exploratory thought experiments. PhD thesis (J.B. Schiere), Wageningen Agricultural University, The Netherlands.

Seetharam, A., Subba Rao, A. and Schiere, J.B. (eds) (1995) *Crop Improvement and its Impact on the Feeding Value of Straw and Stovers of Grain Cereals in India.* Proceedings of a Workshop, 21 November 1994, Krishi Bhavan, Indian Council of Agricultural Research (ICAR), New Delhi, India.

Sen, K.C. and Ray, S.N. (1941) Animal husbandry and crop planning in India, *Science and Culture* 6: 684-689.

Singh, C.B., Rao, S.V.N. and Jain, D.K. (1995) *Farming Systems Research for Improving Livestock Production and Crop Residue Utilization.* Proceedings of a National Seminar, 24-26 November 1994, National Dairy Research Institute (NDRI), Karnal, Haryana, India.

Thole, N.S., Joshi, A.L. and Rangnekar, D.V. (1988) Feed availability and nutritional status of dairy animals in western Maharashtra. In: Kiran Singh and Schiere, J.B. (eds), *Fibrous Crop Residues as Animal Feed*, Indian Council for Agricultural Research (ICAR), New Delhi, pp. 207-212.

Wanapat, M. and Devendra, C. (eds) (1985) *Relevance of Crop Residues as Animal Feeds in Developing Countries.* Funny Press, Bangkok, Thailand.

Whyte, R.O. and Mathur, M.L. (1965) An analysis of feed and fodder resources for livestock populations in India. *Indian Dairyman* 17.

Yazman J.A., Singh, C.B., Patil, B.R. and Udo, H.M.J. (1994) *On-farm Research for Testing of Appropriate Technologies in Crop/Livestock Production Systems.* Technical Publication No. 3. Division of Animal Sciences, Indian Council of Agricultural Research (ICAR), New Delhi, India, and Department of Tropical Animal Production, Wageningen Agricultural University, The Netherlands.

Zemmelink, G., Brouwer, B. and Ifar Subagiyo (1992) Feed utilization and the role of ruminants in farming systems. In: Ibrahim, M.N.M., De Jong, R., van Bruchem, J. and Purnomo, H. (eds), *Livestock and Feed Developement in the Tropics.* Proceedings of the International Seminar, 21-25 October 1991, Brawijaya University, Malang, Indonesia.

Acknowledgements

The author gratefully acknowledges the comments and assistance provided by J.B. Schiere of the Department of Animal Production Systems at Wageningen Agricultural University, The Netherlands.

7. Dynamics of Feed Resources in Mixed Farming Systems of West/Central Asia-North Africa

Thomas L. Nordblom, Anthony V. Goodchild, Farouk Shomo and Gustave Gintzburger
International Center for Agricultural Research in the Dry Areas,
PO Box 5466, Aleppo, Syria

Abstract

Small ruminants predominate in the mixed farming systems of the vast semiarid region that stretches from Morocco to Mongolia. Rising human population and demand for animal products have brought the region's rangelands under increasing pressure. Crop residues have gained in importance in terms of their share of animal diets over the past two decades. They appear certain to play a still greater role in the future. Feed grains and other concentrates have also become more important.

We need to find ways of improving both the quantity and quality of crop residues available for animal feed, and of increasing the efficiency with which these seasonal products can be used. The International Center for Agricultural Research in the Dry Areas (ICARDA) and its national research partners are conducting research to this end.

Introduction

While diverse in natural resources and human populations, the countries of West/ Central Asia-North Africa have several features in common that justify considering them together in a discussion of crop residues for livestock feeds. The cultural similarities across the vast semiarid "steppe" zones (100-400 mm annual rainfall) stretching from Morocco to Mongolia are matched by similarities in the zones' mixed farming systems. Small ruminants predominate, and there is highly seasonal use of rangelands and crop residues in animal diets. Rainfall occurs mainly in the cool to cold winter and early spring months, while summers are almost invariably hot and dry.

For the purposes of research, the International Center for Agricultural Research in the Dry Areas (ICARDA) divides the region's countries into five groups. The first four of these, which constitute the Center's "traditional" mandate area of West Asia-North Africa (WANA), are: (i) northern West Asia, which includes Afghanistan, Iran, Iraq, Jordan, Lebanon, Pakistan, Syria and Turkey; (ii) the Maghreb, including Algeria, Libya, Morocco and Tunisia; (iii) the Arabian Peninsula, comprising Kuwait, Oman, Qatar, Saudi Arabia and the United Arab Emirates; and (iv) the Nile Valley-Red Sea region, grouping Egypt, Ethiopia, Eritrea, Somalia, Sudan and Yemen. To these four groups the center has recently added a fifth: Central Asia, consisting of the five republics of Kazakhstan, Kyrgyzstan, Tajikistan, Turkmenistan and Uzbekistan. Long-time partners with ICARDA in germplasm exchange while part of the Soviet Union, these newly independent countries are seeking to expand their collaborative research with the center. Together the five groups constitute what could now be called the West/Central Asia-North Africa region.

Rising human populations and the associated rise in the demand for animal products have brought the region's rangelands under increasing pressure. More animals than ever graze shrinking areas of range for shorter periods each year (Jones and Darkenwald, 1967; Nordblom *et al.*, 1996). Central Asia, however, constitutes an exception: Kerven *et al.* (1995) report that pastoralists in Kazakhstan, Kyrgyzstan and Turkmenistan have recently extended the time and distance of their migratory cycles to make greater use of rangeland and reduce dependence on now expensive cultivated fodder.

Accompanying the improvement in transport and communication throughout the region is a shift from subsistence to market-oriented farming, with producers able to respond more swiftly and flexibly to changing opportunities.

The Role of Crop Residues

Animal Populations and Feed Supplies

Table 7.1 shows the populations of the major animal species for each country in the subregions identified above. It also shows the estimated proportions of national feed resources falling into the region's three major aniaml feed categories— rangeland grazing, feed grain and other concentrates, and crop residues and forage crops.

With a combined flock of 449 million sheep and goats, the West/Central Asia-North Africa region is home to 26% of the world's total small ruminant population of 1719 million. The region also contains 110 million of the world's 1283 million cattle.

Table 7.1. Livestock population and feed consumption in West/Central Asia-North Africa (5-year averages, 1990-1994).

Subregion/ country	Livestock (million head)			Feed composition (%)			
	Sheep	Goats	Cattle	Range grazing	Feed grain and other concentrates	Crop residues and forage crops	Total feed (million t)
Maghreb:							
Algeria	17.6	2.7	1.3	58	15	27	18.08
Libya	4.6	0.9	1.0	76	18	6	5.92
Morocco	15.2	4.9	3.0	33	18	49	18.36
Tunisia	6.6	1.4	0.6	46	15	39	6.07
Northern West Asia:							
Afganistan	14.2	2.2	1.5	76	3	21	19.92
Iran	45.0	24.0	7.0	44	11	45	50.17
Iraq	6.8	1.1	1.2	32	15	53	11.45
Lebanon	0.2	0.5	0.1	33	22	45	1.39
Jordan	1.9	0.5	0.0	40	50	10	1.69
Pakistan	27.1	38.5	17.8	39	7	54	85.44
Syria	13.5	1.1	0.8	14	20	66	11.64
Turkey	40.3	10.9	11.9	7	19	74	64.47
Arabian Peninsula:							
Kuwait	0.1	0.0	0.0	57	38	5	0.32
Oman	0.1	0.7	0.1	73	17	10	0.82
Qatar	0.1	0.1	0.0	8	69	23	0.08
Saudi Arabia	6.8	3.9	0.2	11	42	47	15.31
United Arab Emirates	0.3	0.8	0.1	7	79	14	0.35
Nile Valley/Red Sea:							
Egypt	3.4	2.8	2.9	4	16	80	49.24
Ethiopia/Eritrea	13.8	10.7	18.2	54	7	39	41.34
Somalia	11.0	13.1	3.8	91	1	8	15.35
Sudan	21.8	15.7	21.3	77	1	22	67.65
Yemen	3.7	3.3	1.1	71	3	26	11.31
Total WANA	250.4	136.5	92.8	42	11	47	485.06
Central Asia (CA):							
Kazakhstan	34.3	0.7	9.5	50	12	38	74.17
Kyrgystan	8.9	0.3	1.1	36	18	47	4.90
Tajikistan	2.2	0.8	1.3	37	39	24	1.91
Turkmenistan	5.6	0.2	0.9	68	16	15	9.04
Uzbekistan	8.2	0.9	4.9	34	38	28	12.07
Total Central Asia	59.2	2.9	17.7	49	16	35	102.09
Grand total WANA+CA	309.6	139.4	110.5	43	12	45	587.15

Source: FAO (1995).

These and other livestock (buffalo, camels, horses, mules, asses, chickens and turkeys) were enumerated for WANA by Nordblom and Shomo (1995) and for Central Asia by Nordblom *et al.* (1996), using data from the Food and Agriculture Organization of the United Nations (FAO). We then updated our calculations, on the basis of FAO (1995). Next, we aggregated the animal populations for each country by taking the number of each kind and multiplying by a factor for conversion to a standard tropical livestock unit (TLU), in this case a 250-kg bovine. Sheep, goats and cattle were assigned 0.1, 0.1 and 0.7 TLU per head respectively, following Jahnke (1982). For other animals we took conversion factors from Kosilla (1988), except for buffalo, which we counted as 1.0 TLU.

Estimating the quantities of feed available in each country was more problematic. Crop residues and some forage crops are not reported in the statistical series of any country in the region, nor in FAO statistics. Quantitative estimates for crop residues can be derived, with more or less confidence, by taking the reported statistics for the grain harvested and multiplying by a "residue-yield factor" specific to each. The factors used in our analysis were 1.3 for wheat, barley and rice straws and stubbles, 3.0 for maize residues, and 4.0 for sorghum and millet stalks and leaves and for cotton residues. As regards forage crops, data on beerseem clover were obtained from Abu Akkada (1984) for Egypt, from Slayman *et al.* (1986) for Lebanon, from Wardeh *et al.* (1982) for Oman, and from Noor (1987) for Pakistan. Data on teff in Ethiopia and Eritrea were taken from Kategile *et al.* (1981). We have not so far been able to find any data at all for some potentially important crop residues and forage crops in some countries and subregions (e.g. cotton leaves and alfalfa in Central Asia, and olive leaves and cactus throughout WANA).

Native pasture or rangeland is still the most important feed source for several countries in the region. It accounts for the largest share of the land surface of the Central Asian republics, totaling about 260 million ha (FAO, 1995). This compares with 272 million ha of steppe land (100-400 mm rainfall) in all of WANA (El-Beltagy, 1996). Our data source for pasture areas was FAO (1995). However, grazing offtakes from these lands are not reported in any regular series. We estimated offtakes by multiplying the reported native pasture area of each country by a country-specific assumption of dry-matter (DM) offtake per hectare. For ten countries our assumptions were based on ACSAD/AOAD (1985) values, as follows: Algeria, 300 kg ha^{-1}; Egypt, 90 kg ha^{-1}; Iraq, 110 kg ha^{-1}; Jordan, 79 kg ha^{-1}; Libya, 100 kg ha^{-1}; Morocco, 500 kg ha^{-1}; Saudi Arabia, 20 kg ha^{-1}; Sudan, 500 kg ha^{-1}; Syria, 200 kg ha^{-1}; and Tunisia, 600 kg ha^{-1}. For Pakistan, we took Noor's (1987) estimate of 660 kg ha^{-1}. For Lebanon, we used Slayman *et al.*'s (1986) figure of 470 kg ha^{-1}. For Afghanistan, Ethiopia, Iran, Turkey and Yemen we assumed a figure of 500 kg ha^{-1}; in the case of Yemen, this figure is corroborated by Hassan *et al.* (1982). In the case of Oman, we used the estimate of 218 000 t of total annual pasture offtake found in Wardeh *et al.* (1982). Central Asian rangelands were assumed to produce only 200 kg ha^{-1} (Gilmanov, personal communication; Nordblom *et al.*, 1996).

Feed grains and other concentrates, including barley grain, maize grain, cottonseed (whole, meal and cake), both domestic and imported, still comprise the smallest feed category for most countries in the region. Exceptions are Saudi Arabia, Jordan, Kuwait, Libya and the United Arab Emirates, where spectacular increases in feed imports have been reported over the past 20 years. Relatively high levels of feed grain and other concentrates have also been used over the past 20 years in Uzbekistan and Tajikistan.

Past Trends

The overall 20-year trends and balances of all feeds (crop residues, natural grazing and feed grains) and all livestock units (small ruminants, bovines, draught animals and poultry) are expressed in millions of metric tonnes of DM and millions of aggregate TLUs in Fig. 7.1.

In most countries there has been considerable growth in livestock numbers over the 20-year period. The exceptions are Afghanistan, Ethiopia and Iraq, doubtless reflecting the conflicts suffered in these countries. Feed supplies have risen in all countries.

The figure shows the close match between TLUs and feed across the small and large populations of livestock in both the WANA and Central Asia regions. Indeed, if we assume feeding levels of between 3 and 6% of liveweight per day, the range of annual dry-matter consumption obtained fits neatly with the middle range of the regions' feed and livestock balances. In this part of the world, we do not observe animal populations greatly in excess of the total feed base. Little of the domestic supply of cheap feeds such as crop residues and rangeland grazing goes unused, any available "niche" being immediately being filled up with animals. And where market demand for livestock products has outstripped what can be satisfied using low-cost feed sources, it becomes profitable to use feed grains, domestic or imported.

For each country, the three feed categories should add up to 100% of feed use at any given point in time. Data for the three categories in the 1990-1994 period have already been presented in Table 7.1. We have made similar calculations for 1970-1974 and are thus are able to plot 20-year trend lines for each of the countries in a composition chart (Fig. 7.2). This works on the same principle as a soil chart showing fractions of sand, silt and clay. That is, every point within the figure represents a percentage combination of the three feed sources, summing to 100%. Again, each arrow joins two points, 20 years apart in time.

As the figure clearly shows, crop residues have gained in importance over the past two decades. The gain is not merely relative—a higher share of animal diets—but also absolute, reflecting increased crop areas and yields.

Feed grains and other concentrates have also gained in importance. In most countries this is associated with the explosive growth of poultry production and the steady expansion of dairying. The most spectacular shifts towards feed grains have occurred in Jordan, Lebanon, Libya and Saudi Arabia, which have moved into feeding configurations that have long been typical of countries such as Uzbekistan

TLU (million)

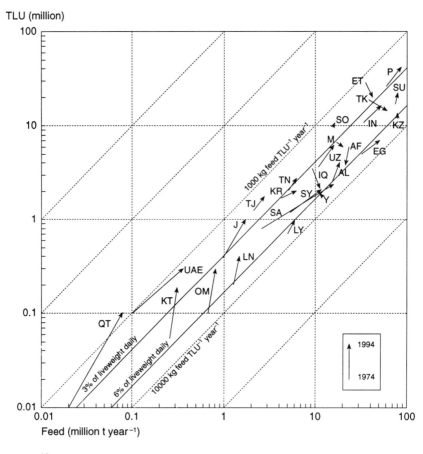

Feed (million t year⁻¹)

Key:

AF = Afghanistan, AL = Algeria, EG = Egypt, ET = Ethiopia/Eritrea,
IN = Iran, IQ = Iraq, J = Jordan, KZ = Kazakhstan, KT = Kuwait,
LN = Lebanon, LY = Libya, M = Morocco, OM = Oman, P = Pakistan,
QT = Qatar, SA = Saudi Arabia, SO = Somalia, SU = Sudan,
SY = Syria, TJ = Tajikistan, TN = Tunisia, TK = Turkey, TR = Turkmenistan,
UAE = United Arab Emirates, UZ = Uzbekistan, Y = Yemen.

Fig. 7.1. Feed availability per TLU, 1974 and 1994 (Nordblom *et al.*, 1996).

and Tajikistan, making considerable use of feed grains and agroindustrial by-products.

Except for Afghanistan, Kazakhstan and Yemen, all countries show diminishing proportions of range grazing in livestock diets. This result was obtained without accounting for rangeland degradation, for which we have no reliable quantitative estimates; it is simply the outcome of increased livestock populations, increased

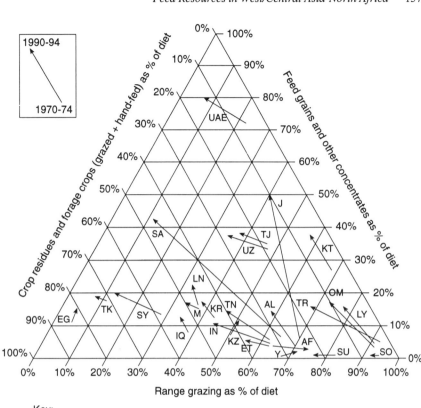

Key:

AF = Afghanistan, AL = Algeria, EG = Egypt, ET = Ethiopia/Eritrea, IN = Iran,
IQ = Iraq, J = Jordan, KZ = Kazakhstan, KR = Kyrgyzstan, KT = Kuwait,
LN = Lebanon, LY = Libya, M = Morocco, OM = Oman, P = Pakistan,
QT = Qatar, SA = Saudi Arabia, SO = Somalia, SU = Sudan, SY = Syria,
TJ = Tajikistan, TN = Tunisia, TK = Turkey, TR = Turkmenistan,
UAE = United Arab Emirates, UZ = Uzbekistan, Y = Yemen

Fig. 7.2. Trends in the composition (%) of livestock diets by feed category in West/Central
Asia-North Africa, 1970-1974 to 1990-1994.

cropping activities (generating crop residues) and increased use of feed grains. If
we could have quantified rangeland degradation, which has occurred across the
region (UNEP, 1992), the reduced role of this feed source would have appeared
still more pronounced.

Future Outlook

Domestic feed supplies in most countries of the region are likely to continue to
grow in response to developments in crop and sown pasture production. According

to a study by the United States Department of Agriculture (USDA), farmers in most of the former Soviet Union are cutting back on areas planted to less profitable crops. Over the next 10 years, grain yields are projected to rise, while the area devoted to cereals is likely to stabilize, resulting in higher output and more room for other crops (Shend, 1995). Thus, in this subregion at least, the quantities of crop residues and feed grains available should increase, as also should the rotation of cereal crops with forage and food legume crops. This trend is likely to prevail in many other parts of the region (Osman *et al.*, 1990), although the speed and degree of intensification will vary greatly from country to country.

The trend toward greater use of feed grains and concentrates is also likely to continue. These feeds will be used to supplement diets based primarily on crop residues and natural grazing. However, rising amounts will be directed towards the expanding commercial dairy, poultry and intensive meat production subsectors (Khaldi, 1984; Sarma, 1986; Glenn, 1988).

Better management of livestock, combined with investment in revegetation, are essential to rehabilitate the range (Cooper and Bailey, 1991). Possible interventions include substituting open-access range grazing at the time of flowering and seed-set for range plants with a combination of hand-feeding of conserved crop residues and other feeds, and restricted-access grazing of specially sown pastures. Revision of the property rights institutions controlling access will be necessary if such long-term private or corporate (group) investments in range improvement are to take place (Young, 1987, 1992; Bromley and Cernea, 1989).

Overall, in 22 of the 27 WANA and Central Asian countries, domestic feed supplies are unlikely to rise fast enough to keep up with demand. Possible exceptions are Ethiopia, Kazakhstan, Kyrgyzstan, Sudan and Turkey. Elsewhere, even substantial increases in crop production (3% annually) will not turn the tide of decline in the number of TLUs per capita of human population that can be supported by domestic feed production between now and the year 2020 (Nordblom and Shomo, 1995; Nordblom *et al.,* 1996).

The few oil-rich countries, such as Libya, Saudi Arabia and the other Gulf states, will probably increase their imports of livestock products and feeds substantially in the coming decades as their populations grow. But the poorer countries, which are home to the vast majority of the region's population, and particularly the poorer segments of this population, will have access to less and less animal protein per capita, beginning from an already low base (Nordblom and Shomo, 1995). Pulses, which substitute for livestock products at lower incomes, are likely to gain in importance in absolute (but not relative) terms.

Seasonal Shifts in Livestock Diets

The large seasonal differences in temperature and rainfall characteristic of the West/Central Asia-North Africa region are strongly reflected in cropping and feeding calendars. This is best illustrated by a case study from Syria.

Ten Bedouin group heads, who had signed contracts for 1-month (April) grazing rights on the 6500-ha Maragha rangeland reserve near Aleppo, were interviewed to understand how this grazing fitted in with their annual feeding calendars (Nordblom *et al.*, 1995a). The interviews covered the timing and relative proportions of hand-fed and grazed materials in sheep diets over the course of the previous year, beginning in April 1994 and ending in April 1995.

The results are shown in Fig. 7.3, which demonstrates the distinctive seasonal feeding and grazing pattern typical of the Bedouin in Syria. The figure distinguishes between two feed classes: hand-fed feeds (cereal straw, legume straw and energy feeds), shown in capital letters; and grazed materials (cereal stubble, other crop residues, unharvested barley, rangeland and shrub plantations), shown as lower case. Hand-feeding of cereal and legume straws, together with energy feeds such as barley grain, dominates sheep diets in winter and early spring, when lambing and early lactation occur on the range. Rangeland grazing in March and April is normally supplemented with the grazing of unharvested barley, sown by the Bedouin for this purpose. Under the contract, grazing of the range shrub plantation was allowed only in April. Cereal stubble grazing predominates from May to August, when flocks are moved to the higher-rainfall zones 50 to 100 km from the winter range. Finally, residues of irrigated crops (cotton and vegetables) are grazed in the autumn months, before the animals return to the range.

These results were consistent with records of hand-feeding practices at other, nearby sites (Thomson *et al.*, 1989; Leybourne, 1993; Nordblom *et al.*, 1995b). They are also consistent with the nationwide pictures of feed use given in Fig. 7.2 and Table 7.1, which show a predominant role for crop residues and minor roles for feed grains and range grazing in Syria.

The role of fodder shrub reserves and their potential for rangeland rehabilitation are the subject of ongoing research at ICARDA (Leybourne *et al.*, 1994; Osman and Shalla, 1994). Grazing of *Atriplex* spp. is usually complemented with hand-feeding of cereal residues and/or other feeds to avoid digestive problems.

Making Better Use of Crop Residues

While crop residues, particularly cereal straws, provide the bulk of livestock feed, their nutritive value is often so low at present that farmers must supplement them with feed grains and other concentrates. Improving the nutritional value of straws and the efficiency of their use in mixed diets is an important option for increasing livestock production in the region.

We define straws as the residue (leaves, awns, stems) remaining after the mature grain crop has been harvested. Straw is collected and removed from the field, then chopped and stored for use in periods when range or stubble grazing is unavailable. The total standing crop may be grazed at maturity in dry years when low yields do not justify harvesting, while in good years crops may be grazed green in winter before stem elongation, then allowed to mature for harvest (Nordblom, 1983a,b).

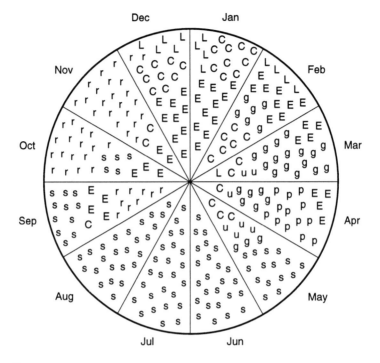

Key:

Upper-case symbols denote hand-feeding:
C = cereal straw (wheat, barley)
L = legume straw (lentil, chickpea, faba bean)
E = energy feeds (barley grain, cottonseed cake, cottonseed and trash,
    wheat bran, dry bread, sugarbeet (and sugarbeet pulp, leaves, tails),
    cottonseed hull, sunflower discs).

Lower-case symbols denote that animals were grazing:
s = cereal stubble (wheat, barley)
r = crop residues (sugarbeet, vegetable, maize, soya, sunflower, cotton)
u = unharvested barley
g = rangeland grazing (rangeland, fallow land, mountains)
p = shrub plantations (*Atriplex* spp., *Salsola vermiculata* with local native
    grasses and other plants)

Each character is 4% of monthly diet

Fig. 7.3. Year-round small ruminant diets among the Bedouin of Syria (Nordblom
et al., 1995a).

Farmers who do not keep livestock often collect their cereal straw for sale.
Straws may have high market values in poor years when roughages are scarce and
grains have to be imported. Farmers sell grazing rights in their fields to livestock
owners, who graze their animals on cereal stubble and on cotton and vegetable
residues where available. However, in intensively farmed areas where there are

few resident livestock or where irrigation allows a summer crop, straw and stubbles may be ploughed under or burned off. The wholesale removal of straw and stubble (whether through grazing or burning) is unquestionably one of the causes of the low organic matter content typical of the region's soils (White, 1992). ICARDA is examining this aspect of crop residue management through its own long-term on-station trials, as well as through collaborative trials with national research groups.

In the right conditions, cereal crop residues provide energy for ruminants in the form of digestible fibre. It is generally accepted that they should be accompanied by small amounts of a suitable nitrogen supplement, such as oilseed cakes. If their nutritive value is low or the desired level of animal production is well above maintenance, farmers must, in addition, feed an energy supplement such as cereal grain to ensure biological and economic efficiency. Such supplementary feeds are usually more expensive than crop residues.

One alternative to using supplementary feeds is to improve the quality of cereal crop residues by chemical treatment, but this requires additional labour and materials and reduces flexibility in use (Ceccarelli, 1993). Another is to replace the oilseed cake with the residues of leguminous feeds grown in rotation with cereals on the farm. These residues are relatively rich in nutrients, but their quantity is generally so low as to make them little more than a minor complement to cereal residues. Their value varies according to their content not only of nutrients but also of so-called antinutritional factors, some of which (e.g. tannin) can have *beneficial* effects on protein digestion if present in moderate quantities (Barry *et al.*, 1986). Cereal straws likewise vary in their response to legume haulms.

Plant Breeding: A Tool with Unexplored Potential

Food crop breeders have traditionally considered straws as costly seed-bearing structures. For the past 100 years, cereal breeders have bred for a decreased straw:grain ratio, selecting against straw yield and leaving the total biological yield unchanged (Riggs *et al.*, 1981). Lipton and Longhurst (1989) argued that differences in the utilization of a crop, and especially of its by-products under different input levels, are one reason why the modern varieties of the Green Revolution have contributed little to the low-input agricultural systems characteristic of many developing countries. Traxler and Byerlee (1993) have shown that the adoption of modern cereal varieties with high grain yields but lower straw yields has been slow in some developing countries where straw is a valuable source of animal fodder for small-scale farmers.

These observations are borne out by the limited adoption of higher-yielding grain varieties of barley in the rainfed farming systems of WANA. Although modern short-strawed varieties are often leafier and respond to high fertilizer levels (Khush and Kumar, 1987), farmers in WANA generally prefer the quality of the straw from their traditional landraces. Many of these landraces are tall, making them easier to harvest and improving straw yields in times of drought (Ceccarelli, 1993).

Plant breeders have tended not to select for straw quality not just because they have considered this trait to be an unnecessary sidetrack but also because appropriate tests for straw quality have been lacking. According to Ceccarelli (1993), such tests must fulfil a tall order: to predict the heritable component of nutritive value with sufficient accuracy, to be practical for large plant populations and to be statistically independent from measures of other desirable characters such as grain yield, plant height and adaptation to specific stresses. In particular, tests should down-weight the straw quality improvements that normally accompany terminal drought stress. Uncritically selecting selecting straw for nutritive value under stress conditions tends to favour late-maturing leafy plants susceptible to drought. Such tests have given rise to the belief that straw quality is always inversely related to grain yield (Reed *et al.*, 1988; Goodchild *et al.,* 1992), shown by recent work to be false (ICARDA, 1996). Inappropriate tests of this kind risk discrediting the entire concept of selecting for straw quality.

Traditionally, straw quality tests have been of two broad types: conventional laboratory tests having uncertain correlations with animal performance, or animal performance trials requiring considerable resources. Animal trials, which on cost grounds can be only few in number, require quantities of straw that are impossible to obtain from early generations of new genotypes. For all these reasons, genetic progress has been slow and erratic.

ICARDA's Research on Breeding for Straw Quality

ICARDA began its work on cereal straw quality in the early 1980s, convinced that (i) the relative importance of straw in ruminant diets in WANA would increase, as illustrated earlier in this paper; (ii) there was scope for improvement (voluntary cereal straw intake regularly varies by more than 10%, according to Capper *et al.,* 1988); and (iii) current approaches to increasing the nutritive value of residues through chemical treatment would not be readily adopted by farmers because of their high start-up costs, their complexity and the non-availability of the necessary chemicals. These convictions have been largely borne out for the region as a whole, although urea treatment is now showing promise in Egypt, as is ammonia treatment in mud-covered straw stacks in the Maghreb. ICARDA has therefore focused its research on identifying methods for selecting crop varieties with good straw qual- ity. We have given priority to barley because of its importance as a feed (both grain and straw) in WANA and the relatively advanced state of knowledge of its genome.

Progress has been made despite the problems of testing for straw quality. A quantum leap forward came with our decision to focus on the technology of near infra-red reflectance spectroscopy (NIRS), in which straw samples can be assessed by matching their spectra to those of straws used in past animal trials. This now permits the prediction of intake and other aspects of nutritive value in less than a minute (Goodchild *et al.*, 1994), with the result that selection can be made amongst large numbers of recently developed lines (Ceccarelli, 1993).

There are probably several alternative genetic mechanisms leading to plants with desirable characters. Given good collaboration with animal scientists and a laboratory with a NIRS instrument, plant breeders and physiologists should be able to identify mechanisms for straw quality that are compatible with the plant's human food value. Genetic means for improving residue quality may also be sought directly: delayed leaf senescence (the so-called "stay-green" trait) is one possibility (Thomas *et al.,* 1992), although its effects on residual soil moisture need to be considered.

ICARDA would like to share its experience in selecting for straw quality with researchers in other crops and ecoregions. Although the center studies crops significant in its mandate region, the worldwide potential of methodologies that can increase the contribution of crop residues to the nutrition of livestock is obvious. However, new tests alone will not bring about a revolution in breeding for crop residue quality. Other constraints must be confronted. These include the historically weak links between cereal breeders and animal scientists and the difficulties breeders may face in coming to terms with the need to broaden their selection criteria. On a brighter note, if the selection of crop varieties for straw quality shows signs of providing farmers with something they need, the breeders' perennial desire to see their varieties widely adopted should stimulate their efforts in this direction.

Future Research Needs

A collaborative effort between livestock and plant scientists is needed to identify research priorities and to initiate a global programme of research to simplify the prediction of straw quality. An indicative research agenda might have the following objectives: (i) to define crop residue quality in terms of animal performance; (ii) to create publicly available reference samples with well-defined characteristics; (iii) to develop tests for small samples of crop residues of various species; (iv) to validate these tests in different environments, comparing the results with farmers' perceptions of quality; and (v) to identify opportunities to utilize favourable dietary interactions between different crop residues and other feed resources. The failure to take advantage of such interactions is one problem affecting the use of straw in small ruminant diets in the region at present.

The presence of ICARDA's headquarters practically at the geographic centre of the steppes of West/Central Asia-North Africa, a region with common cultural roots and similar farming systems, provides the opportunity to conduct the more intense and better integrated research that is now urgently needed on this topic.

Conclusion

Crop residues appear certain to play an increasingly important role as feeds in the future, as human and animal populations expand in the newly defined "WCANA" region. We need to find ways of increasing the efficiency with which they are used.

References

Abu Akkada, A.R. (1984) *Evaluation of Present Status and Potential Development of Animal Feed Resources in the Arab Countries,* vol. 10: Arab Republic of Egypt. Arab Center for the Study of Arid Zones and Dry Lands (ACSAD) and Arab Organization for Agricultural Development (AOAD), Khartoum, Sudan.

ACSAD/AOAD (1985) *Evaluation of Present Status and Potential Development of Animal Feed Resources in Arab Countries.* Arab Center for the Study of Arid Zones and Dry Lands (ACSAD) and Arab Organization for Agricultural Development (AOAD), Khartoum, Sudan.

Barry, T.N., Manley, T.R. and Duncan, S.J. (1986) The role of condensed tannins in the nutritional value of *Lotus pedunculatus* for sheep, 4: Sites of carbohydrate and protein digestion as influenced by dietary reactive tannin concentration. *British Journal of Nutrition* 55: 123-127.

Bromley, D.W. and Cernea, M.M. (1989) *The Management of Common Property Resources: Some Conceptual and Operational Fallacies.* Discussion Paper No. 57, World Bank, Washington DC, USA.

Capper, B.S., Thomson, E.F. and Herbert, F. (1988) Genetic variation in the feeding value of barley and wheat straw. In: Reed, J.D., Capper, B.S. and Neate, P.J.H. (eds), *Plant Breeding and the Nutritive Value of Crop Residues.* Proceedings of a Workshop, 7-10 December 1987, International Livestock Centre for Africa (ILCA), Addis Ababa, Ethiopia, pp. 177-192.

Ceccarelli, S. (1993) Plant breeding technologies relevant to developing countries. In: Gill, M., Owen, E., Pollott, G.E. and Lawrence, T.L.J. (eds*), Animal Production in Developing Countries.* British Society of Animal Production Occasional Publication No. 16, pp. 37-55.

Cooper, P.J.M. and Bailey, E. (1991) Livestock in Mediterranean farming systems: A traditional buffer against uncertainty is now a threat to the agricultural resource base. In: Holden, D., Hazell, P. and Pritchard, A. (eds), *Risk in Agriculture.* Proceedings of the Tenth Agriculture Sector Symposium, 9-10 January 1990, World Bank, Washington DC, USA.

El-Beltagy, A. (1996) Reversing desertification trends in West Asia and North Africa. Keynote address to the International Conference on Desert Development in the Arab Gulf Countries, 23-26 March 1996, Kuwait. Food and Agriculture Organization of the United Nations (FAO), Rome, Italy.

FAO (1995) FAOSTAT. PC Data Base. Food and Agriculture Organization, Rome, Italy.

Glenn, J.C. (1988) *Livestock Production in North Africa and the Middle East: Problems and Perspectives.* Discussion Paper No. 39, World Bank, Washington DC, USA.

Goodchild, A.V., Ceccarelli, S., Grando, S., Hamblin, J., Treacher, T., Thomson, E. and Rihawi, S. (1992) Breeding for straw quality. In: El-Shazly, K. (ed.), *Proceedings of the International Conference on Manipulation of Rumen*

Microorganisms, 20-23 September 1992, Alexandria, Egypt, pp. 317-335.

Goodchild, A.V., Jaby El-Haramein, F. and Treacher, T.T. (1994) Predicting the voluntary intake of barley straw with near infra-red reflectance spectroscopy. *Animal Production* 58: 455.

Hassan, N.I., Abou-Raya, A.K., Abou-Akkada, A.R. and Wardeh, M.F. (1982) *Evaluation of Present Status and Potential Development of Animal Feed Resources in Arab Countries,* vol. 17: Yemen Arab Republic. Arab Center for the Study of Arid Zones and Dry Lands (ACSAD) and Arab Organization for Agricultural Development (AOAD), Damascus, Syria (Arabic).

ICARDA (1996) Relationship between barley straw quality and yield. In: *Annual Report 1995,* International Centre for Agricultural Research in the Dry Areas, Aleppo, Syria, pp. 30-31.

Jahnke, H.E. (1982) *Livestock Production Systems and Livestock Development in Tropical Africa.* Kieler Wissenschaftsverlag Vauk, Kiel, Germany.

Jones, C.F. and Darkenwald, G.G. (1967) *Economic Geography.* Macmillan, New York, USA.

Kategile, J.A., Said, A.N. and Sundst, F. (eds) (1981*) Utilization of Low-quality Roughages in Africa.* Proceedings of a Workshop, 18-22 January 1981, Arusha, Tanzania. Agricultural Development Report No. 1, Agricultural University of Norway, Aas.

Kerven, C., Channon, J. and Behnke, R. (1995) *Planning and Policies on Extensive Livestock Development in Central Asia.* Overseas Development Institute (ODI), London, UK.

Khaldi, N. (1984) *Evolving Food Gaps in the Middle East-North Africa: Prospects and Policy Implications.* Research Report No. 47, International Food Policy Research Institute (IFPRI), Washington DC, USA.

Khush, G.S. and Kumar, I. (1987) Genetic selection for improved nutritional quality of fibrous crop residues of cereal crops: A plant breeder's viewpoint. In: Dixon, R.M. (ed.), *Proceedings of the Sixth Annual Workshop of the Australasian-Asian Fibrous Agricultural Residues Research Network,* 1-3 April 1986, Los Baños, The Philippines, pp. 25-32.

Kosilla, V. (1988) The availability of crop residues in developing countries in relation to livestock populations. In: Reed, J.D., Capper, B.S. and Neate, P.J.H. (eds), *Plant Breeding and the Nutritive Value of Crop Residues.* Proceedings of a Workshop, 7-10 December 1987, International Livestock Centre for Africa (ILCA), Addis Ababa, Ethiopia.

Leybourne, M. (1993) Links between the steppe and cultivated areas through migration: The socioeconomic organization of production of the semi-nomadic agropastoral society of the Syrian steppe. Diplôme de Recherche No. 78, Institut Universitaire d'Etudes du Développement, Geneva, Switzerland.

Leybourne, M., Ghassali, F., Osman, A.E., Nordblom, T.L. and Gintzburger, G. (1994) The utilization of fodder shrubs (*Atriplex* spp., *Salsola vermiculata*) by agropastoralists in the northern Syrian steppe. In*: Pasture, Forage and Livestock Program Annual Report 1993.* International Center for Agricultural Research in the Dry Areas (ICARDA), Aleppo, Syria, pp. 142-160.

Lipton, M. and Longhurst, R. (1989) *New Seeds and Poor People*. Unwin Hyman, London, UK.

Noor, M. (1987) *Rangeland Management in Pakistan*. International Center for Integrated Mountain Development (ICIMOD), Katmandu, Nepal.

Nordblom, T.L. (1983a) *Livestock/Crop Interactions: The Case of Green Stage Grazing of Barley*. Discussion Paper No. 9, International Center for Agricultural Research in the Dry Areas (ICARDA), Aleppo, Syria.

Nordblom, T.L. (1983b) *Livestock/Crop Interactions: The Decision to Harvest or to Graze Mature Grain Crops*. Discussion Paper No. 10, International Center for Agricultural Research in the Dry Areas (ICARDA), Aleppo, Syria.

Nordblom, T.L. and Shomo, F. (1995) *Food and Feed Prospects to 2020 in the West Asia-North Africa Region*. Social Sciences Paper No. 2. International Center for Agricultural Research in the Dry Areas (ICARDA), Aleppo, Syria.

Nordblom, T.L., Arab, G., Gintzburger, G. and Osman, A.E. (1995a) April 1995 survey of Bedouin groups with contracts to graze the government rangeland plantation at Maragha, Aleppo Province, Syria. Paper presented at the Regional Symposium on Integrated Crop/Livestock Systems in the Dry Areas of West Asia and North Africa, 6-8 November 1995, Amman, Jordan.

Nordblom, T.L., Goodchild, A.V. and Shomo, F. (1995b) Livestock, feeds and mixed farming systems in West Asia and North Africa. In: Wilson, R.T., Ehui, S. and Mack, S. (eds), *Livestock Development Strategies for Low-income Countries*. Proceedings of the Joint FAO/ILRI Round Table, 27 February-2 March 1995, International Livestock Research Institute (ILRI), Addis Ababa, Ethiopia, pp. 101-123.

Nordblom, T.L., Shomo, F. and Gintzburger, G. (1996) Food and feed prospects for resources in Central Asia. Paper presented at the SR-CRSP Regional Assessment Workshop on Central Asian Animal Production, 27 February-1 March 1996, Tashkent, Uzbekistan. International Center for Agricultural Research in the Dry Areas (ICARDA), Aleppo, Syria.

Osman, A.E. and Shalla, M.A. (1994) Use of edible shrubs in pasture improvement under a Mediterranean environment in northern Syria. In: Squires, V.R. and Ayoub, A.T. (eds), *Halophytes as a Resource for Livestock and for Rehabilitation of Degraded Lands*. Kluwer Academic Publishers, Dordrecht, The Netherlands, pp. 255-258.

Osman, A.E., Ibrahim, M.H. and Jones, M.A. (1990) *The Role of Legumes in the Farming Systems of the Mediterranean Areas*. Proceedings of a Workshop, 20-24 June 1988, United Nations Development Programme (UNDP) and International Center for Agricultural Research in the Dry Areas (ICARDA), Tunis, Tunisia. Kluwer Academic Publishers, Dordrecht, The Netherlands.

Reed, J.D., Capper, B.S. and Neate, P.J.H. (eds) (1988) *Plant Breeding and the Nutritive Value of Crop Residues*. Proceedings of a Workshop, 7-10 December 1988, International Livestock Centre for Africa (ILCA), Addis Ababa, Ethiopia.

Riggs, T.J., Hanson, P.R., Start, N.D., Miles, D.M., Morgan, C.L. and Ford, M.A.

(1981) Comparison of spring barley varieties grown in England and Wales between 1880 and 1980. *Journal of Agricultural Science* (Cambridge) 97: 599-610.

Sarma, J.S. (1986) *Cereal Feed Use in the Third World: Past Trends and Projections to 2000.* Research Report No. 57, International Food Policy Research Institute (IFPRI), Washington DC, USA.

Shend, J.Y. (1995) Grain imports continue to fall as FSU agricultural sectors contract. In: Foster, C.J. and Liefert, W.M (eds), *Former USSR.* Situation and Outlook Series, International Agriculture and Trade Reports, United States Department of Agriculture (USDA), Washington DC, USA.

Slayman, F., Daghir, N. and Saoud. N. (1986) *Encyclopedia of Animal Resources in the Arab Countries,* vol. 14: Lebanon. Arab Center for the Study of Arid Zones and Dry Lands (ACSAD) and Arab Organization for Agricultural Development (AOAD), Damascus, Syria (Arabic).

Thomas, H., Ougham, H.J. and Davies, T.G.E. (1992) Leaf senescence in a non-yellowing mutant of *Festuca pratensis:* Transcripts and translation products. *Journal of Plant Physiology* 139: 403-412.

Thomson, E.F., Bahhady, F.A. and Martin, A. (1989) Sheep husbandry at the cultivated margin of the northwest Syrian steppe. Mimeo, International Center for Agricultural Research in the Dry Areas (ICARDA), Aleppo, Syria.

Traxler, G. and Byerlee, D. (1993) A joint-product analysis of the adoption of modern cereal varieties in developing countries. *American Journal of Agricultural Economics* 75: 981-989.

UNEP (1992) *World Atlas of Desertification.* United Nations Environment Program and EdwardArnold/Hodder & Stoughton, London, UK.

Wardeh, M.F., Abou-Raya, A.K., Abou-Akkada, A.R. and Hassan, N.I. (1982) Evaluation of present status and potential development of animal feed resources in Arab countries, vol. 20: The Sultanate of Oman. Arab Center for the Study of Arid Zones and Dry Lands (ACSAD) and Arab Organization for Agricultural Development (AOAD), Damascus, Syria (Arabic).

White, P. (1992) Nitrogen balances in integrated crop/livestock systems. In: *Pasture, Forage and Livestock Program Annual Report 1990/1991.* International Centre for Agricultural Research in the Dry Areas (ICARDA), Aleppo, Syria, pp. 97-108.

Young, M.D. (1987) Land tenure: Plaything of governments or an effective instrument? In: Chisholm, A. and Dumsday, R. (eds), *Land Degradation, Problems and Policies.* Cambridge University Press, London, UK.

Young, M.D. (1992) *Sustainable Investment and Resource Use: Equity, Environmental Integrity and Economic Efficiency.* Vol. 9, Man and the Biosphere series. United Nations Educational, Scientific and Cultural Organization (UNESCO), Paris, France.

Acknowledgements

The authors gratefully acknowledge critical comments by Simon Chater and Eva Weltzien on an earlier draft of this paper. Opinions expressed are those of the authors and do not necessarily represent the policy of ICARDA.

8. Dynamics of Feed Resources in Mixed Farming Systems of Latin America

Roberto A. Quiroz[1], Danilo A. Pezo[2], Daniel H. Rearte[3] and Felipe San Martín[4]

[1]*International Potato Center/Consortium for the Sustainable Development of the Andes, PO Box 1558, Lima 12, Peru*
[2] *Universidad de Costa Rica, PO Box 235-2300, San José, Costa Rica*
[3] *Instituto Nacional de Technología Agropecuaria, CC 276-7620 Balcarce, Buenos Aires, Argentina*
[4] *Instituto Veterinario de Investigaciones Tropicales y de Altura, Apartado 41-0068, San Borja, Lima, Peru*

Abstract

This paper discusses the utilization of feed resources by ruminants in the four most important ecoregions of Latin America: the tropics, the Andes, the Southern Cone, and the arid and semiarid areas. Common feeding practices within the region's predominant mixed crop/livestock production systems are addressed, together with research findings on feed resources, including non-conventional resources. Research and development needs are projected, based on the production required to meet the region's increasing domestic demand for animal products and on the expected additional demand for exports.

Introduction

Livestock production is important in the economies of all Latin American countries. In most of them, resource-poor farmers rely on animal agriculture for their subsistence, while in the better endowed areas there are profitable commercial enterprises. However, the vast majority of farms are small, and most of these are mixed crop/livestock farms. Pastures, both native and introduced, are the principal feed resource for ruminant animals, and are complemented with crop residues and by-products.

This paper describes the salient features of feed resources utilization in the mixed crop/livestock systems of Latin America. Due to the considerable variability encountered, it is difficult to generalize. To simplify the task, we divide the region into four ecoregions: the tropics, the Andes, the Southern Cone, and the arid and semiarid areas. Temperate regions in Mexico are not comprehensively covered; the feed resources and production constraints and opportunities here are considered broadly similar to those found in the temperate areas of South America.

The Four Ecoregions

The Tropics

About 77% (1683 million ha) of the land in Latin America and the Caribbean is tropical. Its population is around 389 million and is growing at an average annual rate of 2%. On average, less than 50% of the population lives in rural areas, but this figure varies from 8% in Trinidad and Tobago to 61% in Haiti. The contribution of agriculture to the gross domestic product is relatively low in the region's main oil-producing countries (5, 9 and 9% for Trinidad and Tobago, Mexico and Venezuela respectively), but greater in other countries (27, 27, 24 and 23% for Honduras, Paraguay, Bolivia and Nicaragua respectively). The sector's contribution to exports varies from lows of 1% in Trinidad and Tobago and 6% in Venezuela to highs of 85, 91 and 93% for Cuba, Paraguay and Nicaragua respectively (Riesco, 1992).

Mixed crop/livestock production systems are common in the small and medium-sized farms of tropical America, but the contribution of crop residues and by-products to the animal diet is directly related to the length and harshness of the dry season (CATIE, 1983). The type of livestock system also affects the magnitude of crop residue utilization in animal feeding. In general, crop residues are more important for dual-purpose cattle than in specialized dairy operations. However, a common characteristic uniting this ecoregion's diverse livestock production systems is their strong dependence on the use of native and/or introduced pastures (Pezo *et al.*, 1992). Examples include: (i) the specialized dairy cattle systems of the intermediate plateaux and valleys of Central and South America; (ii) the ranching beef production systems of the Colombian and Venezuelan savannas, the Amazon and Orinoco valleys, the semiarid Chaco, and the lowlands of the Pacific and Atlantic basins in Central America; and (iii) the dual-purpose cattle systems of the subhumid and humid lowlands and hillsides.

Despite the important role of pastures in livestock production, more than 50% of the pasture land has symptoms of degradation (Pezo *et al.,* 1992). Much has been allowed to fallow, giving rise to secondary forest. This situation has been attributed to several factors, including: (i) lack of adaptation of the first generation of introduced grasses (namely *Digitaria decumbens, Hyparrhenia rufa, Panicum maximun* and *Cynodon nlemfuensis*) to the predominant low-fertility acid soils; (ii) biotic pressures (pests and diseases) on introduced grasses and legumes; and (iii) inappropriate management practices, such as lack of fertilizer inputs and overgrazing.

There is clearly a need to reclaim degraded pastures so as to reduce the pressure on primary forest. Several approaches to reclamation have been developed for different ecosystems and socioeconomic conditions, including: (i) the introduction of grass-legume pastures after slash-and-burn of secondary vegetation in the rainforest margins of the Peruvian Amazon Valley (Loker *et al.*, 1991); (ii) the establishment of improved pastures in association with acid-tolerant rice varieties in the Colombian savannas (Toledo, 1994), and with maize or soyabean in the more fertile soils of the Costa Rican humid tropics (Duarte *et al.*, 1995); (iii) the strip over-seeding of legumes in degraded grass monoculture pastures (Hurtado *et al.*, 1988); and (iv) the use of acid-tolerant fast-growing herbaceous legumes (e.g.*Vigna unguiculata*) in a green manuring system (Ara and Ordoñez, 1993). These approaches have proved ecologically sustainable and economically attractive to farmers (Lascano and Pezo, 1994).

The successful application of these approaches requires the use of germplasm adapted to the major soil and biotic constraints, and pasture technologies geared to the rational use of these genotypes, the aim being not only to increase livestock production but also to preserve or enhance soil fertility (Toledo, 1994). Research by the Centro Internacional de Agricultura Tropical (CIAT) and its national partners in the Red Internacional de Evaluacion de Pasturas Tropicales (RIEPT) has been instrumental in identifying the most promising genotypes of grasses (such as *Brachiaria decumbens, B. dictyoneura, B. brizantha, B. humidicola* and *Andropogon gayanus*) and legumes (such as *Arachis pintoi* and *Stylosanthes guianensis*), which are now commercially available in several Latin American countries (Miles and Lapointe, 1992).

There are several distinct agroecosystems in tropical America, the most important of which for our purposes are the intermediate valleys and plateaux, the humid tropics and the subhumid areas.

Intermediate Valleys and Plateaux

The tropical American intermediate valleys and plateaux are located at altitudes of 1200 to 2800 m, with rainfall varying from 1600 to 2800 mm year^{-1} distributed over 7 to 12 months of the year. In most cases they have deep fertile soils (e.g. the volcanic Andosols of Colombia, Costa Rica and Guatemala). In these areas, specialized milk production, based on the use of European breeds (mainly Holstein), is the most common livestock production system (Table 8.1). Frequently, this is the only large farm enterprise, with small areas only cultivated to vegetables, fruits and potato, the latter usually in rotation with pastures. When cheese is the main livestock product, a whey-based pig fattening enterprise is often a complementary activity. During the past decade, many dairy farms have reduced their area in pastures in order to introduce some non-traditional crops (including strawberries, flowers and vegetables), mainly for export.

Table 8.1. Major agricultural systems in small and medium-sized farms located in different agroecological zones of tropical America.

Agroecological zones	Major agricultural systems	Main livestock products	Predominant feeding system
Intermediate plateaux (1200-2800 m)	Dairy cattle/ vegetables/fruits/ potato	Milk/cheese	Grazing, chopped grasses, conserved forages, commercial concentrates, molasses, banana fruits
Humid hillsides (600-1200 m)	Dual-purpose cattle or dairy goats/ coffee and/or sugarcane	Milk/cheese, weaned animals	Limited grazing, chopped grasses, residues from coffee shade trees (plantains/ bananas, legumes), sugarcane tops
	Dual-purpose cattle/slush-mulch maize or sorghum/ legume cover crops/beans/ cassava	Cheese/milk, weaned or mature animals	Grazing, maize or sorghum straw and legumes, browse
Subhumid/dry hillsides (600-1200 m)	Dual-purpose cattle/maize, sorghum or dryland rice/beans/ bananas	Cheese/milk (mainly during the rainy season), weaned or mature animals	Grazing the whole year; during the dry season, maize, sorghum, rice and/or bean straws, *guatera* and other conserved forages, browse and fruits
Humid lowlands (< 600 m)	Dual-purpose or specialized beef cattle/maize/ paddy rice/ cassava/yams/ bananas	Milk/cheese, weaned animals (in dual-purpose systems)	Grazing, molasses, banana fruits, commercial concentrates, legume tree foliages
Subhumid/dry lowlands (< 600 m)	Dual-purpose or beef cattle/maize or sorghum/paddy rice/cassava/ cowpea/ sugarcane/citrics	Cheese/milk (mainly during the rainy season), weaned or mature animals	Grazing the whole year; during the dry season, maize, sorghum, rice and/or cowpea straws, *guatera* and other conserved forages, chopped sugarcane, browse and fruits

In this zone, pastures are the main component of livestock diets. These are mostly managed under rotational grazing, and in the majority of farms some fertilizers are applied, mainly to the paddocks used by milking cows. The most commonly cultivated grasses vary with altitude, but include African star grass (*Cynodon nlemfuensis*) at < 1500 m, Kikuyu grass (*Pennisetum clandestinum*) from

1500 to 2500 m and ryegrass (*Lolium perenne*) at above 2500 m (these altitude limits may vary with latitude). White clover (*Trifolium repens*) is usually present in pastures at intermediate and high elevations and is well adapted to these environments, spreading naturally without being sown. In many farms, grazing is complemented with chopped grasses (mainly Napier grass, king grass and other tall *Pennisetum* genotypes, but also *Axonopus scoparius* at higher elevations).

Supplementation is a common practice in these farms. Commercial concentrates and molasses are the most frequently used supplements, but brewer's grain, green bananas and plantain pseudo-stalks are also used. Survey data for systems of this kind in Costa Rica (CATIE, 1983; van der Grinten *et al.*, 1992) show that concentrate feeding varies from 1.3 to 10.3 kg cow^{-1} day^{-1}, while milk production ranges from 4.7 to 27.1 kg cow^{-1} day^{-1} (16.3 kg on average). As most concentrate ingredients are imported (including maize, wheat bran and soyabean meal) and their international prices are increasing rapidly, along with those of fertilizers, some farmers are looking for alternatives. Among the options being tried are more productive grasses, improved pasture rotation schemes, higher stocking rates and supplementation with locally produced by-products, such as green bananas rejected for export.

The case of green bananas is particularly interesting, since a great deal of information was generated by CATIE in the 1970s and early 1980s on their nutritional value and utilization in feeding systems for dairy and beef cattle (Villegas, 1979; Cerdas, 1981; Ruiz, 1981; Pérez *et al.*, 1990a,b). At that time the major constraint to commercial application was the cost of transport, given the considerable distance between the banana plantations and the major cattle production areas. In 1987 a new road was built in Costa Rica, halving transportation costs and transit times. Coupled with recent increases in the costs of the imported grain (mainly maize) used in concentrates, the new road has facilitated a rapid increase in the use of green banana as an energy supplement for dairy cows. It is expected that this technology will become more widely applied in the future.

Although many farms experience a pronounced dry period during the year, the use of conserved forages such as silage and/or crop residues is not common. A few enterprises with medium to large dairy herds make silage, but this technology is seldom adopted by small farmers, who consider themselves unable to afford the necessary machinery and infrastructure. Most small farmers get through dry periods by simply modifying the grazing system used in the rainy season. The stocking rate is reduced in paddocks grazed by milking cows, as dry cows and heifers are moved to marginal areas kept in reserve for such times (these are generally unfertilized pastures). In the future, the use of crop residues is likely to increase, especially for dry cows and heifers. The incorporation of forage crops such as oats and maize into the traditional crop-pasture rotation system will probably also increase.

Humid Tropics

Over the past three decades, most of the increase in pasture area observed in tropical America has occurred in the humid tropics (lowlands and hillsides). Large-scale clearance of the rainforest for pasture establishment was promoted by opportunities to increase beef exports in Central America and Mexico (Myers, 1981) and, in addition, by economic subsidies in the Amazon Basin of Brazil (Serrao and Toledo, 1990). In both cases, extensive beef cattle production systems were the main type of enterprise established, and a high proportion of pastures have suffered degradation. In the present decade these policies have been reversed by most governments, with the result that the increase in pasture land has slowed markedly. There is still some deforestation in the forest margins, but mainly for colonization purposes (Leonard, 1987; Hecht, 1993).

In the humid lowlands, most large farms are dedicated either to specialized beef cattle or to plantation crops. Bananas, which are the most common of these crops (others are oilseeds and coconut palms), are grown mainly by national or transnational companies, mostly for export. Approximately 20% of the fruits do not meet quality standards for export and are therefore left next to the packing areas, where truckers collect them free of charge and transport them to farms for sale as animal feed. Fruits not loaded by truckers are thrown into the surrounding area, often beside rivers or streams, where they create serious pollution problems.

Small to medium-sized farms usually have a mixed crop/livestock system. Dual-purpose cattle production is the most widespread livestock enterprise, but some farmers keep specialized dairy cattle. The main crops are maize, rice, cassava and yams. Most farmers also have an orchard with citrus, bananas, plantains and other fruit trees, plus a few pigs and some poultry around the house.

In these systems, cattle feeding is mainly based on the use of grasses (e.g. *Cynodon nlemfuensis, Brachiaria ruziziensis, B. decumbens, Ischaemum indicum* and *Axonopus compressus*), managed in a rotational system. Milking cows may also receive small amounts (<1.0 kg cow^{-1} day^{-1}) of commercial concentrates, together with some sugarcane molasses (0.5 kg day^{-1}) and/or green banana (2-3 kg day^{-1}) as energy supplements, and some mineral supplements (Murillo and Navarro, 1986).

Most milk-producing animals in the humid lowlands are the progeny of crosses between Criollo or zebu cattle and European breeds (mainly Holstein, but also Brown Swiss and Jersey). Their average milk production ranges from 2.0 to 5.2 kg cow^{-1} day^{-1} in dual-purpose systems (CATIE, 1983; Vaccaro, 1992) and from 5.9 to 7.3 kg cow^{-1} day^{-1} in specialized milk production systems (Murillo and Navarro, 1986; Teodoro and Lemos, 1992).

Animals may also consume some of the foliage pruned from live fences, which are made mainly from shrub or tree legumes such as *Gliricidia sepium* or *Erythrina berteroana*. The foliage is high in crude protein (CP) (>20%), but its in vitro dry matter digestibility (IVDMD) is similar to or even lower than the values obtained for grasses (Table 8.2). Considerable variability among genotypes has been detected in digestibility, intake and the content of some anti-quality factors, such as tannins and coumarins in *Gliricidia sepium* and alkaloids in various species and ecotypes of *Erythrina* (Pezo *et al.*, 1990; Ruiz, 1992; Payne, 1993). In the past decade, researchers (Pezo *et al.*, 1990; Kass *et al.*, 1992; Romero *et al.*, 1993) have made considerable efforts to develop agronomic management practices that will allow increased foliage production and hence utilization by animals, under different silvopastoral systems such as protein banks, live fences, barriers on terraces or hedges in alley cropping systems. Emphasis has been placed on persistence under intensive management and hence on one aspect at least of system sustainability. Although more attention has been given to the genuses *Erythrina* and *Gliricidia*, some information has also been obtained on other non-legume woody perennials, such as *Morus alba* (24.2% CP, 79.3% IVDMD) and *Malvaviscus arboreus* (21.0% CP, 68.3% IVDMD), which are mostly used for feeding goats (Benavides *et al.*, 1992; Lopez *et al.*, 1994; Rojas and Benavides, 1994).

Table 8.2. Crude protein (CP) content and in vitro dry matter digestibility (IVDMD) of some tree foliages used for animal feeding in the Latin American humid lowland tropics.

Latin name	CP (%)	IVDMD (%)
Erythrina poeppigiana	24.2	51.4
Gliricidia sepium	24.8	62.2
Leucaena leucocephala	22.0	-52.7
Pithecellobium dulce	24.1	59.6
Enterolobium cyclocarpum	21.7	68.8
Morus spp.	24.2	79.3
Cnidoscolus acutinifolium	41.7	84.4
Sambucus mexicana	24.3	75.8
Hibiscus rosa-sinensis	19.9	71.2
Verbesina myriocephala	20.3	69.8
Verbesina turbacensis	20.2	68.4
Diphysa robinoides	26.9	69.8
Malvaviscus arboreus	21.0	68.3
Cestrum baenitzii	37.1	65.8
Spondias purpurea	16.5	56.6
Guazuma ulmifolia	15.6	54.1
Cecropia peltata	19.8	51.7
Brosimum alicastrum	16.1	59.0
Cassia siamea	13.9	60.6
Acacia angustissima	19.9	23.2
Albizia falcataria	20.3	42.2
Calliandra calothyrsus	20.2	21.0
Inga spp.	21.8	23.2

Sources: Adapted from Valerio (1990), Benavides *et al.* (1992) and Araya *et al.* (1994).

Turning to the humid hillsides, one of the major crop/livestock systems of this zone is perennial crops (coffee and/or sugarcane) combined with either dual-purpose cattle or dairy goats. This type of system is most common in relatively fertile areas that were subject to intensive, early colonization, where adequate roads and at least some processing facilities are available. In these systems, animals usually have limited access to a grazing area, but the forage ration is complemented either with chopped grasses, sugarcane tops and banana/plantain pseudo-stems or with tree foliage (CATIE, 1978). The latter two are harvested from coffee plantations, where trees are grown for shade.

In some areas, sugarcane tops are left in the field, but in others they are collected to feed animals. Sugarcane tops represent almost 20% of the total biomass, are highly digestible (69% IVDMD) and are readily consumed by cattle (San Martín *et al.*, 1983) despite their low nitrogen (N) content (0.86%). Farmers leaving sugarcane tops in the field argue that their costs of collection and transportation are too high (Caielli, 1988).

On coffee-growing farms, banana, plantain and/or legume trees such as *Erythrina poeppigiana* and *Inga* spp. are commonly intercropped with coffee as shade crops. The legume trees are pruned once or twice a year, and the leaves are usually left on the field as mulch. However, some farmers collect part of the foliage and offer it to their animals. Banana leaves have 12% CP and 46% IVDMD (see Table 8.4, below), while tree foliage usually has over 20% CP and IVDMD values ranging between 50 and 65% (Table 8.2). Banana and plantain pseudo-stems are cut, carried, chopped and fed to animals, usually to milking cows. They are highly digestible (77% IVDMD), but poor in CP (2.5-3.5%) and high in moisture (93%), which limits their voluntary intake (Martínez, 1980).

As regards the coffee itself, the main by-products with some potential for animal feeding are the pulp and hulls. These are available where the coffee is processed, so have to be carried back to the farms if they are to be used. They have a high moisture content and poor quality. Caffeine, a secondary compound present in coffee pulp, has a depressing effect on feed intake, as well as a stimulatory effect on urinary N excretion (Braham *et al.*, 1973; Ruiz and Ruiz, 1977). All these factors have restricted use. However, when coffee pulp is not either used as animal feed or processed to produce an organic fertilizer, it can create serious pollution problems in watersheds, since it is often thrown into rivers or streams.

A second major crop/livestock mixed farming system found in the humid hillsides includes dual-purpose cattle as the main livestock component, in combination with annual crops such as maize or sorghum, beans and cassava. Maize and sorghum are managed in a "slush-mulch" system with legume cover crops. In this system legume cover crops are sown into an initial cereal crop and left to mature after the cereal crop is harvested. They are then cut and left in the field as a mulch, into which a second cereal crop is sown. *Mucuna deeringiana* and *Canavalia ensiformis* are the legumes most frequently used. They contribute to system sustainability by fixing atmospheric nitrogen and transferring it to companion grain crops, improving soil water retention, preventing soil erosion, increasing soil fertility and reducing pest attacks (Buckles *et al.*, 1993; Zea, 1993; Barreto *et al.*, 1994; García *et al.*, 1994). They also markedly improve the quality of crop residues.

The residues of maize and sorghum grown in monoculture are low in N and rich in fibre, making them poorly digested by animals (Table 8.3). However, when these cereals are intercropped with legumes, the latter provide additional ferment-

Table 8.3. Changes in cultivated area and availability of residues and by-products for the major crops of Central America and Mexico, 1970-1992.

Residue	Cultivated area ('000 ha)				Crop residue availability ('000 t)			
	1970	1981	1992	Rate of change (% year⁻¹)	1970	1981	1992	Rate of change (% year⁻¹)
Maize stalk	8 965	8 378	8 569	– 0.20	10 828	12 332	14 087	+ 1.37
Sorghum straw	1 199	1 797	1 758	+ 2.12	2 877	3 306	3 426	+ 0.87
Rice	363	411	354	– 0.11				
Straw	–	–	–	–	302	448	429	+ 1.91
Polishings	–	–	–	–	100	152	146	+ 2.09
Cassava	20	21	25	+ 1.44	–	–	–	–
Leaves	–	–	–	–	39	41	55	+ 1.86
Tubers	–	–	–	–	42	47	68	+ 2.81
Bean haulm	2 094	1 875	1 663	– 2.45	1 256	1 170	1 109	– 0.53
Sugarcane	734	871	924	+ 1.18				
Tops	–	–	–	–	11 185	12 956	16 248	+ 2.06
Bagasse	–	–	–	–	15 650	18 139	22 747	+ 2.06
Molasses	–	–	–	–	1 785	2 073	2 600	+ 2.08
Banana	630	720	740	+ 0.79				
Fruits	–	–	–	–	1 010	1 139	1 234	+ 1.01
Pseudo-stems	–	–	–	–	6 380	7 200	7 600	+ 0.87
Leaves	–	–	–	–	1 253	1 425	1 568	+ 1.14
Coffee pulp	994	1 205	1 345	+ 1.61	337	482	619	+ 3.80

Sources: Escobar and Parra (1984), FAO (1986, 1994) and Ruiz (1993).

able N, which in these feeding systems is the most limiting nutrient not only for the animal but also for rumen microorganisms. The atmospheric N fixed by the legumes also results in slightly higher N content in the cereal crop residues. Substantial positive effects on a range of parameters including feed intake, digestibility, liveweight gain and milk production were observed when forage legumes such as *Lablab purpureus, Pueraria phaseoloides, Centrosema plumieri* and *C. pubescens* were intercropped with maize, either in relay or sown simultaneously. In addition, grain yields were not negatively affected (Espinoza, 1983; Sinclair *et al.,* 1991).

Subhumid Tropics

Most of the flat and relatively fertile land in the subhumid lowlands of tropical America is occupied by large farms. Small ones tend to have poorer-quality land, located mainly up the hillsides (Thiesenhusen, 1994). The majority of large farmers do not use their land intensively—in many cases almost a third of it lies fallow. In contrast, small farmers must use their land as intensively as possible. Increasingly, they are compelled to farm previously uncultivated steep areas, with serious soil degradation effects (Leonard, 1987).

Large farms tend to have specialized crop (including sugarcane, citrus, paddy rice, maize, soyabean and sorghum) or livestock systems (mainly beef production in an extensive ranching type of system), whereas on small and medium-sized farms mixed crop/livestock systems are more common. Large crop farms generate substantial amounts of crop residues and by-products with a potential for animal feeding. A high proportion of these resources are not used, but some are sold either to beef cattle feedlot owners or to farmers facing feed shortages due to drought or other unforeseen events.

Feedlot feeding systems based on the use of crop residues as the forage component have been designed by researchers, but few operate commercially. Examples worth mentioning are systems based on untreated (Ruiloba *et al.,* 1978) and treated rice straw (Preston and Leng, 1990), black bean straw (Lozano *et al.,* 1980a), sweet potato vines (Backer *et al.,* 1980) and sugarcane tops (Ruiz, 1984). In all of these, either urea or poultry litter provides the fermentable nitrogen required for adequate rumen microbial activity (Ruiz and Ruiz, 1978), while sugarcane molasses alone or with green bananas provides readily fermentable carbohydrates (Ruiz, 1981). However, in recent years more attention has been paid to rice polishings as a source of fermentable carbohydrates, long-chain fatty acids and by-pass nutrients (Preston and Leng, 1990), and to the use of tree foliage as a locally produced source of by-pass protein (Pezo *et al.,* 1990; Kass *et al.,* 1992; Preston and Murgueitio, 1992).

The sugarcane industry provides a wide variety of products and by-products for animal feeding. The research on this topic done by the Cubans deserves special mention. In this field, Cuba has led the way in tropical America, in terms not only of technology development but also of extensive commercial application. Some 24 to 35% of the diet of the national cattle herd is now derived from the sugar industry

(Pérez, 1993). The feedstuffs produced range from highly fibrous materials, such as untreated and steam-treated bagasse, bagasse pith and dry leaves, through energy-rich products (sugarcane juice) and by-products (molasses), to protein-rich by-products such as *Torula* and *Saccharomyces* yeasts. Several composite products have been developed, including mel-urea, molasses-urea blocks, pajumel, garamber and saccharine. Comprehensive reviews of the use of sugarcane products and by-products for livestock feeding and other purposes can be found in Sansoucy *et al.* (1988), Elías *et al.* (1990), Preston and Leng (1990), Preston and Murgueitio (1992) and Pérez (1993).

On small and medium-sized farms, dual-purpose cattle are the main component of the livestock enterprise, but farmers also raise some poultry and swine in the backyard. In the drier areas, goats may also be important. In the subhumid lowlands, maize or sorghum, paddy rice, cassava, cowpea and yams are the most frequent crops, whereas in the hillsides dryland rice and common bean replace paddy rice and cowpea (Table 8.1). Pigeonpea (*Cajanus cajan*) is also important for some farmers located in the subhumid hillsides, especially in the Caribbean.

The dual-purpose cattle production systems prevalent in this zone vary more than in the humid zone in terms of animal genetic makeup (zebu, Criollo *Bos taurus* and zebu x European crossbreds), system orientation (relative contribution of milk and beef to total income), milking intensity and length of the milking period (Wadsworth, 1992). Flexibility is the main advantage of the dual-purpose over the specialized system. The dual-purpose system can be oriented to optimizing net revenues according to changes in beef:milk price ratios. It can also respond to seasonality by adjusting nutrient demands to changes in the availability and quality of feed resources (many farmers stop milking during the dry season).

These factors explain the considerable variability of milk production from different dual-purpose systems in this zone. For example: a monitoring study carried out by CATIE (1983), covering more than 30 small farms in the subhumid lowlands and hillsides of Central America, found milk yields varying from 370 to 1083 kg lactation^{-1} and lactation lengths from 148 to 285 days; Wadsworth (1992) observed a range of 142 to 1736 kg lactation^{-1} for 12 farms monitored in Costa Rica; Vaccaro *et al.* (1992), monitoring nine medium-sized dual-purpose cattle herds in Venezuela, obtained milk yields varying from 675 to 1102 kg lactation^{-1} and lactation lengths from 217 to 279 days; lastly, Anderson *et al.* (1992) monitored five herds in Yucatan (Mexico), obtaining on average 2.3 to 6.3 kg of milk cow^{-1} day^{-1}, with the number of cows milked ranging from 48 to 87% of the herd.

Throughout the subhumid tropics, cattle feeding systems are based mainly on pastures, but during the dry season (4 to 7 months) animals also graze some crop residues (e.g. maize and/or sorghum stalks, rice stubble), browse trees and shrubs and, in some cases, are fed other crop residues (e.g. maize husks and cobs, bean straw) and by-products (e.g. molasses, citrus pulp and cottonseed hulls). Some conserved forages may also be used, including whole sugarcane (*Saccharum officinarum*) as a strategic fodder crop or an energy bank for the dry season (Preston and Murgueitio, 1992).

The pasture species traditionally used in the subhumid tropics are *Hyparrhenia rufa* and *Panicum maximum*, as well as the so-called natural grass (a complex community composed mainly of several genotypes of *Axonopus* spp. and *Paspalum* spp.), which dominates degraded pastures. A wide diversity of herbaceous legumes is also found, mostly as components of native pastures. The species found include *Aeschynomene, Calopogonium, Cassia, Centrosema, Desmodium, Galactia, Macroptilium, Rhynchosia, Stylosanthes* and *Vigna*). Other common sown pastures in this zone are *Cynodon nlemfuensis, Dichantium aristatum, Brachiaria dyctioneura, B. decumbens* and *Andropogon gayanus* (Pezo *et al.*, 1992).

Turning to trees and shrubs, several multipurpose species play an important role in the subhumid tropics, providing tangible benefits such as live fences, shade, fodder, fuelwood, timber and fruits, but also contributing to soil fertility through nutrient cycling and to soil conservation when used as live barriers on hillsides (Lascano and Pezo, 1994). The contribution of these species to animal diets tends to be greater in the subhumid than in the humid zone, since dry-season feed shortages are more acute here. They are also more important for goats than for cattle (Benavides *et al.*, 1992). The most promising woody perennials, in terms of both foliage and fruit production, in the subhumid zone include *Gliricidia sepium, Guazuma ulmifolia, Brosimum alicastrum, Spondias* spp., *Enterolobium cyclocarpum, Acacia* spp., *Cratylia argentea, Morus* spp., *Malvaviscus arborescens, Hibiscus rosa-sinensis* and *Leucaena leucocephala* (Benavides *et al.*, 1992; Araya *et al.*, 1994). Data on CP content and IVDMD values for most of these species are shown in Table 8.2.

Among crop residues, maize and sorghum stalks are the most commonly used in the subhumid zone. They are mainly utilized by direct grazing, usually during the second half of the dry season, when pasture grazing is scarce. In spite of the very poor content of fermentable N found in both crop residues (Table 8.4) and pastures, most farmers do not provide any protein or mineral supplements. In this type of system the use of multinutrient blocks is one option that has worked well for growing animals eating sorghum stover (Combellas, 1994; Ordoñez, 1994) and for dual-purpose cows fed poor-quality hays (Combellas and Mata, 1992; Domínguez, 1994).

Sorghum and maize residues are used in the traditional *guatera* system of Guatemala, El Salvador and Honduras, in which these crops are cultivated exclusively for fodder purposes. In this system, cereals are planted late in the rainy season at a very high density and in some cases fertilized with up to 120 kg N ha^{-1} (Roldán and Soto, 1988). Due to drought stress, plants usually reach only a vegetative or early flowering stage (no grains are formed). After cutting, they are dried in the field for 6 days, then tied in bunches and stored in the shade to prevent nutrient losses (IICA, 1995). Farmers consider *guatera* to be a high-quality forage, suitable for feeding milking cows. It has about 70% IVDMD, but marginal CP contents (< 7.0%). Small increases (13.6%) in milk production resulted when dual-purpose cows receiving *guatera* were supplemented with a simple mixture of urea, common salt and minerals (IICA, 1995).

Table 8.4. Nutritive values (%) of the residues and by-products of crops commonly cultivated in Central America[1].

Crop residue or by-product	DM	N	CWC	IVDMD	Ca	P
Maize stalk	83.2	0.67	76.4	40.4	0.23	0.19
Sorghum straw	82.6	0.83	67.4	54.0	0.40	0.14
Rice						
Straw	89.0	0.69	78.2	42.3	0.21	0.31
Polishings	89.3	3.66	6.4	87.6	0.23	0.72
Cassava						
Foliage	26.1	3.36	43.5	52.9	0.62	0.42
Tubers	32.0	0.58	5.3	79.0	0.09	0.12
Sweet potato foliage	14.6	2.32	29.5	74.6	0.86	0.40
Bean haulm	92.0	0.61	26.7	45.6	–	–
Cottonseeds (whole)	91.6	3.66	20.5	76.6	0.15	0.73
Sugarcane						
Tops	24.0	0.86	68.3	69.3	0.31	0.28
Bagasse	48.41	0.39	90.8	30.3	–	–
Molasses	75.0	0.65	–	>95.0	0.73	0.11
Banana						
Fruits	21.2	0.72	2.7	96.0	0.23	0.09
Pseudo-stems	6.8	0.38	38.7	77.4	0.18	0.14
Leaves	21.8	1.95	53.2	45.9	0.51	0.19
Coffee						
Pulp	20.0	2.13	21.0	61.2	0.55	0.11
Hulls	89.6	0.48	45.7	48.0	0.15	0.02

[1] DM = dry matter; N = nitrogen; CWC = cell wall constituents; IVDMD = *in vitro* dry-matter digestibility; Ca = calcium; P = phosphorus.

Common black bean (*Phaseolus vulgaris*) is another important crop on small farms in Central America, but the area cultivated to it has fallen over the past decade (Table 8.3). The crop is harvested when plants are mature and the pods are almost dry, but the above-ground part is cut and stored under shade to dry completely. After the grains have been collected, the haulm (consisting mainly of stems and empty pods) has to be disposed of. In many cases it is burnt rather than being returned to the field. A series of experiments was conducted at the Centro Agronómico Tropical de Investigación y Enseñanza (CATIE) at Turrialba, Costa Rica, to estimate the availability of bean residues under different cropping systems, to evaluate its nutritive value (Lozano *et al.*, 1980b; Ruiz *et al.*, 1980a) and to develop feeding systems for beef fattening enterprises (Lozano *et al.*, 1980a). In the past few years some technicians and participating farmers have demonstrated an interest in the utilization of black bean haulm as the roughage component of dual-purpose cattle diets during the dry season.

The major root and tuber crops in the subhumid zone are cassava (*Manihot esculenta*), sweet potato (*Ipomoea batatas*) and yams (*Colocasia esculenta*). However, when used for animal feeding, they are mostly offered to pigs. In the early 1980s a few experiments were conducted on the management and utilization of cassava (Pezo *et al.*, 1984) and of sweet potato (Ruiz *et al.*, 1980b) as dual-purpose crops (roots and foliage). Feeding systems based on the use of sweet potato vines and non-marketable roots were developed for milking and growing cattle (Backer *et al.*, 1980). Although biological and economic analysis indicated that dual-purpose cultivation was viable, extension efforts to transfer these systems to farmers have had poor results.

The Andes

The Andean ecoregion covers around 2 million km^2 and includes territory in six countries. It extends from the northern coasts of Venezuela and Colombia (latitude 11° north) to Argentina (55° south). Several characteristics differentiate this ecoregion from other mountainous parts of the world (Gastó, 1993), especially its very steep slopes and extreme climatic variations, associated with considerable regional diversity.

Production systems are, therefore, highly varied. Table 8.5 represents an attempt to group the commodities associated with different agroecological zones. Further differences exist within zones, due mainly to water availability, frost risk, slope, access to markets and market demands.

Weather conditions in the inter-Andean valleys are similar to those of temperate areas. These are favourable environments for milk production, but feed resources are not optimally utilized at present. Appropriate pasture management and forage conservation techniques, combined with better use of agricultural by-products, are required to achieve higher levels of production at competitive prices.

The availability and quality of feed resources in the other four zones described in the table differ considerably from those of the inter-Andean valleys. Seasonality is more evident, both in quantity and quality. Crop residues are a buffer that keeps animals alive through the dry season, which may last 4 to 6 months (Ruiz-Canales

Table 8.5. Predominant mixed crop/livestock production systems of the Andes.

Agro-ecological zone	Altitude (m) Rainfall (mm) Slope (%)	Crops/livestock	Main livestock products	Predominant feeding system
Inter-Andean valleys	200 - 2700 250 - 700 5 - 40	Maize, faba bean, ryegrass, white clover, alfalfa, cows	Milk/cheese	Grazing, conserved forages, agricultural by-products, commercial concentrates
Hillsides	2700 - 3500 500 - 800 20 - 90	Maize, wheat, barley, potato, other root and tuber crops, triple-purpose cattle, sheep and goats	Milk/cheese, weaned or mature animals, wool, dung	Grazing, crop residues, conserved forages
Suni	3400 - 4000 500 - 600 0 - 80	Potato, oat, barley, wheat, quinoa, faba bean, sheep, camelids, triple-purpose cattle	Milk/cheese, wool/fibre, weaned or mature animals, dung	Grazing, crop residues, conserved forages
Jalca	3400 - 4000 700 - 1300 30 - 90	Potato, barley, oat, sheep, triple-purpose cattle	Milk/cheese, wool, weaned or mature animals, dung	Grazing, crop residues, conserved forages
Puna	3800 - 4500 600 - 1200 20 - 60	Bitter potato, quinoa, qañiwa, cereals, sheep, camelids, triple-purpose cattle	Wool/fibre, weaned or mature animals, milk/cheese, dung	Grazing, crop residues, conserved forages

Source: Tapia (1996).

and Tapia-Núñez, 1987; Quiroz *et al.*, 1991). It is important to note that animals constitute not only the main source of income but also the savings account of most rural households (Li Pun and Paladines, 1993). This highlights the importance of using crop residues to reduce mortality rates.

Crop residues and by-products are fed mainly to large ruminants, but after a good harvest they are also given to small ones (Quiroz *et al.*, 1991). With the introduction of new pastures such as ryegrass, alfalfa and white clover, practices are changing. Cattle and sheep are now favoured, with lower-quality feedstuffs being left to camelids, which are more able to digest them (San Martin and Bryant, 1987).

Livestock production in the Andes is based on pasture utilization, complemented with crop residues or agricultural by-products. Native pastures adapted to different altitudes and soils dominate grasslands across the ecoregion. Both native and introduced pastures vary greatly according to the agroecological zones described in Table 8.5.

The humid inter-Andean valleys have the most favourable combination of climate and soils for pasture growth and development. Conditions in this zone resemble those of temperate regions. Predominant native species include *Axonopus affinis, A. compressus, A. micay, Bromus inermis, Holcus lanatus, Paspalum notatum, Desmodium* spp., *Medicago hispida* and *Trifolium dubium.* Among the introduced species are *Dactylis glomerata, Festuca arundinacea, Lolium multiflorum, L. perenne, Pennisetum clandestinum, Phalaris tuberosa, Trifolium pratense and T. repens* (Ruiz-Canales and Tapia-Núñez, 1987; Florez-Martínez, 1993; León-Velarde and Izquierdo-Cadena, 1993; Lotero, 1993)

Dry-matter (DM) production of native pastures at farm level varies from 2 to 5 t ha^{-1} year^{-1}, while introduced pastures produce between 4 and 10 t ha^{-1} year^{-1}. Production varies according to season, year type, water availability and management. It has been suggested that these levels of production could be doubled (Lotero, 1993).

The CP content of native pastures varies from 8 to 15% and their IVDMD from 45 to 65%. For introduced pastures, CP is between 11 and 25% and IVDMD 60 to 80%.

In the other four zones, conditions are less favourable for the production of high-yielding, good-quality pastures. Many native forage species are encountered, clustered according to their adaptation to different environments. Species of the Graminae, Bromeliaceae, Juncaceae, Iridaceae, Portulacaceae, Ranunculaceae, Cruciferae, Rosaceae, Leguminosae, Malvaceae, Cactaceae, Umbelliferae, Gentianaceae, and Compositae families are all encountered (Ruiz-Canales and Tapia-Núñez, 1987; Alzérreca, 1992; Florez-Martínez, 1993).

The production of DM is highly variable, according to environmental and management constraints. It ranges from < 1 to 5 t ha^{-1} year^{-1}. The content of CP varies from 5 to 10% and IVDMD from 40 to 60%. Where animal selectivity increases ingesta quality, the figures rise to 15 and 75% respectively.

Several introduced pasture species have adapted to these adverse environmental conditions. *Medicago sativa, Festuca arundinacea, Phalaris tuberosa, Lolium perenne* and *Lolium multiflorum* are now widespread (Ruiz-Canales and Tapia-Núñez, 1987; Alzérreca, 1992; Florez-Martínez, 1993; León-Velarde and Izquierdo-Cadena, 1993). The DM production of these species is from 2 to 10 t ha^{-1} year^{-1}, with a CP content of 11 to 20% and IVDMD of 60 to 85%.

The *suni, puna* and *jalca* zones are the bleak tablelands of the high Andes. The *suni,* found in the southern part of the range, is the most arid of the three, with average temperatures of 7°C. The average temperature in the puna, also in the south, is 3.6°C. Here pasture growth is favoured by large naturally irrigated areas whose water supply is melted snow from the high peaks. The *jalca,* located in the central and northern Andes, is a cloudier zone with higher rainfall and an average temperature of 8°C. In parts of all three zones, shrubs play an important part in animal feeding throughout the dry season (4 to 6 months), when they are browsed as a valued protein complement to lignified grasses (Alvarez-Vilcapaza, 1993).

The dominant shrub species include *Parastrephia lepidophyla, P. phylicaeformis, Ephedra americana, Suaeda foliosa* and *Atriplex cristata. Parastrephia* is the most widespread, but its contribution to the diet is seldom above 5% (Villca and Genin, 1995) due to its high content of anti-nutritional factors. *S. foliosa* is a good alternative for areas with saline soils. Under farmer management, this species produces 1 to 3 t ha^{-1} of edible DM with 12-16% CP and 70-85% IVDMD (Alzérreca, 1992).

The Southern Cone

Most livestock production in this ecoregion is located in temperate humid areas where the cattle diet is based on pasture grazing. There is some beef and dairy production in the subtropical and semiarid areas, based on natural grasslands, cultivated pastures, crop residues and agroindustrial by-products. Beef production in semiarid areas is considered to be of minor importance.

In temperate areas, beef animals graze high-quality temperate pastures strategically supplemented with cereal grains and agroindustrial by-products. Farmers constantly look for more efficient feeding systems that will increase liveweight gain and/or fat deposition. Aberdeen Angus and Hereford are the main breeds used. Productivity is around 80-120 kg ha^{-1} year^{-1} in cow-calf systems (located mainly in lowland areas not suitable for crop production) and around 280-300 kg meat ha^{-1} year^{-1} in rearing and fattening systems. Obviously, crop residues are of minor importance in these production systems.

In subtropical areas, beef production is also based on the grazing of pastures and natural grasslands, but here crop residues are more important. Chopped sugarcane, molasses-urea mixtures, rice bran and soyabean meal are the main by-products available. Pure zebu and zebu x Hereford or Angus crosses are the most widespread breeds. Meat productivity is about 30-70 kg ha^{-1} year^{-1}.

Dairy production in temperate areas is based on the grazing of temperate perennial pastures and on annual crops fed as fresh forage supplemented with maize silage and concentrates. Cereal grains such as maize, oat, sorghum and barley and by-products such as sunflower meal, wheat bran, whole cottonseed and brewer's grain constitute the major sources of concentrates. Holstein is the main breed, with milk productivity ranging from 100 to 200 kg butterfat ha^{-1} year^{-1} (15-20 l cow^{-1} day^{-1}).

Tropical pastures are also the basis for dairy production in subtropical areas, but they are supplemented with chopped forages, concentrates and crop residues such as sugarcane, molasses and rice bran. Pure zebu and Holstein x zebu crosses are the commonest breeds here and milk productivity is around 30-50 kg butterfat ha^{-1} year^{-1} (8-10 l cow^{-1} year^{-1}). Dual-purpose beef-milk production systems are also common.

Beef production in semiarid areas is very extensive, again based on natural grasslands. Crop residues and by-products are not very much used because no crops are grown in these areas. Criollos and Angus x Criollo crosses are used and meat productivity is very low, ranging from 5 to 30 kg ha^{-1} year^{-1}.

For years, livestock production systems in this ecosystem were extensive, with animal feeding based on forage from natural grasslands and cultivated pastures. However, economic stability and free-market policies have recently changed the relationship between fixed and operational costs. Over the past few years some operational costs, such as energy, seeds, fertilizers, machinery and equipment, have decreased or remained more or less constant, while others, such as taxes and labour, have risen sharply. Systems are being intensified as a way of increasing productivity while cutting costs (Rearte, 1994).

Formerly, beef and milk production were based mainly on forages provided by natural grassland, cultivated pasture and annual crops, whereas concentrates were strategically used only when the grain:beef/milk price ratio was favorable. In these extensive systems, applying N fertilizers to pasture was uncommon, while phosphorus was used only at pasture establishment. Today, a better fertilizer:animal product price ratio has made regular N fertilizer application profitable (Darwich, 1994).

In the traditional system, the stocking rate was adjusted according to the availability of forages during months with lower biomass availability (in the temperate zone, winter). This meant that the highest pasture growth, which occurred in early spring, could not be efficiently used because animal numbers were not high enough. The surplus forage was often lost because forage conservation practices were either little used or ineffective. Hay making was the most common conservation practice, but quality tended to be poor. Hay was made at the end of the growing season when rainfall decreased, allowing rapid drying under field conditions. But by that time of the year grasses had a low nutritive value due to maturity and the consequent increase in cell wall lignification. Silage was occasionally used in dairy systems, but was not used at all for beef production. The main crop for silage making was maize, which was fed to dairy cows during the winter, when green pastures became insufficient.

Pasture utilization is the main area in which improvements can be sought in the intensification process that is now taking place. Formerly, the proportion of pasture grazed by animals was no higher than 35-40%. A higher stocking rate is needed to graze spring regrowth. Consequently, more conserved forage is required to feed animals through the winter (Rearte and Elizalde, 1993). Although grazed forage continues to be the main component of cattle diets, more silage, hay and concentrates are now being offered. In the Argentina of 10 years ago, concentrates comprised no more than 20% of the total ration for dairy cows. Today, a typical dairy ration contains 30-35% pastures, 30-35% silage and 35% concentrates. A similar trend is occurring in beef production systems, in which higher levels of fertilizer are now applied to pastures and more concentrate is fed to steers.

The way in which feedstuffs are fed to animals is also changing. In previous years, pasture hays and even maize silage were considered as feed reserves, made when surpluses occurred and offered to animals when pasture biomass availability became a limiting factor. Today, these forages are no longer considered as a reserve but as important ingredients in the cow's ration. They are offered to animals not just when forage is scarce but when they are needed to supplement nutritional components in which pasture is deficient. Maize silage is a good example: it used to be fed to dairy cows during the winter, when pastures were not available. Nowadays, it is a vital supplement for overcoming the energy/protein imbalance that characterizes pastures during the autumn season. Its starch content provides the energy required to utilize the large amounts of nitrogen released in the rumen when animals graze a pasture with a high content of easily degradable protein (Rearte *et al.*, 1990).

Another important point concerning intensification is the requirement that conserved forage should be of high quality. Hay making allows the farmer to obtain a large amount of relatively cheap conserved forage, but its nutritive value at present is generally low—inadequate to satisfy the requirements of high-producing animals. One way to improve the nutritive value of conserved spring pasture would be to cut the forage earlier, at a less mature stage. But earlier hay making is not possible because of weather conditions, so silage seems to be the only alternative (Hargreaves, 1994). The amount of pasture land allocated to silage is steadily growing, although hay continues to be an important forage source for animals with lower requirements, such as dams in cow-calf systems.

In the quest for increased productivity, higher levels of concentrates and by-products are also being used. Concentrates based on cereal grains such as maize, sorghum, barley or oat are often considered too expensive for the low nutrient quantities they provide (Nicholson, 1984). The residues of these crops are good alternatives for the semiarid or marginal zones of the ecoregion, where forage availability from pastures and natural grasslands is seldom sufficient, but in the region as a whole they are not very important. By-products are more important. Their higher nutritive values make them excellent ingredients for concentrates (Manterola and Cerda, 1994).

Horticulture and fruit production are important sources of both crop residues and by-products in some areas of this ecoregion. In Chile, they play a major role in livestock feeding. The high water content of these materials makes storage and handling difficult. They are also susceptible to yeast and bacterial attack, which can affect their nutritive value. These storage and marketing problems mean that horticultural residues and by-products can be used only on farms close to where they are produced.

The favourable prospects for livestock production in the temperate areas of the Southern Cone suggest that both crop residues and by-products should become more important in the future. Several factors support this hypothesis:

1. More cattle will need to be fed in the more marginal areas, leaving higher-potential areas for cash crop production. Livestock with lower nutrient requirements, such as the dams in cow-calf systems, will be moved to areas where crop residues can compensate for lower pasture productivity.

2. Due to the need for more intensive production systems, higher levels of concentrates will be used to supplement grazing cows.

3. Analysis of the international meat market shows that the highest prices for beef at present are paid by Southeast Asia. The type of meat preferred by that market is high-quality beef produced in feedlots using high levels of concentrate supplementation to obtain carcasses with a higher fat content. The implications are that new feedlot systems will be launched in the Southern Cone, even in areas where livestock production is currently based mainly on grazing.

4. Higher demand for concentrates, combined with the higher price of cereal grains, will mean that agroindustrial by-products will provide the best alternative for reducing costs.

Arid and Semiarid Areas

The arid and semiarid areas of Central and South America occur mainly in Mexico and Brazil (Shelton and Figuereiro, 1990). Much of the Mexican area has access to irrigation, making year-round crop production feasible (Ramos-Sánchez and Galomo-Rangel, 1991). Alfalfa/maize-based intensive dairy systems are pervasive. Alfalfa is the forage crop favored for use under irrigated conditions because of its ability to colonize and improve desert soils and its high quality, which makes it particularly suitable for supplementing ruminant rations based on straws of low digestibility (Seré *et al.*, 1995).

In non-irrigated areas, goat and sheep production are the main sources of income for resource-poor farmers. The animals graze rangeland, which is supplemented with crop residues. Predominant vegetation includes *Opuntia, Flourensia cernua, Parthenium incanum, Dalea tuberculata, Mimosa, Prosopis, Atriplex* and *Larrea tridentata*. Crop residues (maize stover and bean haulm) are utilized by 90% of farmers (Salinas *et al.*, 1991).

The arid and semiarid area of Brazil is mainly found in the northeast, a vast region accounting for approximately 20% of the country and including all or parts of ten separate states (Malechek and Queiroz, 1989). The typical crops are maize, cotton, common bean and cassava. Goats and hair sheep are the most frequently kept domestic animals.

The principal year-long forage resource for ruminant animals is the sparse forest growth known as *caatinga,* which includes herbaceous plants, the live leaves and shoots of low or coppiced trees and shrubs and the fallen leaves of taller trees that become available during the dry season (Malechek *et al.*, 1989).

Crop residues are vital for the subsistence of ruminant animals throughout northeast Brazil. The most important ones used by farmers are: bean leaves, stalks and pods; cotton hulls and leaves; and maize cobs, husks, leaves, stalks, stover and tassels. The quality of these residues is variable, but all of them are high in cell wall contents. This limits the productivity of small ruminants, which cannot digest fibrous crop residues efficiently because of their limited rumen capacity (Johnson *et al.*, 1990).

Future Research Needs

Forecasting is a dangerous game, since projections are based on past trends. In spite of this limitation, different scenarios answering "What if?" questions are useful for designing and implementing research and development policies. Using the global food projections of Rosegrant *et al.* (1995) (Table 8.6) and their expected impact on Latin America, it is possible to outline the research required to meet the demand for livestock products.

Table 8.6. Projected annual per capita demand and net trade in livestock and crops, 1990-2020.

Commodity	Per capita demand (kg)		Net trade ('000 t)	
	1990	2020	1990	2220
Beef	21	24	959	2 002
Sheep meat	1	1	- 111	- 83
Total meat	40	50	864	1 724
Wheat	59	65	- 3 588	- 1 383
Rice	26	32	- 889	- 2 815
Maize	133	153	- 5 116	- 1 740
Other grains	41	41	- 3 715	- 2 255
Soyabean	60	80	6 678	9 809

Source: Rosegrant *et al.* (1995).

Out of the various ruminant species in the region, only cattle are expected to continue to contribute substantially to the region's trade. This is because of the expected continuing strength of demand for beef and, to a lesser extent, dairy products.

The basic questions, therefore, are: given the current population of beef cattle and existing trends in growth rates, is it biologically possible to meet the projected demands? Are new technologies required? If so, what research topics should be emphasized?

On the basis of the current calving rates of less than 60% and mortality rates of around 10%, it will be difficult to meet the demands of the region's growing population, especially if rising incomes increase per capita consumption from the current 21 to 24 kg year^{-1}. However, there are some possible improvements in beef cattle operations that will propel the sector in the right direction.

In cow-calf systems, nutritional interventions could improve the cow's reproductive performance. As mature cows will be increasingly maintained in marginal areas, they will greatly benefit from improved utilization of crop residues and by-products. For growing animals, the negative effects of periods of feed shortage could be minimized through the strategic budgeting of crop residues, by-products and improved pastures. Lastly, during fattening a combination of good-quality pastures and by-products is required to improve the cost-effectiveness of current feeding systems.

Milk production in dual-purpose and specialized dairy systems is expected to increase, covering the existing regional deficit. In the humid and subhumid tropics, both animal numbers and yields are projected to rise. The demand for additional nutrients can be satisfied through a combination of good-quality pastures, tree foliage and agricultural by-products. In contrast, for the high-producing specialized dairy cows raised in the temperate areas, new energy sources will be needed, capable of substituting expensive maize grain.

Free-trade policies will drive down prices, forcing the whole sector to become more competitive. This will compel farmers to pay more attention to the utilization of feed resources produced on the farm, such as green bananas and fodder trees. Such resources are also becoming more attractive because of price increases in grain and conventional protein sources. Changes in this direction are already taking place in Cuba, where sugarcane by-products are replacing inputs from outside the region (Pérez, 1993), following a dramatic increase in prices caused by the blockade and by the elimination of subsidies from the former Soviet Union. An additional driving force behind the increasing use of non-conventional feed resources is protection of the environment. In many places in tropical America, by-products such as citrus and coffee pulp and green banana are major pollutants.

Projections for small ruminant products suggest that there will not be a surplus for export. Production could be increased by more efficiently combining local feed resources such as native grasslands, woody perennials and crop residues. This will ensure that good-quality pastures will be made available to animals producing goods with higher economical value.

American camelids are not included in the projections made by Rosegrant *et al.* (1995), but one of two possible outcomes may be expected. If the fibre only continues to command good prices, current production systems will not need to be substantially changed, although the carrying capacity of rangeland will need to increase. However, if meat products from these species become more highly valued, so that the current practice of moving animals down from the altiplano to the inter-Andean valleys continues, new measures will need to be taken, since these species will compete with cattle for crop residues.

References

Alvarez-Vilcapaza, J.B. (1993) Composición botánica y valor nutricional de las dietas de alpacas (*L. pacus*), llamas (*L. glama*) y ovinos (*O. aries*) al pastoreo libre, durante el período de secano en puna seca. MSs thesis, Universidad Nacional Agraria La Molina, Lima, Peru.

Alzérreca, H. (1992) *Producción y Utilización de los Pastizales de la Zona Andina de Bolivia.* Red de Pastizales Andinos (REPAAN), La Paz, Bolivia.

Anderson, S., Santos, J., Boden, R. and Wadsworth, J. (1992) Characterization of cattle production systems in the State of Yucatán. In: Anderson, S. and Wadsworth, J. (eds), *Dual-purpose Cattle Production Research.* Proceedings of the IFS/FMVZ-UADY International Workshop, 23-27 March 1992, Mérida, México. International Foundation for Science, Stockholm, Sweden, pp. 150-161.

Ara, M.A. and Ordoñez, J.H. (1993) Establecimiento de pasturas en Ucayali: Estado del arte. *Revista de Investigaciones Pecuarias* (IVITA) 6: 67-79.

Araya, J., Benavides, J.E., Arias, R. and Ruiz, A. (1994) Identificación y caracterización de árboles y arbustos con potencial forrajero en Puriscal, Costa Rica. In: Benavides, J.E. (ed.), *Arboles y Arbustos Forrajeros en América Central.* Technical Report No. 236, vol. 1. Centro Agronómico Tropical de Investigación y Enseñanza (CATIE), Turrialba, Costa Rica, pp. 31-63.

Backer, J., Ruiz, M.E., Muñoz, H. and Pinchinat, A.M. (1980) El uso de la batata (*Ipomoea batatas* L.) en la alimentación animal, 2: Producción de carne de res. *Producción Animal Tropical* 5: 166-175.

Barreto, H.J., Pérez, C., Fuentes, M.R., Queme, J.L. and Larios, L. (1994) Efecto de dosis de urea-N en el redimiento de maíz bajo un sistema de rotación con leguminosas de cobertura. *Agronomía Mesoamericana* 5: 88-95.

Benavides, J., Ramlal, H. and Pezo, D.A. (1992) Feeding resources for goats in Central America and the Caribbean region. In: Acharya, R.M. (ed.), *Fifth International Conference on Goats, Invited Papers,* vol. 2. Indian Council of Agricultural Research (ICAR), New Delhi, India, pp. 134-142.

Braham, J.E., Jarquín, R., González, J.M. and Bressani, R. (1973) Pulpa y pergamino de café, 3: Utilización de la pulpa de café en forma de ensilaje. *Archivos Latinoamericanos de Nutrición* 23: 379-388.

Buckles, D., Ponce, I., Saín, G. and Medina, G. (1993) Uso y difusión del frijol de abono (*Mucuna deeringiana*) en las laderas del Litoral Atlántico de Honduras. *Agronomía Mesoamericana* 5: 15-29.

Caielli, E.L. (1988) Case study: Brazil sugarcane as feed. In: Sansoucy, R., Aarts,

G. and Preston, T.R. (eds), *Sugarcane as Feed.* Proceedings of an FAO Expert Consultation, 7-11 July 1986, Santo Domingo, Dominican Republic. FAO Animal Production and Health Paper No. 72. Food and Agriculture Organization of the United Nations, Rome, Italy, pp. 100-105.

CATIE (1978) Milk and Beef Production Systems for Small Farmers Using Crop Derivatives. Progress Report, Centro Agronómico Tropical de Investigación y Enseñanza, Turrialba, Costa Rica.

CATIE (1983) Investigación Aplicada en Sistemas de Producción de Leche. Final Report 1979-1983, CATIE-BID Project. Centro Agronómico Tropical de Investigación y Enseñanza, Turrialba, Costa Rica.

Cerdas, R. (1981) Banano de desecho (*Musa acuminata*) como suplemento a vacas lecheras en pastoreo, en diferentes estados de lactancia. MSc thesis. Universidad de Costa Rica (UCR) and Centro Agronómico Tropical de Investigación y Enseñanza (CATIE), Turrialba, Costa Rica.

Combellas, J. (1994) Influencia de los bloques multinutrientes sobre la respuesta productiva de bovinos pastoreando forrajes cultivados. In: Cardozo, A.F. and Birbe, B. (eds), *Memorias de la Primera Conferencia Internacional sobre Bloques Multinutricionales,* 29-31 July 1994, Universidad Experimental de los Llanos Ezequiel Zamora (UNELLEZ), Guanare, Venezuela, pp. 67-70.

Combellas, J. and Mata, D. (1992) Suplementación estratégica en bovinos de doble propósito. In: Fernández-Baca, S. (ed.), *Avances en la Producción de Leche y Carne en el Trópico Americano.* Food and Agriculture Organization of the United Nations (FAO), Santiago, Chile, pp. 99-130.

Darwich, N.A. (1994) Los sistemas mixtos y la fertilidad de los suelos. In: *Memorias de lo Segundo Simposio Tecnologico de AACREA,* 29-30 September 1994, Asociación Argentina de Consorcios Regionales de Experimentación Agropecuaria, Buenos Aires, Argentina, pp. 1-9.

Domínguez, C. (1994) El uso de los bloques multinutricionales en el Estado Guárico, 1: Efectos sobre producción de leche, reproducción y crecimiento en ganado de doble propósito. In: Cardozo, A.F. and Birbe, B. (eds), *Memorias de la Primera Conferencia Internacional sobre Bloques Multinutricionales,* 29-31 July 1994, Universidad Experimental de los Llanos Ezequiel Zamora (UNELLEZ), Guanare, Venezuela, pp. 97-116.

Duarte, J.M., Pérez, H.E., Pezo, D.A., Arze, J. and Argel, P. (1995) Producción de maíz (*Zea mays* L.), soya (*Glycine max* L.) y caupí (*Vigna unguiculata* L.Walp) sembrados en asociación con pastos en el trópico húmedo. *Pasturas Tropicales* 17: 12-19.

Elías, A., Lezcano, O., Lezcano, P., Cordero, J. and Quintana, L. (1990) A review of the development of a protein-sugarcane enrichment technology through solid-state fermentation: "Saccharina". *Cuban Journal of Agricultural Sciences* 24: 1-13.

Escobar, A. and Parra, R. (1984) Procesamiento y tratamiento físico-químico de los residuos de cosecha con miras al mejoramiento de su valor nutritivo. In: Ruiz, M.E., Ruiz, A. and Pezo, D. (eds), *Estrategias para el Uso de Residuos de Cosecha en la Alimentación Animal.* Proceedings of a Workshop, 19-21 March 1980, Centro Agronómico Tropical de Investigación y Enseñanza (CATIE), Turrialba, Costa Rica. International Development Research Centre (IDRC), Ottawa, Canada, pp. 93-128.

Espinoza, J.R. (1983) Consumo y parámetros de digestión en rastrojos de maíz cultivado sólo o en asocio con leguminosas. MSc thesis. Universidad de Costa Rice (UCR) and Centro Agronómico Tropical de Investigación y Enseñanza (CATIE), Turrialba, Costa Rica.

FAO (1986) *FAO Production Yearbook,* vol. 40. Food and Agriculture Organization of the United Nations, Rome, Italy.

FAO (1994) *FAO Production Yearbook,* vol. 48. Food and Agriculture Organization of the United Nations, Rome, Italy.

Flórez-Martínez, A. (1993*) Producción y Utilización de los Pastizales Alto Andinos del Perú.* Red de Pastizales Andinos (REPAAN), Lima, Peru.

García, R., Quiroga, R. and Granados, A. (1994) Agroecosistemas de productividad sostenida de maíz en las regiones cálido-húmedas de México. In: Thurston, H.D. (ed.), *Los Sistemas de Siembra con Coberturas.* Centro Agronómico Tropical de Investigación y Enseñanza (CATIE) and Cornell International Institute for Food, Agriculture and Development (CIIFAD), Cornell University, Ithaca, New York, USA, pp. 65-79.

Gastó, J. (1993) Aproximación agroecosistémica. In: *El Agroecosistema Andino: Problemas, Limitaciones, Perspectivas.* Proceedings of a Workshop, 30 March-2 April 1993, Centro Internacional de la Papa (CIP), Lima, Peru, pp. 31-49.

Hargreaves, A. (1994) Estrategias de suplementacion con ensilajes en pastoreo. In: Gonzales, M. and Bortolameolly, G. (eds), *II Seminario Produccion y Utilizacion de Ensilajes de Pradera para Agricultores de la Zona Sur,* 8-9 June, Instituto Nacional de Investigación Agropecuaria (INIA), Osorno, Chile, pp. 77-95.

Hecht, S.B. (1993) The logic of livestock and deforestation in Amazonia. *BioScience* 43: 687-695.

Hurtado, J.A., Pezo, D., Chaves, C. and Romero, F. (1988) Caracterización de una pradera degradada de pasto estrella africana *(Cynodon nlemfuensis)* bajo el efecto del pastoreo y la introducción de leguminosas en el trópico húmedo. In: Pizarro, E.A. (ed.), *Memorias de la Primera Reunión RIEPT/CAC,* 17-18 November 1988, Veracruz, México. Centro Internacional de Agricultura Tropical (CIAT), Cali, Colombia, pp. 341-347.

IICA (1995) Mejoramiento de Sistemas de Producción Bovina de Doble Propósito en Guatemala. Final Report, Third Phase. Instituto Interamericano de Cooperación para la Agricultura (IICA), Instituto de Cienca y Technología Agrícolas (ICTA), Dirección General de Servicios Pecuarios (DIGESEPE) and Universidad de San Carlos (USAC), Guatemala.

Johnson, W.L., Barros, N.L. and Oliveira, E.R. (1990) Supplemental feed resources and their utilization by hair sheep. In: Shelton, M. and Figuereiro, E.A.P. (eds*), Hair Sheep Production in Tropical and Subtropical Regions.* Small Ruminant Collaborative Research Support Program (SR-CRSP), University of California-

Davis, Berkeley, USA, pp. 79-95.

Kass, M., Benavides, J., Romero, F. and Pezo, D. (1992) Lessons from main feeding experiments conducted at CATIE using fodder trees as part of the N-ration. In: Speedy, A. and Pugliese, P. (eds), *Legumes and Other Fodder Trees as Protein Sources for Livestock.* Proceedings of the FAO Expert Consultation, 14-18 October 1991, Malaysian Agricultural Research and Development Institute (MARDI), Kuala Lumpur, Malaysia. FAO Animal Production and Health Paper No. 102, Food and Agriculture Organization of the United Nations, Rome, Italy, pp. 161-175.

Lascano, C.E. and Pezo, D.A. (1994) Agroforestry systems in the humid forest margins of tropical America from a livestock perspective. In: Copeland, J.W., Djajanegara, A. and Sabrani, M. (eds), *Agroforestry and Animal Husbandry for Human Welfare.* Proceedings of an International Symposium, 11-16 July 1994, Bali, Indonesia. ACIAR Proceedings No. 55, Australian Centre for International Agricultural Research, Canberra, pp. 17-24.

Leonard, M.J. (1987) *Recursos Naturales y Desarrollo Económico en América Central: Un Perfil Ambiental.* Technical Report No. 127, Centro Agronómico Tropical de Investigación y Enseñanza (CATIE), Turrialba, Costa Rica.

León-Velarde, C. and Izquierdo-Cadena, F. (1993) *Producción y Utilización de los Pastizales de la Zona Andina: Compendio.* Red de Pastizales Andinos (REPAAN), Quito, Ecuador.

Li Pun, H. and Paladines, O. (1993) Funcion de las pasturas y la ganaderia en la sostenibilidad de los sistemas de producción Andina. In: *El Agroecosistema Andino: Problemas, Limitaciones, Perspectivas.* Proceedings of an International Workshop, 30 March-2 April 1993, Centro Internacional de la Papa (CIP), Lima, Peru, p. 15.

Loker, W.M., Hernández, E. and Rosales, J. (1991) Establecimiento de pasturas en el trópico húmedo: Experiencias en la Selva Peruana. In: Lascano, C.E. and Spain, J.M. (eds), *Establecimiento y Renovación de Pasturas: Conceptos, Experiencias y Enfoque de Investigación.* Proceedings of the Sixth Meeting of the Steering Committee of the Red Internacional de Evaluación de Pastos Tropicales (RIEPT), 14-16 November 1988, Veracruz, México. Centro Internacional de Agricultura Tropical (CIAT), Cali, Colombia, pp. 321-346.

López, G.Z., Benavides, J.E., Kass, M. and Faustino, J. (1994) Efecto de la suplementación con follaje de amapola *(Malvaviscus arboreus)* sobre la producción de leche de cabras estabuladas. In: Benavides, J.E. (ed.), *Arboles y Arbustos Forrajeros en América Central.* Technical Report No. 236, Centro Agronómico Tropical de Investigación y Enseñanza (CATIE), Turrialba, Costa Rica, pp. 321-339.

Lotero, J. (1993) *Producción y Utilización de los Pastizales de las Zonas Alto Andinas de Colombia.* Red de Pastizales Andinos (REPAAN), Medellín, Colombia.

Lozano, E., Ruiz, M.E. and Ruiz, A. (1980a) Desarrollo de sistemas de alimentación de bovinos a base de rastrojo de frijol *(Phaseolus vulgaris* L.), 3: Producción de carne. *Turrialba* 30: 153-159.

Lozano, E., Ruiz, A. and Ruiz, M.E. (1980b) Desarrollo de sistemas de alimentación de bovinos con rastrojo de frijol (*Phaseolus vulgaris* L.), 2: Balance metabólico a varios niveles de energía y proteína suplementaria. *Turrialba* 30: 63-70.

Malechek, J.C. and Queiroz, J.S. (1989) Ecological characteristics of the semiarid tropics of northeast Brazil. In: Johnson, W.L. and Oliveira, E.R. (eds), *Improving Meat Goat Production in the Semiarid Tropics*. Small Ruminant Collaborative Research Support Program (SR-CRSP), University of California-Davis, Berkeley, USA, pp. 1-11.

Malechek, J.C., Schacht, W.H., Pfister, J.A., Kirmse, R.D., Hardesty, L.H. and Provenza, F.D. (1989) Improving the productivity of grazing lands in the semiarid tropics. In: Johnson, W.L. and Oliveira, E.R. (eds), *Improving Meat Goat Production in the Semiarid Tropics*. Small Ruminant Collaborative Research Support Program (SR-CRSP), University of California-Davis, Berkeley, USA, pp. 49-66.

Manterola, H. and Cerda, D. (1994) Recursos forrajeros, estrategias y metodologias para la alimentacion de rumiantes menores en la zona arida y semi-arida de Chile. In: Iñiguez, L. and Tejada, E. (eds), *Producción de Rumiantes Menores en los Valles Interandinos de Sudamerica*. Proceedings of a Workshop, 16-21 August 1993, Tarija, Bolivia, pp. 33-73.

Martínez, L. (1980) Consumo voluntario, digestibilidad y balance metabólico en novillos alimentados con seudo-tallo de banano (*Musa acuminata* cv. Cavendish) y suplementos energéticos. MSc thesis, Universidad de Costa Rica (UCR) and Centro Agronómico Tropical de Investigación y Enseñanza (CATIE), Turrialba, Costa Rica.

Miles, J.W. and Lapointe, S.L. (1992) Regional germplasm evaluation: A portfolio of germplasm options for the major ecosystems of tropical America. In: *Pastures for the Tropical Lowlands: CIAT's Contribution*. Centro Internacional de Agricultural Tropical (CIAT), Cali, Colombia, pp. 9-28.

Murillo, O. and Navarro, L.A. (1986) Validación de prototipos de producción de leche en la Zona Atlántica de Costa Rica. Technical Report No. 90, Centro Agronómico Tropical de Investigación y Enseñanza (CATIE), Turrialba, Costa Rica.

Myers, N. (1981) How Central America's forest became North America's hamburger. *Ambio* 10: 3-8.

Nicholson, J.W.G. (1984) Digestibility, nutritive value and feed intake. In: Sundstøl, F. and Owen, E. (eds), *Straw and Other Fibrous By-products as Feed*. Elsevier, Amsterdam, The Netherlands, pp. 340-372.

Ordónez, J.A. (1994) Levante de toros en soca de sorgo durante la temporada seca: Bloques multinutricionales vs cama de pollo, sorgo molido. In: Cardozo, A.F. and Birbe, B. (eds), *Memorias de la Primera Conferencia Internacional sobre Bloques Multinutricionales,* 29-31 July 1994, Universidad Experimental de los Llanos Ezequiel Zamora (UNELLEZ), Guanare, Venezuela, pp. 79-83.

Payne, L.D. (1993) Chemical constituents of *Erythrina*: Historical perspectives and future prospects. In: Westley, S.B. and Powell, M.H. (eds), *Erythrina* in the New and Old Worlds. Nitrogen Fixing Tree Association (NFTA), Hawaii, USA, pp. 314-321.

Pérez, E., Ruiz, M.E. and Pezo, D. (1990a) Suplementación de bovinos con banano verde, 3: Efecto sobre la degradación ruminal del banano. *Agronomía Costarricense* 14: 61-66.

Pérez, E., Ruiz, M.E. and Pezo, D. (1990b) Suplementación de bovinos con banano verde, 4: Efecto sobre algunos parámetros de fermentación ruminal. *Agronomía Costarricense* 14: 67-72.

Pérez, R. (1993) Experiencias cubanas en el uso de subproductos agroindustriales para el ganado en la seca. In: *Memorias del Seminario Internacional Estrategias de Alimentación en Verano para Ganaderías Tropicales,* 2-4 June 1993, Centro Internacional de Capacitación para el Desarrollo Pecuario (CICADEP), Santafé de Bogotá, Colombia, pp. 31-51.

Pezo, D., Benavides, J. and Ruiz, A. (1984) Producción de follaje y raíces de yuca (*Manihot esculenta* Crantz) bajo diferentes densidades de plantación y frecuencias de poda. *Producción Animal Tropical* 9: 251-262.

Pezo, D.A., Kass, M., Benavides, J., Romero, F. and Chaves, C. (1990) Potential of legume tree fodders as animal feed in Central America. In: Devendra, C. (ed.), *Shrubs and Tree Fodders for Farm Animals.* Proceedings of a Workshop, 24-29 July 1989, Denpasar, Bali, Indonesia. International Development Research Centre (IDRC), Ottawa, Canada, pp. 163-175.

Pezo, D.A., Romero, F. and Ibrahim, M. (1992) Producción, manejo y utilización de los pastos tropicales para la producción de leche y carne. In: Fernández-Baca, S. (ed.), *Avances en la Producción de Leche y Carne en el Trópico Americano.* Food and Agriculture Organization of the United Nations (FAO), Santiago, Chile, pp. 47-98.

Preston, T.R. and Leng, R.A. (1990) Ajustando los Sistemas de Producción Pecuaria a los Recursos disponibles: Aspectos Básicos y Aplicados del Nuevo Enfoque sobre Nutrición de Rumiantes en el Trópico. Consultorías para el Desarrollo Rural Integrado en el Trópico (CONDRIT), Cali, Colombia.

Preston, T.R. and Murgueitio, E. (1992) *Strategy for Sustainable Livestock Production in the Tropics.* Centro para la Investigación en Sistemas Sostenibles de Producción Agropecuaria (CIPAV) and Swedish Agency for Research Cooperation with Developing Countries (SAREC), Cali, Colombia.

Quiroz, R., Mamani, G., Revilla, R., Guerra, C., Sánchez, J., Gonzalez, M. and Pari, G. (1991) Perspectivas de investigación pecuaria para el desarrollo de las comunidades de Puno. In: Arguelles, L. and Estrada, R.D. (eds), *Perspectivas de la Investigación Agropecuaria para el Altiplano.* Proyecto de Investigación de los Sistemas Agropecuarios Andinos (PISA), Centro Internacional de Investigaciones para el Desarrollo (CIID), Lima, Peru, pp. 357-406.

Ramos-Sánchez, A. and Galomo-Rangel, T. (1991) Sistemas de producción agropecuarios y forestales en México. In: Ruiz, A. and Ruiz, M. (eds), *Informe de la Novena Reunión General de la Red de Investigación en Sistemas de*

*Producción Animal en Latinoamérica (RISPAL),*18-26 April 1990, Instituto Interamericano de Cooperación para la Agricultura (IICA), San José, Costa Rica, pp. 209-234.

Rearte, D.H. (1994) La tecnologia futura en la produccion ganadera. In: *Memorias del Segundo Simposio Tecnologico de AACREA,* 29-30 September 1994, Asociación Argentina de Consorcios Regionales de Experimentación Agropecuaria, Buenos Aires, Argentina, pp. 20-35.

Rearte, D.H. and Elizalde, J. (1993) Suplementacion de vacunos en pastoreo. In: *Resumenes de Jornadas de Actualizacion Tecnica en Invernada,* 13-14 May 1993, Centro Regional Buenos Aires Sur (CERBAS) and Instituto Nacional de Tecnología Agropecuaria (INTA), Mar del Plata, Argentina, pp. 46-58.

Rearte, D.H., di Berardino, J. and Melani, G. (1990) Performance of dairy cows grazing pasture and supplement with corn silage. *Journal of Dairy Science* 73 (suppl. 1): 240.

Riesco, A. (1992) La ganadería bovina en el Trópico Americano: Situación actual y perspectivas. In: Fernández-Baca, S. (ed.), *Avances en la Producción de Leche y Carne en el Trópico Americano.* Food and Agriculture Organization of the United Nations (FAO), Santiago, Chile, pp. 13-46.

Rojas, H. and Benavides, J.E. (1994) Predución de leche en cabras alimentadas con pasto y suplementadas con altos niveles de morera *(Morus* sp.). In: Benavides, J.E. (ed.), *Arboles y Arbustos Forrajeros en América Central.* Technical Report No. 236, Centro Agronómico Tropical de Investigación y Enseñanza (CATIE), Turrialba, Costa Rica, pp. 305-320.

Roldán, G. and Soto, R.A. (1988) Características del manejo de la *guatera* en dos municipios de Oriente. In: Proyecto Mejoramiento de Sistemas de Producción Bovina de Doble Propósito en Guatemala. Final Technical Report, First Phase, Instituto Interamericano de Cooperación para la Agricultura (IICA), Guatemala, pp. 112-116.

Romero, F., Montenegro, J., Chana, C., Pezo, D. and Borel, R. (1993) Cercas vivas y bancos de proteína de *Erythrina berteroana* manejados para la producción de biomasa comestible en el trópico húmedo de Costa Rica. In: Westley, S.B. and Powell, M.H. (eds), *Erythrina in the New and Old Worlds.* Nitrogen Fixing Tree Association (NFTA), Hawaii, USA, pp. 205-210.

Rosegrant, M.W., Agcaoili-Sombilla, M. and Perez, N.D. (1995) Global Food Projections to 2020: Implications for Investment. Food, Agriculture and the Environment Discussion Paper, 2020 Vision. International Food Policy Research Institute (IFPRI), Washington DC, USA.

Ruiloba, M.H., Ruiz, M.E. and Pitty, C. (1978) Producción de carne durante la época seca a base de sub-productos, 3: Integración de componentes y validación de sistemas de alimentación de engorde. *Ciencia Agropecuaria* (IDIAP) 1: 87-92.

Ruiz, A. and Ruiz, M.E. (1977) Efecto del consumo de pasto verde sobre el consumo de pulpa de café y la ganancia de peso en novillos. *Turrialba* 27: 23-28.

Ruiz, A. and Ruiz, M.E. (1978) Utilización de gallinaza en la alimentación de bovinos, 3: Producción de carne en función de diversos niveles de gallinaza y almidón. *Turrialba* 28: 215-223.

Ruiz, C. (1992) Aceptabilidad por ovinos de la biomasa comestible de procedencias, familias e individuos de *Gliricidia sepium* en Guápiles, Costa Rica. MSc thesis, Centro Agronómico Tropical de Investigación y Enseñanza (CATIE), Turrialba, Costa Rica.

Ruiz, M.E. (1981) The use of green bananas and tropical crop residues for intensive beef production. In: Smith, A.J. and Gunn, R.G. (eds), *Intensive Animal Production in Developing Countries*. Occasional Publication No. 4, British Society of Animal Production, pp. 371-383.

Ruiz, M.E. (1984) Estrategia para la intensificación de la producción de carne. In: Ruiz, M.E., Ruiz, A. and Pezo, D. (eds), *Estrategias para el Uso de Residuos de Cosecha en la Alimentación Animal.* Proceedings of a Workshop, 19-21 March 1980, Centro Agronómico Tropical de Investigación y Enseñanza (CATIE), Turrialba, Costa Rica. International Development Research Centre (IDRC), Ottawa, Canada, pp. 39-80.

Ruiz, M.E. (1993) Experiencias en el manejo de residuos de cultivos para la alimentación de rumiantes. In: *Memorias del Seminario Internacional Estrategias de Alimentación en Verano para Ganaderías Tropicales,* 2-4 June 1993, Centro Internacional de Capacitación para el Desarrollo Pecuario (CICADEP), Santafé de Bogotá, Colombia, pp. 61-74.

Ruiz, M.E., Olivo, R., Ruiz, A. and Fargas, J. (1980a) Desarrollo de subsistemas de alimentación de bovinos con rastrojo de frijol (*Phaseolus vulgaris* L.), 1: Disponibilidad, composición y consumo de rastrojo de frijol. *Turrialba* 30: 49-55.

Ruiz, M.E., Pezo, D. and Martínez, L. (1980b) El uso del camote (*Ipomoea batatas* L.) en la alimentación animal, 1: Aspectos agronómicos. *Producción Animal Tropical* 5: 157-165.

Ruiz-Canales, C. and Tapia-Núñez, M.E. (1987) *Producción y Manejo de Forrajes en los Andes Altos del Perú.* Proyecto de Investigación de los Sistemas Agropecuarios Andinos (PISA), Lima, Peru.

Salinas, H., Quiroga, P., Martínez, M., Guerrero, A., Espinoza, J., Cano, J.F. and Avila, J.L. (1991) Sistemas de producción caprina en México. In: Ruiz, A. and Ruiz, M. (eds), *Informe de la Novena Reunión General de la Red de Investigación en Sistemas de Producción Animal en Latinoamérica* (RISPAL), 18-26 April 1990, Instituto Interamericano de Cooperación para la Agricultura (IICA), San José, Costa Rica, pp. 81-98.

San Martín, F. and Bryant, F.C. (1987) Nutrición de los Camélidos Sudamericanos: Estado de Nuestro Conocimiento. Technical Report T-9-505. College of Agricultural Sciences, Texas Technical University (TTU), Lubbock, USA.

San Martín, F., Pezo, D., Ruiz, M.E., Vohnout, K. and Li Pun, H.H. (1983) Suplementación de bovinos con banano verde, 2: Efecto sobre el consumo de punta de caña. *Producción Animal Tropical* 8: 240-246.

Sansoucy, R., Aarts, G. and Preston, T.R. (1988) *Sugarcane as Feed.* Proceedings of an FAO Expert Consultation, 7-11 July 1986, Santo Domingo, Dominican Republic. FAO Animal Production and Health Paper No. 72, Food and Agriculture Organization of the United Nations, Rome, Italy.

Seré, C., Steinfeld, H. and Groenewold, J. (1995) World livestock production systems: Current status, isues and trends. In: Gardiner, P. and Devendra, C. (eds), *Global Agenda for Livestock Research.* Proceedings of a Consultation, 18-20 January 1995, International Livestock Research Institute (ILRI), Nairobi, Kenya, pp. 11- 38.

Serrao, E.A.S. and Toledo, J.M. (1990) The search for sustainability in Amazonian pastures. In: Anderson, A.B. (ed.), *Alternatives to Deforestation: Steps Towards Sustainable Use of the Amazonian Rainforest.* Columbus University, New York, USA, pp. 195-214.

Shelton, M. and Figuereiro, E.A.P. (1990) *Hair Sheep Production in Tropical and Subtropical Regions.* Small Ruminant Collaborative Research Support Program (SR-CRSP), University of California-Davis, Berkeley, USA.

Sinclair, R., Wedge, L. and Romero, A. (1991) Utilización de rastrojos en la alimentación de animales. *Pasturas Tropicales* 13: 20-22.

Tapia, M.E. (1996) *Ecodesarrollo en los Andes Altos.* Fundación Friedrich Ebert, Lima, Peru.

Teodoro, R.L. and Lemos, A. de Matos (1992) Cruzamientos de bovinos para producción de leche y carne. In: Fernández-Baca, S. (ed.), *Avances en la Producción de Leche y Carne en el Trópico Americano.* Food and Agriculture Organization of the United Nations (FAO), Santiago, Chile, pp. 209-260.

Thiesenhusen, W.C. (1994) The relation between land tenure and deforestation in Latin America. In: Pezo, D., Homan, J.E. and Yuill, T.M. (eds), *Animal Agriculture and Natural Resources in Central America: Strategies for Sustainability.* Proceedings of a Workshop, 7-12 October 1991, Centro Agronómico Tropical de Investigación y Enseñanza (CATIE) and University Group for International Animal Agriculture (UGIAAG), San José, Costa Rica. CATIE, Turrialba, pp. 243-256.

Toledo, J.M. (1994) Ganadería bajo pastoreo: Posibilidades y parámetros de sostenibilidad. In: Homan, E.J. (ed.), *Memorias Simposio-Taller sobre Ganadería y Recursos Naturales en América Central: Estrategias para la Sostenibilidad,* 7-12 October, Centro Agronómico Tropical de Investigación y Enseñanza (CATIE) and University Group for International Animal Agriculture (UGIAAG), San José, Costa Rica. CATIE, Turrialba, pp. 141-162.

Vaccaro, L. (1992) Evaluación y selección de bovinos de doble propósito. In: Fernández-Baca, S. (ed.), *Avances en la Producción de Leche y Carne en el Trópico Americano.* Food and Agriculture Organization of the United Nations (FAO), Santiago, Chile, pp. 171-208.

Vaccaro, L., Vaccaro, R., Verde, O., Alvarez, R., Mejías, H., Ríos, L. and Romero, E. (1992) Genetic improvement of dual-purpose herds: Some results from Venezuela. In: Anderson, S. and Wadsworth, J. (eds), *Dual-purpose Cattle Production Research.* Proceedings of the IFS/FMVZ-UADY International

Workshop, 23-27 March 1992, Facultad de Medicina Veterinaria y Zootecnia, Universidad Autónoma de Yucatán, Mérida, México. International Foundation for Science, Stockholm, Sweden, pp. 133-149.

Valerio, S. (1990) Efecto del secado y métodos de análisis sobre estimados de taninos y la relación de estos con la digestibilidad in vitro de algunos forrajes tropicales. MSc thesis, Centro Agronómico Tropical de Investigación y Enseñanza (CATIE), Turrialba, Costa Rica.

van der Grinten, P., Baayen, M.T., Villalobos, L., Dwinger, R.H. and 't Manneteje, L. (1992) Utilization of kikuyu grass (*Pennisetum clandestinum*) pastures and dairy production in a high-altitude region of Costa Rica. *Tropical Grasslands* 26: 255-262.

Villca, Z. and Genin, D. (1995) Uso de los recursos forrajeros por llamas y ovinos, 1: Comportamiento alimenticio. In: Genin, D., Picht, H.J., Lizarazu, R. and Rodríguez, T. (eds), *Waira Pampa: Un Sistema Pastoral Camélidos-ovinos del Altiplano Arido Boliviano*. Institut Français de Recherche pour le Développement en Coopération (ORSTOM), Proyecto de Auto Desarrollo Campesino, Fase de Consolidación (CONPAC), Oruro and Instituto Boliviano de Tecnología Agropecuaria (IBTA), La Paz, Bolivia, pp. 117-130.

Villegas, L.A. (1979) Suplementación con banano verde a vacas lecheras en pastoreo. MSc thesis, Universidad de Costa Rica (UCR) and Centro Agronómico Tropical de Investigación y Enseñanza (CATIE), Turrialba, Costa Rica.

Wadsworth, J. (1992) Dual-purpose cattle production: A systems overview. In: Anderson, S. and Wadsworth, J. (eds), *Dual-purpose Cattle Production Research*. Proceedings of the IFS/FMVZ/UADY International Workshop, 23-27 March 1992, Facultad de Medicina Veterinaria y Zootecnia, Universidad Autónoma de Yucatán, Mérida, México. International Foundation for Science, Stockholm, Sweden, pp. 2-27.

Zea, J.L. (1993) Efecto residual de intercalar leguminosas sobre el rendimiento de maíz (*Zea mays* L.) en nueve localidades de Centro América. *Agronomía Mesoamericana* 4: 18-22.

9. Crop Residues as a Strategic Resource in Mixed Farming Systems

Marc Latham

International Board for Soil Research and Management, PO Box 9-109, Bangkok 10900, Thailand

Abstract

Despite their other uses as animal feed, fuel and construction materials, crop residues are extremely important for managing soils. They are needed to improve soil tilth and structure, and to reduce soil erosion. They help enhance water availability to plants by promoting infiltration and increasing the soil's water retention capacity. And they are a vital source of nutrients in systems where few inputs are used.

The alternative uses of crop residues can give rise to conflicts in rural communities. As a result there is a need to discuss their use in a participatory manner with all stakeholders. Participatory on-farm research on socioeconomic issues is vital, since few of the results available from research have yet been applied by farmers. Strategic research on biophysical aspects may also be necessary, in such areas as the levels of organic matter necessary to maintain soil structure, the means of reducing nutrient losses, and the development of crop varieties that can simultaneously meet the need for grain and residues. New approaches to extension, such as the involvement of farmers' self-help groups and other non-government organizations (NGOs), can complement the traditional approaches used by formal government extension services.

For resource-poor farmers in developing countries, the key to making more rational use of crop residues lies in escaping poverty. Given higher prices for their produce, farmers could afford other materials for animal feed, fuel and construction.

Introduction

The term "residue", with its connotations of something left over that nobody wants, gives a false impression of the value of the straws, stubbles and other vegetative parts of crops that remain after harvest, especially since many farmers burn them or otherwise dispose of them.

The main reason for burning crop residues is to eradicate insect pests and rodents—but some farmers persist in burning out of sheer habit. In the developed world, burning has recently gone out of fashion, as farmers have realized that valuable nutrients and organic matter are permanently lost in this way. Indeed, according to Unger (1990), crop residues represent about half the nutrients exported through the production of grain, fibre or nut crops. They should not, therefore, be wasted.

In the developing world, and especially in the semiarid zone, crop residues are very much in demand. Pastoralists make arrangements with crop farmers to graze their cattle directly on fields after harvest, exchanging feed for the provision of manure. Mixed farmers store residues for later feeding. Villagers use them as fuel or construction materials. Farmers everywhere—whether traditional or modern—use residues as sources of organic matter and nutrients to be returned to the soil.

Unfortunately, these different uses often conflict. In addition, crop residues are often used inefficiently (Bonfils, 1987). The consequences are lost crop and livestock production, soil erosion and declining soil fertility, while the residues themselves are also often wasted. This process is part of the spiral of decline in rural societies associated with poverty, environmental degradation and population growth, as described by Greenland *et al.* (1994). Degradation is particularly severe in the semiarid and subhumid zones of the tropics, where vegetation is less prolific. Here the situation is aggravated by the expansion of agriculture into former rangelands, which can create conflicts between farmers and pastoralists during dry years when there is not enough feed for the animals (Lericollais, 1990). But in the humid tropics too, removing organic residues can lead to nutrient depletion, acidification and deteriorating soil structure. In the temperate zone, the virtues of retaining crop residues have recently been rediscovered following the damaging effects of wind erosion, particularly in North America (NRC, 1989; Rosaasen and Lokken, 1993).

The management of crop residues is thus a major issue affecting the sustainability of rural development, both in the tropics and in the temperate zone. Arguably, it is as much a socioeconomic issue as a biophysical one. It is therefore worthwhile asking how best to approach it and what place research has in its resolution.

The Biophysical Context

Crop residues are important in boosting soil tilth conditions and the development of crop roots. They also enhance soil moisture and restore to the soil nutrients that would otherwise be lost.

Root Development

Crops need good soil tilth to develop their roots. This requirement can only be met if the topsoil is properly managed, including erosion control where necessary, and if root penetration in the subsoil is enhanced.

Soil structure also affects the rooting pattern of plants. The structure of a soil can be observed, described and measured, although the latter operation is rather complex (Henin *et al.,* 1958; Emerson, 1967; Le Bissonnais and Le Souder, 1995). The links between structure and organic matter content and quality, although subject to continuing debate, have long been established (Charreau and Nicou, 1971; Tisdall and Oades, 1982; Albrecht *et al.*, 1992). But there is still no general understanding of how much (or how little) organic matter is needed to protect the soil structure. Pieri (1989) describes the relationship between organic matter and the percentage of fine particles required to maintain proper soil physical properties in the savannas of West Africa (Fig. 9.1). Similar descriptions for other agroecologies are few and far between, and not very convincing.

One of the most important consequences of the structural degradation of a soil is surface crusting. This impedes both the emergence of seedlings and the infiltration of water (Valentin *et al.*, 1992). Crusting is pronounced in the semiarid zone, but can also be found in the subhumid tropics wherever the soil surface is not properly managed. We now know quite a lot about the process of crusting and the typology of different crusts (Casenave and Valentin, 1989). But we still know very little about how to combat crusting practically.

Erosion, whether by water or wind, is another consequence of a deterioration in soil structure. Erosion removes part or all of the topsoil, leaving a crust which is

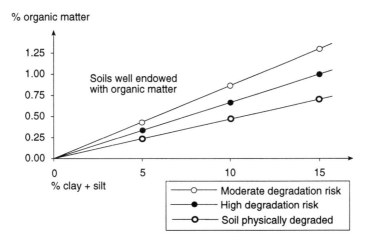

Fig. 9.1. Organic matter content (%) required to maintain soil structure for soils of varying clay and silt percentages in West Africa (Pieri, 1989).

difficult for roots to penetrate. It also entails the loss of soil fertility (Roose, 1989; Lal, 1990). Leaving crop residues on the soil surface is a good way of reducing soil erosion (Table 9.1). Yet this is not always practical, especially when there are other demands for these residues.

Table 9.1. Effect of different treatments on soil loss (t ha^{-1}).

Location	Residues removed/burned	Residues left on the soil (DM)	Alley cropping
Chiang Mai (Thailand)	32.0	3.8	0.7
Kuamang Kuning (Indonesia)	91.1	40.7	20.0
Dac Lac (Vietnam)	6.7	5.4	4.9

Source: Sajjapongse (1995).

Water Availability

Of the various means of increasing the availability of water to plants, the most efficient are those that promote its infiltration and storage in the soil profile and slow down its evaporation. Here, porosity and water absorption are the keys. Crop residues can help to improve these two parameters.

Porosity, particularly through macropores associated with cracks or soil biological activity, is essential for the proper flow of water into the soil profile. The role of macrofauna in this regard should be emphasized (Lavelle, 1994). Practices that increase soil biological activity have been developed on the research station but have not so far been widely implemented by small-scale farmers. Often, they depend on a conjunction of circumstances, including the availability of sufficient labour and of suitable natural resources. As such, they may imply major changes in the farming system for which farmers are not yet prepared.

Water absorption and storage in the soil are also linked to soil organic matter and hence to the return of crop residues (Parr et al., 1990). Greb (1983) showed that, in the Great Plains of the USA, the application of mulch gave an almost linear increase in water storage. Similar results have been obtained at the International Institute of Tropical Agriculture (IITA), in the forest-savanna transition zone of Nigeria (Fig. 9.2).

It is difficult, in a field experiment, to separate the effect of water storage as such from that of the reduced evaporation that can also be achieved by applying a mulch. In the case of the transitional forest-savanna environments around IITA in

Nigeria, mulch has proved quite efficient in reducing evaporation. However, Gregory (1989) points out that it may not be practical where it is most needed—in the semiarid tropics, where evaporation is the highest.

Nutrients Store

The main reason for advocating the return of crop residues to the soil is that this provides the soil with the nutrients needed to grow future crops. Fernández and Sanchez (1990), Swift *et al.* (1992) and Feller (1995) have all shown how important nutrient cycling has been and still is in tropical agriculture. As already noted, crop residues contain about half the nutrients exported from the soil through crop production. Returning them to the soil in systems where zero or few inputs are used is essential in slowing down nutrient losses. However, crop residues by themselves are not enough to offset what has been termed "nutrient mining". New nutrients need to be added if the system is to be sustained (Woomer and Swift, 1994).

The status of poor tropical soils, with their very low nutrient reserves, is well documented. It contrasts with that of European or North American soils, where the overuse of fertilizers in the recent past has built up large surpluses of nutrients. The return of crop residues alone, without the importation of nutrients from outside the field, is not a feasible option for tropical soils. This is especially the case in Africa,

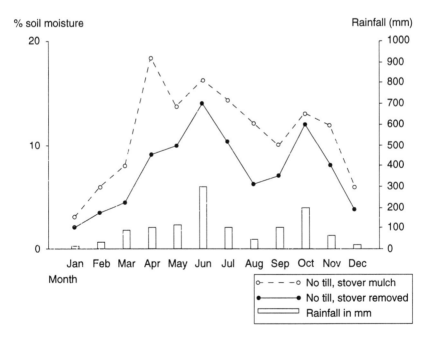

Fig. 9.2. Effect of organic residue mulch on soil moisture storage (Juo, 1990).

where soils have been extensively mined by resource-poor farmers who have no other way of providing for themselves and their families.

Crop residues and organic matter should not be seen solely as direct sources of nutrients. Feller (1995) has shown that organic matter can increase the cation exchange capacity of non-acid tropical soils in West Africa. Similar results were obtained on acid sandy soils in northeast Thailand (Willett, 1995). The results here may be linked to the effect of the organic matter in building exchangeable aluminium complexes (Hargrove and Thomas, 1981), which reduces aluminium toxicity (Bell and Edwards, 1987). The experiments of the International Board for Soil Research and Management (IBSRAM) in Southeast Asia have confirmed these findings (Nguyen tu Siem *et al.*, 1994).

Interactive Effects of Organic Matter

It is difficult to separate the various effects of returning crop residues to the soil. A good soil tilth combined with the proper provision of water and nutrients promotes good yields. If only one of these components is missing, the whole system may be disrupted. For this reason it is better, at least at first, to try to improve the whole system than to focus too narrowly on individual components. Component research should be undertaken only when it is certain that the component in question constitutes a critical constraint.

As already pointed out, returning crop residues restores only a proportion of the nutrients exported through crop production from what are, in many cases, already severely impoverished soils. Obviously, crop residues cannot replace rainfall either. But they can help to enhance these two production factors and may serve as a buffer during periods of temporary shortage. In contrast, crop residues are the only components which can protect the soil structure and reduce soil erosion.

The question here is how much or how little organic matter is really necessary to protect the physical environment. The levels given by Pieri (1989) for West Africa can be assumed to be the minimum ones below which soil structural conditions are seriously affected. The data obtained by Albergel and Valentin (1990) suggest that these thresholds have already been crossed in many places in semiarid Africa. Replenishing the soil's organic matter is thus crucial for the future of agriculture in this region.

The Socioeconomic Context

Returning crop residues to the soil is not common practice for most farmers, who prefer to use them for other purposes—livestock feed, fuel and building materials—or to burn them or remove them from their fields. The rationale for this behaviour lies in the fact that residues are an obstacle to tillage operations and that, in certain cases, they can sustain pests, whose populations build up to levels that threaten the next season's harvest. The burning of crop residues is not peculiar to the tropics, but became common in Europe and North America when tractors replaced animal draught power.

Renewed Interest in Crop Residue Management

Farmers in temperate zones began changing their farming practices when they perceived that depleting the soil's organic matter led to soil degradation. Indeed, the change from animal draught power to mechanized agriculture in the developed world and the accompanying intensification of cropping patterns have caused serious degradation problems. The famous Oklahoma Dust Bowl of the 1930s led not only to the creation of the US Soil Conservation Service but also to the realization on the part of farmers themselves that they must take action. Whereas burning was widespread in the 1960s, the use of mulch, conservation tillage and other conservation practices is now quite common in agriculture in developed countries. Yet land degradation is still an issue in Europe and North America (NRC, 1989). Conservation practices are more complex to implement than conventional ones. If they are to be implemented in the context of modern, intensive farming, they also require new equipment and technologies to reduce their labour-intensiveness.

In the tropical world we are still a long way from such changes. Nevertheless, researchers have demonstrated the advantages of making better use of crop residues (Pushparajah and Bachick, 1987; Bationo *et al.,* 1989; Lal, 1995) (Table 9.2), although the technology has yet to reach the farming community. It might be thought that the problem is less urgent in mixed farming systems due to the return of farmyard manure to the soil. But in many rural areas dung is increasingly being used for fuel as the supply of firewood runs out. It may also be used as a building material. In addition, an unknown but possibly quite high proportion of dung is lost during the

Table 9.2. Effect of pearl millet residues on yield of pearl millet and organic matter content of the soil.

Treatment	Grain yield (kg ha^{-1})	Total soil organic matter (%)
Control	56	0.24
Crop residue mulch	743	0.29
Fertilizer	816	0.25
Crop residue mulch and fertilizer	1532	0.33

Source: Bationo *et al.* (1989).

recycling process, due to failure to collect it. And the dung that is returned may have lost valuable nutrients to the air before it is incorporated.

Soil Resilience

Soil degradation occurs slowly, due to the resilience of the soil (Greenland and Szabolcs, 1992). Consequently, the relationship between soil stress and productivity is far from immediately obvious. This is both reassuring and worrying. It is reassuring because a well-managed soil may not suffer too greatly from short-term stresses such as flooding or severe drought. It is worrying because prolonged stress takes time to show its effects, such that crop yields may remain relatively high as stress levels build. As a result, farmers may realize too late the problems they are facing.

Once a soil has become degraded, rehabilitating it is a long and costly process. Experiments by IBSRAM on a degraded soil in northern Thailand have shown that soil conservation practices still had no effect on crop yields after 5 years of use. There was a high incidence of erosion in farmers' practice plots, with more than 100 t ha^{-1} year^{-1} lost in this way, compared to around 10 t ha^{-1} year^{-1} for the other treatments (Anecksamphant *et al.*, 1995) (Table 9.3). Farmers seemed more concerned with their immediate problems, such as pests and diseases or nutrient needs, than with longer-term issues such as preventing soil degradation.

Inappropriate Innovations

Although it makes sense agronomically to return crop residues to the soil, farmers' socioeconomic circumstances often prevent them from accepting this practice. In many cases farmers do not even realize that they have soil degradation problems. This much is obvious from interviews conducted by IBSRAM scientists with farmers

Table 9.3. Effect of soil conservation treatment on rice yield (t grain ha^{-1}) in Chiang Rai, Thailand.

Treatment	Fourth year (1992)	Fifth year (1993)
Farmers' practice	0.87	0.98
Alley cropping	0.90	0.93
Bahia grass strips	0.76	0.81
Hillside ditches	0.77	0.87
Agroforestry	0.92	0.93

Source: Anecksamphant *et al.* (1995).

in northern Thailand. Yet the solutions researchers propose to farmers are mostly partial and are far from attractive to them economically or socially. Fujisaka (1992) lists 13 reasons why farmers do not adopt innovations oriented towards sustainable soil management. Mulching, alley cropping and the return of crop residues emerge as too demanding of labour and not economically viable for resource-poor farmers, who do not receive a high enough price for their products. The real issue here is poverty and how to break free of it.

Other Uses

Poverty is the reason why crop residues are used as feed, fuel and construction materials rather than being returned to the soil. Crop residues usually make poor substitutes for the other materials that should be used for these purposes (Unger, 1990), but they are the only option which farmers and villagers can afford in their present economic situation. Sandford (1989) states that, in semiarid West Africa, crop residues can represent up to 80% of livestock feed during the peak months of the dry season and up to 40% of total feed for the year. Under these circumstances returning crop residues to the soil is well nigh impossible, since it would imply preventing animals from grazing.

Is it futile to seek to manage crop residues to enhance agricultural productivity and sustainability? I hope not. But to change the situation, farmers must experience agriculture more as an economic activity than as a means of survival. In particular, livestock must become a profitable enterprise based on the production of forage, rather than being viewed as an insurance policy against lean years.

The Principles of a New Agenda

The basic principles that should apply to the management of crop residues are simple: to aim for maximum efficiency in their use in both the short and long term, and to preserve the vital interests of the different stakeholders. However, these principles are mutually contradictory, since the sharing of resources in resource-poor farming areas rarely leads to their efficient use. Examples of successful complementarity in crop/livestock mixed farming systems are few and far between and appear to be linked mainly to relatively wealthy farmers (Sandford, 1989). The challenge for research and development is, therefore, first to understand the local situation before seeking to intervene, then to use a participatory approach to planning and implementing local projects.

The following principles should guide discussions with farmers:

1. Developing the production capacity of mixed farming systems depends critically on returning crop residues to the soil, either directly or indirectly (via livestock), so as to contribute to both productivity and sustainability.

2. Returning crop residues alone is not viable; there is a need to add other nutrients through inputs such as inorganic fertilizers, lime, rock phosphate and so on.
3. Soil biological activity should be enhanced without provoking excessive oxidization of the soil organic matter through aggressive tillage practices. Conservation tillage, in particular, should be promoted.
4. The production of forage and wood should be promoted at village level so as to satisfy local demand for these products.

These principles will obviously be easier to follow if the economic circumstances are conducive. Some interventions can be effected with relatively small initial investments. All, however, depend on the full participation of the village community, especially those individuals representing conflicting interest groups. A strong will to overcome obstacles and negotiate solutions will be needed.

Research and Extension

Support from national research and extension groups is essential if successful projects to improve the management of crop residues are to be formulated and implemented, based on the principles outlined above. Rather than conducting more biophysical studies on specific aspects of the processes involved, researchers should focus on the social and economic determinants of successful crop residue management. Such an approach would be in line with the conclusions of Greenland *et al.* (1994), mentioned above.

Participatory Research

Clearly, research on the socioeconomic aspects of crop residue use needs strengthening. If farmers apply so few of the wealth of biophysical research results that are now "on the shelf", it is important to understand why and to see what can be done about it.

The key to obtaining improved understanding and achieving higher success rates in the future is participatory on-farm research. Such research needs to focus first on the conditions determining the use of crop residues. What will make farmers more inclined to use them more efficiently? How can their profits from crop and livestock production be increased so that these activities are practised on a commercial basis rather than primarily for subsistence? What kind of investment and support are required to make this occur?

Following the diagnostic phase of the on-farm research programme, or perhaps at the same time as it, a participatory land use planning exercise should be conducted at village level (Latham, 1993). Farmers, livestock owners (if they are different people) and other villagers should all be included, with researchers on hand to provide support where needed. The planning exercise needs to be firmly geared to what is feasible, avoiding over-ambitious plans that will never work. Feasibility can often be gauged from the success (or otherwise) of previous on-farm research programmes. The implementation of the plan should be monitored and the plan adapted where necessary in the light of experience, again using a participatory approach. The successful design and implementation of innovations in the use of crop residues, as in the management of all natural resources, depends on adopting a bottom-up approach involving a real dialogue among farmers, researchers and other stakeholders.

Strategic Research

If a participatory bottom-up approach is used to define the agenda for socioeconomic research, it is even more important that the same approach be used to define the biophysical research agenda—if biophysical research is needed at all. Again, the needs should be identified through dialogue between researchers, farmers' groups and other interested parties. This dialogue should essentially be an exercise in problem-solving.

Initially, three broad areas of strategic research come to mind:

1. We need to improve our understanding of the relationships between soil organic matter, soil structure and erosion processes, with a view to developing new management practices.

2. Ways must be sought of reducing the loss of nutrients, especially nitrogen. This can be done by devising new cropping systems and crop rotations that are oriented towards the conservation of natural resources. It is important to develop technologies that provide farmers with visible returns in the short term and that are not too demanding in terms of their labour requirements.

3. New crop varieties are required that respond both to the need to produce sufficient human food and to the need for sufficient crop residues to sustain the soil.

These research suggestions are just that, of course. They should not be interpreted as my attempt to pre-empt the research agenda, which will be set by farmers and researchers together.

Extension

Extension has long been left to government officers who are ill-prepared to meet the challenges it presents (Cerna *et al.*, 1984). It is time to view extension differently, especially when it comes to resource management issues. Extensionists should be fully trained in rapid rural appraisal (RRA), in a farming systems approach and in participatory on-farm research, so that they can collaborate on an equal footing with researchers and farmers. A training programme of this kind has been provided over the past 2 years, in a consortium between IBSRAM, the Royal Thai Government and Thai universities (IBSRAM, 1995). Similar programmes need to be developed in other areas of the tropics, particularly in semiarid sub-Saharan Africa.

At the same time as reinvigorating national extension systems, we need to encourage the involvement of new participants, such as farmers' self-help groups and other non-government organizations (NGOs). Often under dedicated leadership, these organizations can usefully complement (and galvanize) government services, which have not always had a dynamic, participatory approach. Interesting experiences with such groups have recently been reported in The Philippines (Cerna *et al.,* 1995). Amidst all the enthusiasm for working with NGOs, we should not forget that these organizations should be firmly rooted in, or at least highly familiar with, the farming community if they are to fulfil their role successfully. In addition, they may need training in the diagnosis of constraints and the application of new technology.

Conclusions

The management of crop residues is arguably the single most important sustainability issue facing the mixed farming systems of the semiarid tropics. Yet it is not primarily a technical issue. Research has demonstrated the advantages of returning crop residues to the soil and of using other materials to meet other needs. To manage crop residues in this way, crop and livestock producers must come to view their activities as commercial rather than subsistence-based. If this is to happen, agriculture will need stronger support from governments and regional or international agencies such as development banks. However, the primary responsibility for this change rests with land users themselves.

To implement a research and development programme aimed at broad-based poverty alleviation and protection of the environement, a good understanding is needed both of farmers' circumstances and of the potential of research to improve those circumstances. Research should be interdisciplinary, with a strong social science component, and it should use a bottom-up approach. All stakeholders must participate in defining the research agenda and in monitoring its outcome. Research should also work hand in hand with extension services, with a view to involving them not just in technology dissemination but in the process of developing and testing technology. NGOs should also be closely involved in the research and development process, particularly when they speak for the interests of resource-poor farmers. Both extension and NGO staff can usefully serve as facilitators in the participatory approach that must underpin all interventions. To play this role, they need to be trained in both biophysical and socioeconomic aspects, so that they can collaborate as equal partners with researchers, policy makers and land users.

Crop residues are primarily soil regenerators, but too often they are either disregarded or misapplied. In fact they could be the catalyst of a new agricultural revolution based on the principles of sustainability and participation. As such our image of crop residues as mere leftovers should be replaced by the more apt notion of a strategic resource that needs to be used with the greatest care.

References

Albergel, J. and Valentin, C. (1990) "Sahélisation" d'un petit bassin versant soudanien, Kongnere-Boulsa, au Burkina Faso. In: Richard, J.F. (ed.), *La Dégradation des Paysages en Afrique de l'Ouest.* Ministère de la Coopération et du Développement, Paris, France, pp. 11-133.

Albrecht, A., Rangon, L., Barret, P. (1992) Effet de la matière organique sur la stabilité structurale et la détachabilité d'un vertisol et d'un ferrisol (Martinique). *Cahiers de l' ORSTOM, Série Pédologique* 25 (1): 121-133.

Anecksamphant, C., Boonchee, S., Inthapan, S., Taejajai, U. and Sajjapongse, A. (1995) The management of sloping lands for sustainable agriculture in northern Thailand. In: *The Management of Sloping Lands for Sustainable Agriculture in Asia.* ASIALAND Network Document No. 12. International Board for Soil Research and Management (IBSRAM), Bangkok, Thailand, pp. 165-204.

Bationo, A., Christianson, C.B. and Mokwunye, U. (1989) Soil fertility management of the pearl millet-producing sandy soils of Sahelian West Africa: The Niger experience. In: Renard, C., van den Beldt, J.F. and Parr, J.F. (eds), *Soil, Crop, and Water Management Systems for Rainfed Agriculture in the Sudano-Sahelian Zone.* International Crops Research Institute for the Semi-Arid Tropics (ICRISAT), Patancheru, India, pp. 159-168.

Bell, L.C. and Edwards, D.G. (1987) The role of aluminium in acid soil infertility. In: *Soil Management under Humid Conditions in Asia.* IBSRAM Proceedings No. 5, International Board for Soil Research and Management (IBSRAM), Bangkok, Thailand, pp. 201-223.

Bonfils, M. (1987) *Halte à la Désertification au Sahel.* Karthala, Paris, France, and Technical Centre for Agricultural and Rural Cooperation (CTA), Wageningen, The Netherlands.

Casenave, A. and Valentin, C. (1989) *Les Etats de Surface de la Zone Sahélienne: Influence sur l'Infiltration.* Office de la Recherche Scientifique et Technique d'Outre-Mer (ORSTOM), Paris, France.

Cerna, L., Moneva, L.A. and Geradino, E.C. (1995) Mag-uumag research and extension for soil, water and nutrient management. In: *The Zschortau Plan for Implementation of Soil, Water and Nutrient Management Research.* Deutsche Stiftung für Internationale Entwicklung (DSE), Zschortau, Germany, and International Board for Soil Research and Management (IBSRAM), Bangkok, Thailand, pp. 191-198.

Cerna, M.M., Coulter, J.K. and Russel, J.F.A. (1984) Building the research-extension-farmer continuum: Some current issues. In: *Research-Extension-Farmers: A Two-way Continuum for Agricultural Development.* World Bank,

Washington DC, USA, pp. 3-10.

Charreau, C. and Nicou, R. (1971) L'amélioration du profil cultural dans les sols sableux et sabloargileux de la zone tropicale sèche Ouest Africaine et ses incidences agronomiques. *Agronomie Tropicale* 26: 209-255, 565-631, 903-978, 1183-1237.

Emerson, W.W. (1967) A classification of soil aggregates based on their coherence in water. *Australian Journal of Soil Research* 5: 47-57.

Feller, C. (1995) La matière organique dans les sols tropicaux à argile: Recherche de compartiments organiques fonctionnels. PhD thesis, Office de la Recherche Scientifique et Technique d'Outre-Mer (ORSTOM), Paris, France.

Fernández, E.C.M. and Sanchez, P.A. (1990) The role of organic inputs and soil organic matter for nutrient cycling in tropical soils. In: Pushparajah, E. and Latham, M. (eds), *Organic Matter Management and Tillage in Humid and Subhumid Africa.* IBSRAM Proceedings No. 10, International Board for Soil Research and Management, Bangkok, Thailand, pp. 169-188.

Fujisaka, S. (1992) Thirteen reasons why farmers do not adopt innovations intended to improve the sustainability of upland agriculture. In: Dumanski, J., Pushparajah, E., Latham, M. and Myers, R. (eds), *Evaluation for Sustainable Land Management in the Developing World.* IBSRAM Proceedings No. 12, International Board for Soil Research and Management, Bangkok, Thailand, pp. 509-522.

Greb, B.W. (1983) Water conservation in the central Great Plains. In: *Dryland Agriculture Agronomy No. 23.* Agronomy Society of America, Madison, Wisconsin, USA, pp. 57-72.

Greenland, D.J. and Szabolcs, I. (eds) (1992) *Soil Resilience and Sustainable Land Use.* CAB International, Wallingford, UK.

Greenland, D.J., Bowen, G., Eswaran, H., Rhoades, R. and Valentin, C. (1994) *Soil, Water, and Nutrient Management Research: A New Agenda.* IBSRAM Position Paper. International Board for Soil Research and Management, Bangkok, Thailand.

Gregory, P.J. (1989) Water use efficiency of crops in the semiarid tropics. In: Renard, C., van den Beldt, R.J. and Parr, J.F. (eds), *Soil, Crop and Water Management in the Sudano-Sahelian Zone.* International Crops Research Institute for the Semi-Arid Tropics (ICRISAT), Patancheru, India, pp. 85-98.

Hargrove, W.L. and Thomas, G.W. (1981) Effect of organic matter on exchangeable aluminum and plant growth in acid soils: In: Stelly, M. (ed.), *Chemistry in the Soil Environment.* Agronomy Society of America, Madison, Wisconsin, USA, pp. 211-227.

Henin, S., Monnier, G. and Combeau, A. (1958) Méthode pour l'étude de la stabilité structurale des sols. *Annales Agronomiques* 9: 73-92.

IBSRAM (1995) *Highlights.* International Board for Soil Research and Management, Bangkok, Thailand.

Juo, A.S.R. (1990) Maintenance and management of organic matter in tropical soils. In: Pushparajah, E. and Latham, M. (eds), *Organic Matter Management and Tillage in Humid and Subhumid Africa.* IBSRAM Proceedings No. 10,

International Board for Soil Research and Management, Bangkok, Thailand, pp. 199-212.

Lal, R. (1990) *Soil Erosion in the Tropics: Principles and Management.* McGraw Hill, New York, USA.

Lal, R. (1995) *Sustainable Management of Soil Resources in the Humid Tropics.* United Nations University Press, Tokyo, Japan.

Latham, M. (1993) Rural development planning for sustainable land management. In: Wood, C. and Dumanski, J. (eds), *Proceedings of the International Workshop on Sustainable Land Management for the Twenty-first Century*, 20-26 June 1993, University of Lethbridge, Alberta, Canada, pp. 261-266.

Lavelle, P. (1994) Faunal activities and soil processes: Adaptive strategies that determine how ecosystems function. In: *Fifteenth World Congress of Soil Science* 50: 189-222.

Le Bissonnais, Y. and Le Souder, C. (1995) Mesurer la stabilité structurale des sols pour évaluer leur sensibilité à la battance et à l'érosion. *Etude et Gestion des Sols* 201: 43-56.

Lericollais, A. (1990) La gestion du paysage: Sahélisation et surexploitation des terroirs Serer au Sénégal. In: Richard, J.F. (ed.), *La Dégradation des Paysages en Afrique de l'Ouest.* Université Cheikh Anta Diop, Dakar, Senegal, pp. 151-169.

Nguyen tu Siem, Sutanto, R., Noor, M.Y.M., Sharifudin, A.H., Duque, C.M., Suthipradit, S., Myers, R.J.K. and Edwards, D.G. (1994) Advances in managing acid upland soils in Southeast Asia. In: *Fifteenth World Soil Congress* 50: 538-550.

NRC (1989) *Alternative Agriculture.* National Research Council and National Academy Press, Washington DC, USA.

Parr, J.I., Papendrick, R.J., Hornick, S.B. and Meyer, R.E. (1990) The use of cover crops, mulches and tillage for soil water conservation and weed control. In: Pushparajah, E. and Latham, M. (eds), *Organic Matter Management and Tillage in Humid and Subhumid Africa.* IBSRAM Proceedings No. 10, International Board for Soil Research and Management, Bangkok, Thailand, pp. 245-262.

Pieri, C. (1989) *Fertilité des Terres de Savanes.* Centre de Coopération Internationale en Recherche Agronomique pour le Développement (CIRAD) and Ministère de la Coopération, Paris, France.

Pushparajah, E. and Bachick, A.T. (1987) Management of acid tropical soils in Southeast Asia. In: Sanchez, P.A., Stoner, E.R. and Pushparajah, E. (eds), *Management of Acid Tropical Soils for Sustainable Agriculture.* IBSRAM Proceedings No. 2, International Board for Soil Research and Management, Bangkok, Thailand, pp. 13-40.

Roose, E. (1989) Gestion conservatoire des eaux et de la fertilité des sols dans les paysages Soudano-Sahéliens de l'Afrique Occidentale. In: Renard, C., van den Beldt, R.J. and Parr, J.F. (eds), *Soil, Crop and Water Management Systems for Rainfed Agriculture in the Sudano-Sahelian Zone.* International Crops Research Institute for the Semi-Arid Tropics (ICRISAT), Patancheru, India, pp. 55-72.

Rosaasen, K.A. and Lokken, J.S. (1993) Canadian agricultural policies and other initiatives and their impacts on prairie agriculture. In: Wood, C. and Dumanski, J. (eds), *Proceedings of the International Workshop on Sustainable Land Management for the Twenty-first Century,* 20-26 June 1993, University of Lethbridge, Alberta, Canada, pp. 343-368.

Sajjapongse, A. (1995) Management of sloping land for sustainable agriculture in Asia. In: Sajjapongse, A. (ed.), *The Management of Sloping Lands for Sustainable Agriculture in Asia.* ASIALAND Network Document No. 12. International Board for Soil Research and Management (IBSRAM), Bangkok, Thailand, pp. 1-35.

Sandford, S.G. (1989) Crop residue/livestock relationships. In: Renard, C., van den Beldt, R.J. and Parr, J.F. (eds), *Soil, Crop, and Water Management Systems for Rainfed Agriculture in the Sudano-Sahelian Zone.* International Crops Research Institute for the Semi-Arid Tropics (ICRISAT), Patancheru, India, pp. 16-182.

Swift, M.J., Kang, B.T., Mulongoy, K. and Woomer, P.W. (1992) Organic matter management for sustainable soil fertility in tropical cropping systems. In: Dumanski, J., Pushparajah, E., Latham, M. and Myers, R. (eds), *Evaluation for Sustainable Land Management in the Developing World.* IBSRAM Proceedings No. 12, International Board for Soil Research and Management, Bangkok, Thailand, pp. 307-326.

Tisdall, J.M. and Oades, J.M. (1982) Organic matter and water-stable aggregates in soils. *Journal of Soil Science* 33: 141-163.

Unger, P.W. (1990) Tillage and residue management in rainfed agriculture: Present and future trends. In: Pushparajah, E. and Latham, M. (eds), *Organic Matter Management and Tillage in Humid and Subhumid Africa.* IBSRAM Proceedings No. 10, International Board for Soil Research and Management, Bangkok, Thailand, pp. 307-340.

Valentin, C., Hoogmoed, W. and Andriesse, W. (1992) The maintenance and enhancement of low fertility soils. In: Dumanski, J., Pushparajah, E., Latham, M. and Myers, R. (eds), *Evaluation for Sustainable Land Management in the Developing World.* IBSRAM Proceedings No. 12, International Board for Soil Research and Management, Bangkok, Thailand, pp. 219-252.

Willett, J. (1995) Role of organic matter in controlling chemical properties and fertility of sandy soils used in lowland rice in northeast Thailand. In: Lefroy, R.B.D., Blair, G.J. and Craswell, E.T. (eds), *Soil Organic Matter for Sustainable Agriculture.* ACIAR Proceedings No. 56, Australian Centre for International Agricultural Research, Canberra, pp. 109-114.

Woomer, P.W. and Swift, M.J. (eds) (1994) *The Biological Management of Tropical Soil Fertility.* John Wiley & Sons, Chichester, UK.

10. Alternatives to Crop Residues as Feed Resources in Mixed Farming Systems

J. Steinbach

Department of Livestock Ecology, Tropical Sciences Centre, Justus-Liebig University, Ludwigstrasse 21, D-35390 Giessen, Germany

Abstract

If better use is to be made of crop residues, either as basal livestock diets or by returning them to the soil, alternative feed resources must be developed. This paper examines those alternatives, concentrating mainly on the role of cultivated forages in the subsistence-oriented mixed smallholder rainfed and irrigated farming systems of the tropics and subtropics. The potential for introducing forages on cropped land is assessed, covering crop rotations, fodder banks, cover crops in plantations and the use of food crops as forages. Uncropped land may also be used to produce forages. Considerable differences in the opportunities for introducing forages in different agroecosystems and regions are found, according to such factors as available arable land per capita, the number of crops that can be grown per year, market access, labour availability and farmers' perceptions of the risks and rewards of investing in their livestock enterprises. Nevertheless, the introduction of forage crops is a promising way of promoting sustainable agriculture in the low-input systems of resource-poor farmers. More research is needed to ensure successful technology development and transfer.

Introduction

Although fibrous crop residues are important feed resources in mixed farming systems in the tropics and subtropics, many of them are characterized by low metabolizable energy and digestible protein contents, which sometimes barely cover animal maintenance requirements. In addition, the supply is frequently seasonal. To satisfy the nutrient needs of ruminant and non-ruminant livestock for the efficient production of food, fibre and power, supplementation strategies are needed. Such strategies seek to utilize all possible feed resources on the mixed farm in an optimized system.

Ruminants are by far the most important group of farm animals in developing countries. The feed resources available to them originate from rangeland, arable areas, plantations and peripheral land (such as roadsides, terrace risers, field bunds and the banks of irrigation channels). Arable areas provide fallow grazing, weeds from crop fields, forage crops and concentrates, in addition to fibrous crop residues. In plantations, the herbaceous cover crops (grasses, legumes) can be used for feeding, while fodder grasses, legumes, shrubs and trees can be grown on peripheral land. Industrial products and the by-products of crop processing can supplement the above resources. In future, the importance of natural rangelands is likely to decrease, due to reduction in area (cropland encroachment, desertification) and reduced productivity (overgrazing). The additional nutrients needed for livestock feeding to meet the expected increase in the demand for animal products will have to come from arable land. However, it is important to realize that for most small-scale crop farmers animal production is of lower priority than crop production. This means that fodder production and feeding strategies must not reduce crop yields or make crop production more risky.

The task of this paper is to review the alternatives to fibrous crop residues as feed resources. The topics that could be discussed under this heading are legion, so I will have to limit them. Firstly, I will focus mainly on subsistence-oriented mixed smallholder rainfed and irrigated farming systems in tropical and subtropical highland, humid and subhumid, arid and semiarid ecosystems, which are deficient in production inputs due to problems of liquidity and availability (resource poverty). Secondly, I will stress cultivated fodders and place less emphasis on concentrates and industrial products, since these tend to be less accessible to smallholders. I will also use large ruminants (cattle and buffalo) as examples wherever possible, since these multipurpose (food, draught, fuel, soil fertility) species contribute more than others to the efficiency of smallholder mixed farming systems (Seré, 1994). This approach is in line with the orientation and agenda of the World Food Summit 1996, which emphasized intensified use of existing cropland. It also offers a means of meeting the considerable deficit of milk, beef and animal power in the tropics recently projected for the year 2020 by the Food and Agriculture Organization of the United Nations (FAO, 1996).

Nutrient supply to livestock is one aspect of mixed farming, crop productivity another. Given rapid population growth and deteriorating soil resources, there is a general trend towards continuous cropping and reduced fallow periods in smallholder production systems, although great regional differences exist. Due to inefficient management and the use of few, if any, inputs, soils are increasingly characterized by inadequate nutrient and organic matter contents, unsatisfactory physical structure, chemical composition and biotic status, unfavourable climates (moisture and temperature) and moderate to severe problems of erosion, leaching, salinity, alkalinity or acidity. Furthermore, many cropping systems suffer from inadequate labour capacity. Using draught animals for cultivation, weeding, harvesting and transport can improve soils and crop yields as well as expand the areas cropped per farming family. I will argue that the efficient integration of crop and livestock production can solve the problems of nutrient supply, declining soil fertility and lack of animal power in mixed farming systems. The means to this end are forage production in crop rotations and as intercrops with permanent crops, and the supplementary feeding of animals with concentrates. Research on the strategic utilization of the feeds available on mixed farms will doubtless be required. The aim will be to develop new and sustainable forage systems which provide feed and maintain soil fertility.

The Forage Potential of Tropical Crop Production Systems

Forages for feeding livestock can be produced on cropped land as well as on uncropped farm or non-farm areas. Herbaceous plants (grasses and legumes) can be grown in crop rotations, on wasteland, as pioneer crops on fallow land, along roadsides, river and canal banks, on terrace risers and field bunds, and as cover crops under tree plantations. Shrubs can be planted on contour lines, along roadsides and on canal banks, or in pastures and cropping fields, while fodder trees can be used almost anywhere on the farm where excessive shading or extensive root systems will not be a problem. The primary purpose of all these forage crops is to close the nutrient gaps specific to the farming system. Broadly speaking, these are:

1. In permanent cropping systems: energy and protein deficits during the cropping season, protein deficit during the dry season.
2. In crop/fallow systems: dry-season protein deficit (fallow grazing during the rainy season).
3. In plantation systems: energy and protein deficits during the dry season (cover crop grazing during the rainy season).
4. In agropastoral systems: dry-season protein deficit (rangeland grazing during the rainy season).

The only types of system in which few deficits may occur are agroforestry systems, in which the presence of trees and shrubs integrated with the production of crops can ensure a year-round supply of forage, and irrigated systems, which have more reliable water supplies for longer periods of the year. Even in these systems, however, sporadic deficits occur.

Thus, in most mixed farming systems, a deficit in digestible crude protein is the most limiting factor affecting the quality and quantity of feed available to domestic livestock, and any forage crop development effort must address this problem. The relative share in diets of herbaceous and lignified fodder plants depends on such factors as the ecoregion, the farming system, the season, and the demand for basic and supplementary feeds. However, in most cases grasses and leguminous forbs will be more productive, while shrubs and trees will have additional ecological and economic advantages.

The ecological and economic advantages of forage production in tropical systems are several. Ecologically, forage crops may increase the primary and secondary productivity of the agroecosystem. They improve soil fertility and water harvesting (infiltration rates, water-holding capacity), suppress weeds, crop diseases and insect pests, control soil erosion and nutrient leaching by wind and water, and help balance animal diets. Economically, forage crops can raise incomes, spread risk through the increased diversity of production, improve the efficiency of crop production (particularly at low to medium input levels), and distribute the demand for labour more evenly throughout the year (draught animal use at times of peak demand, livestock management during the slack dry season).

Forage production adds another dimension to mixed farming systems. It does, of course, require additional investments in terms of land, labour and capital. The competition with food and cash crops for land and labour, the ease with which fodder crops can be integrated into the existing farming system, and the efficiency of production (ratio of inputs to outputs) will determine the adoption rates of any new technology.

Forage Production on Cropped Land

Feed resources from cropped land include forage crops grown in crop rotations, fodder banks to cover specific protein and energy requirements, and cover crops in plantations. In addition, food crops may sometimes be used as feed for livestock.

Crop Rotations

Crop rotations represent a vast underexploited potential for forage production in tropical developing countries. Generally, experience with their introduction through research and development projects is still limited—although more is known now than was the case 20 years ago. Since fodder crops must not reduce food or cash crop yields, land availability is a critical issue. The most obvious arable land resource is fallow land, which varies widely in extent among regions. Surprisingly, very few data exist on fallow periods and extent on a global scale (FAO, 1995). Studies conducted at local level indicate that, despite the continuing reduction in fallow periods, unmanaged fallow land may still amount to 60% or more of total arable land, even in such densely populated areas as the Malawi Lake region (Rischkowsky, 1996).

It is not possible here to discuss individually all the forage crops suitable for managed fallow feed production. A fairly comprehensive list can be found on the Internet (FAO, 1994). Let us turn instead to the criteria used for selecting an appropriate crop for a given location. These criteria include the crop's ameliorative properties, nutrient production, environmental adaptation and agronomic and management requirements.

The ameliorative properties of a forage should improve upon the soil fertility restoring characteristics of the existing fallow. They relate to root depth and distribution, organic matter production, nitrogen fixation (by legumes and grasses), nutrient pumping ability (trees and shrubs), soil cover (shading, temperature reduction, erosion control) and the control of weeds, insect pests and diseases. As regards nutrient production, the important characteristics are above-ground biomass, nutrient contents (metabolizable energy, digestible crude protein, minerals and vitamins, antinutritional factors), phenological parameters (duration and timing of vegetative and generative phases), leaf:stem ratios, the efficiency of soil nutrient and water utilization, and the regeneration potential from the soil seedbank. Depending on location, tolerance to drought, frost and shade, to acid, alkaline, saline or sandy soils, to diseases and pests and to competition from weeds may be important factors affecting the crop's environmental adaptation. Finally, agronomic and management considerations include ease of cultivation, the possibility of undersowing into the preceeding food or cash crop, the fertilizer requirement, harvesting methods (grazing or cut-and-carry) and their frequency, ease of eradication or ploughing for the subsequent food crop, and capacity for regeneration.

It is important to distinguish between annual and perennial forage crops, which differ in their ameliorative and productive efficiencies as well as in their management requirements. Both temporal and spatial intercropping is possible. Mixed cropping systems are easy to establish, but have distinct disadvantages due to shading and the difficulties of weed control, mechanization and harvesting. Relay cropping protects the soil and saves time, while sequential cropping allows mechanized cultivation and avoids competition from intercrops but requires more time and additional cultivation and leaves the soil unprotected for at least part of the year. Strip cropping combats erosion on slopes, but may impede mechanization. An example of the production of short-term intercrop forages is the rice farming system described by Sangakkara (1989), in which *Crotalaria juncea* is grown between two rice crops for 50-60 days each in February-April and August-October, yielding 3.5 and 2.9 t DM ha^{-1} of forage respectively. The wheat-maize system described by Badve (1991), in which sorghum or millet is combined with cowpea during the interseason between April and June, giving a combined yield of 35-40 t ha^{-1} of fresh matter (FM), is another example. Medium-term intercrops usually require 4 to 8 months. The best documented example is the cultivation of berseem (*Trifolium alexandrinum*) between October and February (23 t FM ha^{-1} from 1.5 cuts) or between October and April (62 t FM ha^{-1} from 4 cuts) in the Nile Valley, where it covers 24% of the irrigated area. Berseem can also be grown as a relay crop. Nazir *et al.* (1988) reported that wheat and *Trifolium alexandrinum* grown in multi-row strips yielded 4 t ha^{-1} of wheat grain and 40 t FM ha^{-1} of berseem in four cuts, a combination which had the highest benefit/cost ratio and yielded the greatest net income.

In spite of their undisputed ecological advantages, leys have not attracted the attention they deserve. Leys can be defined as managed fallows, and as such they fit particularly well into cropping systems in which fallow land is still available, as is the case in large parts of Africa and Latin America. Leys have a rejuvenating effect on soil structure and fertility, and are also effective in controlling crop pests. In semiarid regions, however, they may affect soil moisture negatively. The duration of leys required for adequate restoration of soil fertility has been given as 2-3 years (Powell and Mohamed-Saleem, 1987). Leys may be established by sowing grasses and/or legumes into the last food crop of the rotation, such as maize. Alternatively, they may be allowed to regenerate from soil seedbanks, as is practised with annual legumes in the Mediterranean Basin. In humid regions, planting grasses may be prohibitive because of the high labour requirement. Important criteria determining the choice of suitable species or species combinations are persistence, competitiveness and regenerative capacity (resistance to ploughing, etc). Some examples: one grass species often used is *Chloris gayana* (FAO, 1977); a legume recommended recently for northern Nigeria is *Aeschynomene histrix* (Peters, 1992); suitable grass-legume mixtures for the more humid tropics could be *Cynodon nlemfuensis + Centrosema pubescens, Chloris gayana + Stylosanthes* spp., *Panicum maximum + Desmodium intortum* or *Melinis minutiflora + Pueraria phaseoloides*. Ley farming is not necessarily a high-input technology, but suitable animal-drawn soil- and crop-adapted cultivation implements need to be developed if this forage production system is to be widely adopted by resource-poor farmers.

Fodder Banks

Although crop rotations have the greatest potential for increasing forage production in smallholder settings in both the tropics and subtropics, other options do exist. On cultivated land, these options consist of fodder banks and, in tree and shrub plantations, the use of cover crops.

The purpose of fodder banks is to supplement livestock fed from natural pasture or with crop residues with additional protein and/or energy. On sloping land, fodder banks may also serve to control erosion. Banks intended to supply protein can be produced using herbaceous legumes or shrubs and trees. Besides protein yield, selection criteria include phenological characteristics, competitiveness with weeds, regeneration potential and the maintenance of an adequate soil seedbank. Many suitable species have been identified by institutions such as the International Livestock Centre for Africa (ILCA), the International Center for Agricultural Research in the Dry Areas (ICARDA) and the Centro Internacional de Agricultura Tropical (CIAT):*Centrosema brasilianum* stays green in the dry season, *Chamaecrista rotundifolia* regenerates quickly after the first rains, while *Stylosanthes guianensis* is a useful dry-season forage. However, many of the species tested exhibit poor persistence if not weeded regularly, or have low productivity (Peters, 1992). Leguminous forage shrubs and trees occur in great diversity (Gutteridge and Shelton, 1994), and are used extensively as multipurpose species providing the farmer with forage, fuel and more fertile soils. Desirable characteristics are high leaf production and good nutrient content, resistance to ratoon cropping, coppicing ability and fuelwood yield. Typical species used are *Leucaena leucocephala, Gliricidia sepium, Sesbania grandiflora, Calliandra calothyrsus* and *Acacia albida.* Banks designed to increase the supply of energy normally consist of highly productive grasses (e.g. *Pennisetum purpureum, Andropogon gayanus*) or energy crops such as sugarcane (*Saccharum officinarum*). The three-strata forage system developed in India combines the energy grass hybrid Napier with leguminous forage shrubs and trees, such as *L. leucocephala* (Gill and Tripathi, 1991; Patel *et al.*, 1992) or *Casuarina equisetifolia* (Mathew *et al.*, 1992), that have high dry-matter and crude protein yields. However, this system suffers from high labour requirements, while the shading effect can reduce ground cover and fodder productivity by 35-50% up to 10 m from the tree row (Puri *et al.*,1994).

Cover Crops in Plantations

Cover crops in plantations are also multipurpose. Besides producing fodder they assist in weed control, maintain or improve soil fertility and control erosion. Forage production is an efficient way of using cover crops in young and relatively open plantations, where the light is sufficient to satisfy the high light requirements for photosynthesis characteristic of most tropical forage crops. In a young rubber plantation in Indonesia, Siagian and Sumarmadji (1992) harvested around 25-48 t DM ha^{-1} year^{-1} from *Panicum maximum, Pennisetum purpureum* and *Paspalum dilatatum,* with no harmful effect on seedling growth. Jayasundara and Marasinghe (1989) described a model in which an old (40 years) coconut plantation (156 palm trees ha^{-1}) was undercropped with 2500 *Leucaena* shrubs and a ground cover of *Brachiaria miliiformis* and *Pueraria phaseoloides.* The total forage yield was 22.5 t DM ha^{-1} year^{-1}, which translated into marked gains in milk production and liveweight by grazing cows. Soil fertility improved and fertilizer costs were cut by 70%, while the yield of coconuts rose by 11%. Weeding requirements decreased. It is obvious from these examples that fodder production can be combined with plantation agriculture to the benefit of both the cash crop and the subsidiary livestock enterprise. However, inappropriate grazing management may damage the plantation crop, especially while young.

Forage Use of Food Crops

Finally, cultivated land produces food crops, which can be used for feed either in their entirety—if in excess of human requirements or if harvesting for food is considered not worthwhile because of inadequate yields—or in part (leaves, vines). Examples are cassava and banana leaves, groundnut and sweet potato vines, and pigeonpea forage, all of which have high levels of digestible protein (Dixon and

Egan, 1987). In Pakistan, surplus maize plants are harvested continuously during the growing season and fed to cattle. In Mauritius, whole sugarcane or sugarcane tops are used for ruminant feeding, while cassava tubers and leaves are valuable sources of energy and protein respectively for both non-ruminant and ruminant species. In Syria, drought-affected barley is leased out to the Bedouin for direct grazing by sheep. The potential for increasing forage supplies from such sources is considerable, having as yet been only partially realized. For instance, in Indonesia only 3% of cassava leaves and 7% of sugarcane tops are used for feeding livestock (Winugroho, 1996).

Forage Production on Uncropped Land

Depending on farm size, topography, hydrology and infrastructure, uncropped areas may occupy a considerable proportion of total farm land. These areas—roadsides, irrigation channel banks, contour bunds, terrace risers, shelter belts, live fences, ponds and waste lands—can be used for intensive feed production and erosion control. Suitable plants are grasses, such as Napier or sugarcane, for energy production, leguminous shrubs and trees as fodder banks for protein, or water hyacinths (*Eichhornia crassipes*) in ponds.

Utilization

The aim, in feeding livestock, is to provide the animals with a balanced diet throughout the year, with sufficient nutrients to perform as required in terms of milk production, liveweight gain, reproduction and draught work. Where fibrous crop residues provide the basic ration, they must be supplemented quantitatively and qualitatively throughout the different seasons of the year.

The efficiency with which forage resources are used depends on the type of stock as well as on feeding practices. The most efficient conversion of green forages is achieved by dairy cows. However, where cultivation work has top priority, the supplementation of draught oxen may, at least seasonally, be more viable economically. Maximum overall efficiency may be obtained by using multipurpose cows for milk, meat and draught production, but work performance and feed efficiency may be poorer in females than in males (Cole, 1996) and there may be cultural or religious barriers against using cows for draught.

The forages grown should be fed fresh to the animals, since conservation as hay or silage has given disappointing results in the tropics, particularly in smallholder situations. Whether in situ grazing or confinement feeding is practised depends on the crop production system and the importance attached to manure. Where forage resources are derived from intercropping, a cut-and-carry system with stall or kraal feeding is preferable, since it prevents damage to field or plantation crops. Such systems also facilitate the collection of manure required for fuel or fertilizer. However, they require more feed, as an estimated 15-35% of production is wasted (Boodoo, 1991) and a higher labour input in feeding. Leys, three-strata-forage systems and mature plantations can be grazed ad libitum or for a limited time daily, the animals being herded or tethered. Results from the literature on the relative advantages of continuous or rotational grazing are the subject of controversy. However, the cutting or grazing frequency should be sufficiently high to provide the animals with an adequate diet of digestible nutrients.

The establishment of forage calendars for particular cropping systems and environments is a useful way of ensuring that the feed demand of existing livestock populations can be met and of adjusting the nutrient balance at different seasons. Seasonal deficits in certain classes of stock can be covered by redirecting resources. For instance, supplementing grass-fed lambs and dairy cows with *Leucaena leucocephala* leaves improved their nitrogen retention and milk yield respectively (Mtenga and Shoo, 1990; Munga *et al.*, 1992).

In cut-and-carry systems, forage is often collected and fed by women and children. The effects on the workload of these groups should always be considered when seeking to optimize forage production and utilization. Since women are often entitled to the income from milk sales, they can benefit greatly from improvements in the feeding of dairy cattle.

Constraints and Research Needs

Despite the technical options just described, the forage production potential of smallholder mixed farms in the tropics remains largely untapped. For several reasons, the adoption rates of fodder production technologies are poor. This is not simply because extension services are under-resourced and ineffective, but also because of a wide range of real or imagined constraints including increased labour and capital requirements, land shortage (particularly in Asia), poor soil moisture, lack of a market for livestock products (especially milk) and the higher returns to food crops. Traditional rules governing access may also prevent adoption, since in many societies fodder for livestock is considered common property. Improvements may create a heightened potential for conflict, especially if previously communal resources are suddenly privatized. Sometimes the ameliorating properties of forage crops described above are not well understood and appreciated by resource-poor farmers.

There is, in addition, a more pervasive lack of knowledge on the land resources (fallow land, uncropped areas) available for forage production in different parts of the world, on the effects of forage crops and draught animal use on crop yields, and on the complementarity of available feed resources and the optimum structure of feeding calendars for specific situations. Too little is known about indigenous knowledge of local resources—their potential and how best to manage them. Research on sustainable fodder production should be intensified, with the aim of developing strategies to close the feed gap and improve soil fertility in smallholder farming systems. Priority topics for participatory on-farm or (where necessary) station-based research are: (i) low-input means of improving soil fertility and food/fodder crop productivity; (ii) the soil fertility effects of various forage systems (leys, intercropping); (iii) the collection and characterization of forage species; (iv) the evaluation of locally adapted indigenous species; (v) environmental impact assessment (EIA) and economic analysis of intensified forage production; and (vi) socioeconomic studies on indigenous knowledge and problem awareness, labour requirements, gender issues and technology adoption.

In addition, the strategies for technology transfer need overhauling. Participatory on-farm and/or farmer-managed research are promising vehicles for the transfer of improved and locally adapted technologies. They need to be more widely adopted at national level. For technology transfer to succeed, new practices and technologies need to be developed that are adapted to the location and the farmer's goals, manageable within his or her resources, ecologically beneficial, economically viable and socially acceptable.

Concentrates

Most of the feed resources commonly referred to as concentrates can be used equally as human food and animal feed. In a time of food shortages, concentrate feeding to livestock may be ethically questionable. For instance, in the Lake Region of Malawi the food grain supply lasts only for 7 to 9 months of the year, and a surplus for feeding to cattle does not exist (Rischkowsky, 1996). However, in specific situations there may be advantages in supplementing the basic roughage diet with limited amounts of concentrate feeds with a higher nutrient density. Seasonal and location-specific nutrient deficits (protein, energy, minerals) may occur in roughages and crop residues. Animals have only limited nutrient pools, and all classes of livestock need balanced diets with a certain minimum nutrient density for growth, lactation or work. Balancing the diet and increasing the nutrient density will, at least theoretically, increase the intake of fibrous crop residues, raise secondary productivity and improve feed conversion. As a consequence, pollution of air, soil and water resources per unit animal product will be reduced.

There is a wide range of energy- or protein-rich supplements that are useful for animal feeding. These are cereal and grain legumes, roots and tubers as feed grain substitutes, industrial crop by-products, such as milling residues (bran), molasses, brewery grains and oilseed cakes, slaughter and milk processing by-products, industrial supplements (minerals, vitamins, urea), compound feeds and food wastes. After satisfying human needs, non-ruminant production systems (poultry, pigs, fish) should probably take priority in the use of concentrates. Among ruminants, the use of concentrate feeds is most efficient for milk production by cattle, buffalo, goats and sheep. However, concentrates have also been used for smallholder cattle fattening in Malawi, for lamb fattening near Aleppo, Syria, and in draught animal feeding. Cole (1996) demonstrated that supplementing N'dama cattle fed on Guinea grass with 1 kg day⁻¹ of maize meal decreased heat stress by 20% and increased work output by 33%.

Since resource-poor farmers have only limited financial resources, their own farms must be the main source of any concentrated feeds they use. Commonly used by such farmers are cereals, grain legumes, brans, roots and tubers, sugarcane juice, palm oil, copra and household wastes. Market access to commercially available feeds is much more restricted for logistical as well as financial reasons.

Frequently, the theoretical benefits of concentrate feeding are not realized in practice. Economic viability is threatened by poor input:output ratios. More research is required in the areas of on-farm feeding trials, the physiological effects of supplementing fibrous forages, the identification of antinutritional factors, rumen microbes and the complementary effects of supplementation.

Industrially produced feeds such as urea blocks, mineral mixtures and compound feeds are rarely used by resource-poor farmers, who find them unaffordable and only periodically available.

Forage Production Systems

Seré (1994) identified 11 tropical livestock production systems based on grassland, mixed farming or landlessness. Only the mixed farming systems integrating crop and livestock production are relevant here. The six crop/livestock systems of this kind were classified according to three major climatic zones and to the main source of soil moisture—rainfall or irrigation (Table 10.1). A further subdivision of climatic zones into subhumid and semiarid (as distinct from merely humid and arid) and for tropical and subtropical systems might have been desirable.

I will now attempt to apply the principles presented in the first part of this paper to these six production systems, which will first be characterized in slightly greater detail.

Rainfed Systems

Tropical and Subtropical Mountain Areas

These areas are found in the Andean region, in eastern Africa and in the southern Himalayas. They have high population pressure and low land availability. Poor soils are a frequent problem, and growing seasons vary greatly with altitude, latitude and amount of rainfall. Farm sizes permit all types of forage production in sub-Saharan Africa and in Central and South America, but the limited arable land available in Asia (0.3 ha person⁻¹) does not permit ley farming. Poor infrastructure frequently hampers market access and the purchase of inputs. Multipurpose cattle production systems, mostly based on stall feeding, to satisfy the subsistence and

Table 10.1. The forage potential of tropical production systems[1].

| | Production system | | | | | |
| Forage source | Rainfed agriculture | | | Irrigated agriculture | | |
	Mountain	Humid	Arid	Mountain	Humid	Arid
Field crop rotations:						
Ley farming	+	+	+	-	-	-
Annual intercrops	+	+	+	?	-	-
Fodder banks:						
Protein	+	+	+	+	+	+
Energy	+	+	-	+	+	+
Plantations:						
Cover crops	+	+	-	-	-	+
Uncropped land:						
Herbaceous/tree forage	+	+	+	+	+	+
Supplements:						
Farm-grown	+	+	+	+	+	+
Purchased	?	?	?	?	?	?

[1] Potential depends greatly on population density, the availability of arable land, the degree of market integration and farmers' incomes.
Source: Seré (1994).

growing market demand for milk and beef and to provide manure and draught power for crop cultivation appear to be the best way forward.

Humid and Subhumid Tropics and Subtropics

Here, too, human population density tends to be fairly high, with available arable land ranging from 0.2 (Asia) to 1.0 (Latin America) ha person⁻¹. The opportunity to grow two crops a year reduces the pressure somewhat. Soil fertility is generally poor, requiring ameliorative treatment. Again, all the types of forage production described above are possible in sub-Saharan Africa and Latin America, but in Asia the shortage of cropland generally prohibits the use of ley systems and annual forage crops. Multipurpose cattle, buffalo and goat production systems are possible, in addition to the flourishing non-ruminant sector found in Asia and, increasingly, other regions too. Near large cities, specialized smallholder dairy production systems may be found (e.g. around Bombay).

Arid and Semiarid Tropics and Subtropics

These zones are found mainly in Asia, sub-Saharan Africa and West Asia-North Africa. The main climatic constraints are the lack of soil moisture and the short growing season. The situation is further aggravated by the shortage of cropland in many areas (e.g. in Kano State, Nigeria). However, natural grasslands, which are often closely associated with these cropping systems, meet a high proportion of the nutrient requirements of livestock during the short rainy season. Ley farming is possible in sub-Saharan Africa and in most of West Asia-North Africa. Fodder banks for protein supplementation can be established on uncropped land, and supplementation with concentrates is widespread. Multipurpose cattle, buffalo, camels and sheep/goat production systems prevail.

Irrigated Systems

Tropical and Subtropical Mountain Areas

These are typically found on the southern slopes of the Himalayas. They are characterized by extremely high population pressure. Terraced rice production dominates, with two crops possible annually. Due to deforestation these areas are increasingly threatened by devastating droughts and floods. Seasonal forage crops and forage production from uncropped land are promising options here. Stall-fed cattle or buffalo milk production systems combined with draught animal use offer an efficient path for integrating crop and animal production.

Humid and Subhumid Tropics and Subtropics

These are the rice production systems of Southeast Asia. With an average of only 0.16 ha person^{-1} of arable land, these regions are among the most densely populated in the world. However, two or three crops can be grown each year. With respect to forage production, the regions' plentiful supplies of crop residues need to be supplemented with green forage from seasonal intercrops and from fodder banks, and with foliage from leguminous shrubs and trees, grown on uncropped field bunds, roadsides and channel banks. Livestock production systems focus on dairy cattle, milch buffalo and pigs and poultry, as well as on aquaculture and silkworms. For the major domestic species, feeding has to follow a cut-and-carry or zero-grazing system so as to avoid damage to crops. Only in mature plantations is free or tethered grazing possible.

Arid and Semiarid Tropics and Subtropics

Irrigated production systems within these climatic zones refer mainly to the widely distributed oasis and river floodplain systems of Asia and Africa, where available arable land ranges from 0.18 to 2.68 ha person^{-1}. Typical examples are the Tozeur and Turfan oases and the Nile or Huanghe valleys of North Africa and Central Asia, the *fadamas* (depressions) along the Niger River in West Africa, and similar systems on the coast of Peru. The high insolation rates mean that, given adequate amounts of water, primary above-ground productivity is high and two crops can be grown annually. Forage options include seasonal rotation crops, undercropping in date palm plantations, and the use of spent, slightly saline irrigation water to produce salt-tolerant shrubs (e.g. *Atriplex* spp.) or trees (e.g. *Casuarina* spp., *Tamarix* spp.) on uncropped land. Dairy cattle, camels and small ruminants can be kept, under a largely zero-grazing system with supplementary grazing and browsing in the adjacent rangland or steppe areas. Fish and waterfowl may be raised where open water bodies exist.

Conclusions

Underused land resources suitable for various types of fodder production can be identified in most agroecosystems, although such land differs markedly in quantity and quality from region to region. It is important to harness these resources to balance nutrient deficits in feeding systems based on fibrous crop residues, to improve the fertility of cropland and to generate additional income for resource-poor farmers by integrating crop and livestock production. In addition, there is probably still a large pool of underexploited domesticated and wild forage germplasm with varying phenological characteristics and a wide range of ecological adaptations and nutrient productivities. This plant biodiversity should be used to advantage while it is still there.

The assumption underlying this paper is that suitable forage crops can increase soil fertility to such an extent that subsequent food crop yields compensate for the loss in area cropped.

Consequently, the integration of forage production into tropical and subtropical food cropping systems is a promising way of promoting sustainable, low-input agriculture. In addition, producing forage as a pioneer crop on wastelands can allow the return of such land to food crop production. In Europe, these technologies were developed 200 years ago and have been practised successfully for more than a century. Temporarily out of favour as high-input monocropping systems came in after 1945, they are now being re-introduced into temperate agriculture through the organic farming movement. While there may be a need for adaptation to tropical conditions, there are good prospects for success in this climatic zone as well.

However, the integration of forage crops and livestock into food cropping systems requires a holistic approach, taking into account the systems' ecological, economic and social characteristics and the priority attached to different crop and livestock components by the farmers themselves. Shortages of both land and capital dictate systems that rely as much as possible on multipurpose crops and livestock. Although complex, these integrated systems are also less risk-prone and hence more sustainable not only ecologically but also economically than monocropping systems.

Compared with the potential of green feed production from arable and wasteland, the other options discussed—concentrate feeding from own farm or commercial sources—are less important. Concentrate feeding will only be widely adopted if farmers' economic circumstances and improved availability permit.

Obviously, there are still large gaps in our knowledge on how tropical production systems function and how they can be improved through the introduction of forage production. More research is necessary. The priority topics are as follows:

1. The extent of available land resources (fallow, uncropped land, wasteland) in target regions.
2. The collection and evaluation of indigenous forage species, including studies on their adaptation and productivity, their contribution to annual and seasonal nutrient requirements and their effects on soil fertility and subsequent food crop yields.
3. The development of crop rotations (ley farming, intercropping) adapted to specific locations; their contributions to forage production, soil fertility and farm income; environmental impact assessment (EIA).
4. Strategic feed utilization studies on selected combinations of forages with food crop residues and concentrates, and the development of feeding calendars.
5. Socioeconomic studies on resource-poor farmers, assessing their awareness of the problems caused by continuous food cropping, their willingness to try out new solutions, and their adoption of improved forage technologies.

These studies should be carried out as local on-farm research with strong farmer participation, with national research groups taking responsibility and international research institutions providing guidance and support where necessary. The aim is to achieve results that can be turned into a set of recommentations and technologies for a fully fledged technology transfer effort involving non-government organizations (NGOs) alongside conventional government extension services. Efforts of this kind should, at last, lead to significant rates of adoption by the resource-poor farmers of the tropics, who are so greatly in need of improved standards of living through more sustainable forms of agriculture.

References

Badve, V.C. (1991) Feeding systems and problems in the Indo-Ganges plain: Case study. In: Speedy, A. and Sansoucy, R. (eds), *Feeding Dairy Cows in the*

Tropics. FAO Animal Production and Health Paper No. 86, Rome, Italy, pp. 176-183.

Boodoo, A.A. (1991) Milk production from tropical fodder and sugarcane residues: Case study: On-farm research in Mauritius. In: Speedy, A. and Sansoucy, R. (eds), *Feeding Dairy Cows in the Tropics.* FAO Animal Production and Health Paper No. 86, Rome, Italy, pp. 225-235.

Cole, G.O.R. (1996) Comparative study of the use of oxen and heifers as draught animals in northern Sierra Leone. PhD thesis, University of Giessen, Germany.

Dixon, R.M. and Egan, A.R. (1987) Strategies for optimizing the use of fibrous crop residues as animal feeds. Cited by Winugroho, M. (1996).

FAO (1977) *Pasture Handbook of Malawi,* 2nd edn. Food and Agriculture Organization of the United Nations, Lilongwe, Malawi.

FAO (1994) Tropical Feeds. World Wide Web (fao.org/WAICENT/Agricult.htm).

FAO (1995) Agricultural World Census. Gopher (fao02.FAO.ORG:70/00Gopher).

FAO (1996) Towards Universal Food Security. FAO Documents for the World Food Summit 1996. Food and Agriculture Organization of the United Nations, Rome, Italy.

Gill, A.S. and Tripathi, S.N. (1991) Intercropping of *Leucaena* with particular reference to hybrid napier varieties. *Range Management and Agroforestry* 12: 201-205.

Gutteridge, R.C. and Shelton, H.M. (eds) (1994*) Forage Tree Legumes in Tropical Agriculture.* CAB International, Wallingford, UK.

Jayasundara, H.P.S. and Marasinghe, R. (1989) A model for integration of pasture, tree fodder and cattle in coconut smallholdings. *Coconut Bulletin* 6: 15-18.

Mathew, T., Kumar, B.M., Babu, K.V.S. and Umamahesvaran, K. (1992) Comparative performance of four multipurpose trees associated with four grass species in the humid regions of southern India. *Agroforestry* 17: 205-218.

Mtenga, L.A. and Shoo, R.A. (1990) *Leucaena leucocephala* as a supplemental feed for Blackhead Persian lambs. *Bulletin of Animal Health and Production in Africa* 38: 119-126.

Munga, R.W., Thorpe, W. and Topps, J.H. (1992) Voluntary food intake, liveweight change and lactation performance of crossbred cows given ad libitum *Pennisetum purpureum* (napier grass var. Bana) supplemented with leucaena forage in the lowland semihumid tropics. *Animal Production* 55: 331-337.

Nazir, M.S., Habib-ur-Rahman, K., Ali, G. and Ahmad, R. (1988) Inter/relay cropping in wheat planted in multi-row strips at uniform plant populations. *Pakistan Journal of Agricultural Research* 9: 305-309.

Patel, J.R., Sadhu, A.C., Patel, P.C. and Patel, B.G. (1992) Forages from a subabool *(Leucaena leucocephala)*-based agroforestry system under irrigated conditions. *Indian Journal of Agronomy* 37: 630-632.

Peters, M. (1992) Evaluierung von tropischen Weideleguminosen für Fodder Banks im subhumiden Nigeria. PhD thesis, University of Giessen, Germany.

Powell, J.M. and Mohamed-Saleem, M.A. (1987) Nitrogen and phosphorus transfers in a crop/livestock system in West Africa. *Agricultural Systems* 25: 261-277.

Puri, S., Singh Shambhu and Kumar, A. (1994) Growth and productivity of crops in association with an *Acacia nilotica* tree belt. *Journal of Arid Environments* 27: 37-48.

Rischkowsky, B. (1996) Untersuchungen zur Milcherzeugung mit Saanen-kreuzungsziegen in kleinbäuerlichen Betrieben Malawis. PhD thesis, University of Giessen, Germany.

Sangakkara, W.R. (1989) Forage legumes as a component of smallholder rice farming systems: A case study. In: *Proceedings of the Sixteenth International Grassland Congress,* 4-11 October 1989, Association Française pour la Production Fourragère, Nice, France, pp. 1303-1304.

Seré, C. (1994) Livestock and Environment Study: Characterization and Quantification of Livestock Production Systems. Draft Final Report, Food and Agriculture Organization of the United Nations (FAO), Rome, Italy.

Siagian, N. and Sumarmadji (1992) Forage grasses as alternative cover crops for young rubber. *Herbage Abstracts* 62: 225.

Winugroho, M. (1996) Utilization of crop residues in sustainable mixed crop/livestock farming systems in Indonesia. Paper presented at the International Workshop on Crop Residues in Sustainable Mixed Crop/Livestock Farming Systems, 22-26 April 1996, International Crops Research Institute for the Semi-Arid Tropics (ICRISAT), Hyderabad, India.

11. Alternatives to Crop Residues for Soil Amendment

J. Mark Powell[1] and Paul W. Unger[2]
[1]*USDA-ARS Dairy Forage Research Center, 1925 Linden Drive West, Madison, Wisconsin 53706, USA*
[2]*USDA-ARS Conservation and Production Laboratory, PO Drawer 10, Bushland, Texas 79012, USA*

Abstract

Crop residues perform various functions in semiarid farming systems. When left in fields after grain harvest, they enhance soil and water conservation, nutrient cycling and subsequent crop yield. Crop residues are also used for other purposes, such as livestock feed, fuel and construction material. Although retaining crop residues on soil surfaces has numerous beneficial effects, their other uses preclude the return of all residues to the soil. This paper examines the key roles of crop residues in semiarid farming systems and suggests alternative soil amendments to crop residues.

The retention of even small amounts of surface residues can conserve soil organic matter and nutrients, decrease water runoff and increase infiltration, decrease evaporation, and control weeds. Various soil management and related practices allow crop residues to be removed without adversely affecting the soil. These include the introduction of improved crop genotypes that produce both sufficient grain and residues of good quality; the partial rather than total removal of residues; the provision and use of chemical fertilizers and animal manure; substituting forages for crop residues as animal feed; the use of clean or reduced tillage; the use of practices that complement tillage; and the application of surface soil-amending materials. Not all practices are universally adaptable, but one or more should be suitable under most conditions where crop residues are limited or used for other purposes.

Introduction

Millet (*Pennisetum glaucum* [L.] R.Br.) and sorghum (*Sorghum bicolor* [L.] Moench) are the principal cereals cultivated in the semiarid tropics; rice (*Oryza* L. spp.) and maize *(Zea mays* L.) are important in areas of more favourable soil fertility and moisture conditions; and wheat *(Triticum aestivum* L.) is cultivated in cooler areas. Whereas grain is almost exclusively consumed and/or marketed as food, crop residues perform various functions in these agricultural systems. They are used as animal feed, as fuel and construction material, and they provide income through their sale. When left in fields after harvest, crop residues play important roles in nutrient cycling, erosion control and the maintenance of favourable soil physical properties (Pichot *et al.*, 1981; Power *et al.*, 1986; Bationo and Mokwunye, 1991; Unger *et al.*, 1991). In minimum and no-tillage systems, they protect the soil surface from wind and water erosion, provide favourable seedbed conditions and conserve soil water. The magnitude of the beneficial effects associated with returning crop residues to fields depends on the quantity and quality of the residue, the subsequent crop to be grown, edaphic factors, topography, climate and soil management.

Crop residues retained on the soil surface provide soil and water conservation benefits. These benefits result mainly from their physical presence, which moderates the forces of wind and water, reducing the potential for erosion. Surface residues aid water conservation by: (i) protecting soil aggregates against dispersion, thereby reducing the potential for the development of a surface seal that could reduce water infiltration; (ii) slowing water flow across the surface and so providing more time for infiltration; and (iii) reducing evaporation. Conservation of soil and water resources is of paramount importance for sustaining cropland productivity in many semiarid environments. Other benefits of surface residues include greater soil organic matter (SOM) concentrations, soil temperature moderation and increased soil biological activity, all of which are also important for sustaining crop production.

When all crop residues are used as animal feed or removed for other purposes, the above benefits are lost. As a result, sustaining soil productivity becomes more difficult. However, it is possible to sustain crop production by using appropriate alternative practices, such as retaining some residues, replacing the nutrients harvested in grain and crop residues, growing forages that substitute for crop residues, using adapted tillage practices, using appropriate land-forming techniques that complement the tillage practices, and applying materials that stabilize the soil surface. This paper highlights the major roles of crop residues in maintaining cropland productivity and outlines alternative soil amendment and management practices that may permit the removal of some residues.

Role of Crop Residues in Semiarid Farming Systems

Livestock Feed

In many parts of the semiarid tropics, livestock (cattle, sheep and goats) are an integral part of agricultural production and crop residues provide vital feeds during the long dry season (McCown *et al.*, 1979; Sandford, 1989; McIntire *et al.*, 1992). Whereas leguminous crop residues from groundnut *(Arachis hypogaea* L.) and cowpea (*Vigna unguiculata* [L.] Walp.) are harvested and either sold or fed to farmers' livestock, cereal stovers (biomass remaining after grain harvest) are generally left in fields and grazed after grain harvest. More intensive modes of crop and livestock production are adopted in densely populated areas, where all stover may be harvested from fields for stall feeding and/or sale.

Although returning crop residues to soils enhances soil productivity, this is not a viable strategy for many farmers owning livestock. The absence of alternative dry-season feeds creates too great a trade-off in lost livestock production. The stability of agricultural production depends heavily on livestock, so most crop residues must be fed and manure/urine used to fertilize the soil.

Nutrient Cycling

Many farmers in the semiarid tropics rely principally on the recycling of organic matter to sustain agricultural productivity. Crop residues have long played key roles in this process. They are often fed to livestock, with the resulting manure and urine being used to fertilize the soil. Feeding crop residues (and other roughages) to livestock accelerates nutrient cycling (Somda *et al.,* 1995). The release of nutrients from manure applied to soil is more rapid than from crop residues, and exhibits a pattern that coincides more closely with crop nutrient demands (Powell and Ikpe, 1992). However, feeding crop residues results in nutrient losses. When animals graze crop residues, more nutrients are removed than returned via manure (Powell and Williams, 1993). Urine nitrogen (N), which comprises approximately 50% of the N voided by ruminant livestock, is highly susceptible to losses via volatilization and leaching.

Farmers and livestock remove variable amounts of nutrients from croplands (Fig. 11.1). Essentially, all of the grain is used for human food, while most crop residues are used as livestock feed or for fuel or construction material. Any crop residues not removed by farmers or livestock are either incorporated into the soil by insects and microbial activity or burned (resulting in N loss) to facilitate manual cultivation and control pests and diseases. The excessive removal of crop residues from fields can deplete SOM and nutrient reserves and increase the risk of soil erosion and degradation.

Nutrient balances (inputs minus harvests) are negative for many farming systems of semiarid Africa (Stoorvogel and Smaling, 1990). In Mali, cropland nutrient deficits approximately equal the amount of N removed in grain (Table 11.1). The estimated maximum potential amount of N that could be removed by livestock grazing crop residues approximately equals the N returned in trampled fractions of crop residue, in manure, and via rain and soil bacteria. Nitrogen balances probably approach equilibrium (nutrient inputs = harvests) in many farmers' fields, given the widespread application of fertilizers made available for use on cash crops but also used on subsistence crops, and the application of manure to fields between cropping periods.

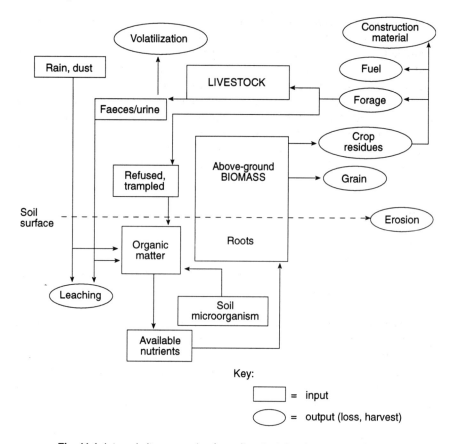

Fig. 11.1. Internal nitrogen cycle of crop/livestock farming systems in Mali.

Some crop residue management strategies conserve and augment SOM reserves, while other strategies deplete them. When more productive cropping technologies are adopted, the return of greater quantities of crop residues can increase SOM and nutrient levels (Jenkinson and Rayner, 1977; Dalal, 1986; Grove *et al.*, 1986). Conservation tillage systems, in which all crop residues are retained on the soil surface, also conserve or increase surface SOM levels. Improvements in crop management, leading to greater crop residue production, may allow sufficient residues to be returned to fields and some to be removed without adversely affecting the soil environment.

Although the SOM of most cultivated soils is highly buffered and maintained at fairly constant levels (Bartholomew, 1972; Magdoff, 1992), dramatic changes in crop residue management can disrupt the SOM equilibrium. When crop residues are removed from high-producing systems, SOM and available nutrient levels decline (Fuller *et al.*, 1956; Larson *et al.*, 1972; Barber, 1979; Powell and Hons, 1992). Cereal stover, which has a high carbon:nitrogen (C:N) ratio, maintains an equilibrium between soil nutrient immobilization and mineralization processes. When normal stover return decreases the equilibrium is disrupted, causing a net shift to SOM mineralization. During the early SOM decay process, labile SOM fractions, which are highly influenced by crop residue management strategies (Sauerbeck and Gonzalez, 1977; Elliott and Papendick, 1986) are the first to mineralize, followed by the more stable SOM components (van Faasen and Smilde, 1985). The effects of stover removal on yields and nutrient cycling are more pronounced in coarser-textured soils having lower SOM and available nutrient reserves and high SOM turnover rates.

Table 11.1. Nitrogen removals, returns and balances (kg ha^{-1}) for croplands in integrated rangeland-cropland systems in Mali.

	Production system			
	Millet- and sorghum-based	Upland rice	Flooded rice	Maize- and cotton-based
Nitrogen removals:				
Grain				
Range	11-28	17-28	26-30	14-19
Mean	18	20	29	16
Crop residues				
Range	8-14	11-12	19-23	9-16
Mean	11	11	22	14
Nitrogen returns:				
Crop residues				
Range	3-5	3-4	4-4	4-5
Mean	4	3	4	5
Manure				
Range	0-6	1-3	0-2	0-2
Mean	1	2	1	1
Crop roots				
Range	2-3	2-2	3-4	2-3
Mean	2	2	4	3
Other (rain/dust, soil microorganisms)				
Mean	4	6	7	5
Nitrogen balance:				
Range	-29 to -12	-19 to -16	-38 to -31	-17 to -11
Mean	-18	-18	-35	-15

Source: Adapted from Powell and Coulibaly (1995).

Soil Conservation

In semiarid West Africa, low soil fertility, high soil temperatures, wind erosion and sand blasting of young plants severely constrain crop establishment and production. When left in the field, cereal stovers provide a physical barrier to soil movement, allow soil and organic matter to accumulate, and enhance soil chemical properties and crop yield (Bationo and Mokwunye, 1991; Geiger *et al.*, 1992). The benefits generally increase with increasing amounts of residues available. However, even small amounts provide some benefits. For example, Mannering and Meyer (1963) reported that maize residues at 2.2 t ha^{-1} resulted in very high infiltration of simulated rain water with essentially no erosion on a silt loam soil with 5% slope. With rates of 0.55 and 1.1 t ha^{-1}, soil losses were 6.7 and 1.1 t ha^{-1} respectively, whereas they were almost 27 t ha^{-1} when no residues were applied. Also, mean soil water storage during fallow was 99 mm with a wheat residue mulch at 1 t ha^{-1}, compared with 72 mm without mulch (Unger, 1978). In the Sahel of West Africa, millet stover applications of 2 t ha^{-1} reduced horizontal soil transport by 59% (Michels *et al.*, 1993) and decreased peak surface soil temperatures (at 1 cm depth) from 48.5 to 39.3°C (Buerkert *et al.*, personal communication). Geiger *et al.* (1992) reported that improved soil fertility following 6 years of continuous millet stover application was due in part to the entrapment of aeolian materials that generally have better fertility characteristics and/or to the protection of the more fertile surface soil from erosion by strong winds. In Texas, USA, Powell (1989) found that annual sorghum stover applications of 5 t ha^{-1} were sufficient to maintain SOM levels, while 2.5 t ha^{-1} was adequate for maintaining available soil phosphorus (P) and potassium (K) levels in a high-producing, continuous grain sorghum-cropping system.

Clearly, the retention of even limited amounts of crop residues on the soil surface can help conserve soil and water and maintain favorable SOM and nutrient levels. Unger *et al.* (1991) discussed various techniques that could help retain more residues on cropland for resource conservation purposes while allowing some use by livestock. Most of the techniques are adaptable to any climate, but some are better suited to more humid regions. In addition, some are more feasible in mechanized farming systems while others work better when animal traction or hand labour is used.

Alternatives to Crop Residues as Soil Amendments

To evaluate alternatives to crop residues as soil amendments, estimates are needed of total crop residue production and the amounts of dry matter (DM) and nutrients harvested and returned by farmers to their fields. Cereal stovers comprise approximately 60 to 75% of total crop biomass production and have lower nutrient concentrations than does grain (Table 11.2). However, considerable caution should be exercised when using harvest indexes to estimate stover production from grain production. Delayed planting of the photosensitive crop genotypes commonly cultivated in semiarid Africa (SAA), combined with a range of other factors such as insufficient rainfall, high temperatures and insect damage, can alter relative grain and stover yields.

Table 11.2. Average and standard error (SE±) of harvest index and nitrogen, phosphorus and potassium concentration (g kg⁻¹) in grain and stover of five major cereals cultivated in semiarid environments.

	Millet	Sorghum	Maize	Rice	Wheat
Harvest index[1]:					
Average	0.26	0.27	0.42	0.44	0.41
SE±	0.08	0.11	0.12	0.08	0.07
N	480	306	344	392	329
Nitrogen:					
Grain					
Average	18.5	16.9	15.5	11.6	21.4
SE±	3.9	2.9	3.0	1.9	4.8
N	125	139	267	184	243
Stover					
Average	7.0	6.5	6.6	6.2	5.1
SE±	3.3	2.2	1.8	1.8	2.6
N	99	105	183	171	200
Phosphorus:					
Grain					
Average	3.1	2.6	2.9	2.0	3.7
SE±	0.5	0.6	0.8	0.6	0.8
N	86	75	101	171	53
Stover					
Average	0.9	0.8	0.8	1.1	0.5
SE±	0.4	0.5	0.4	0.6	0.2
N	86	73	82	158	46
Potassium:					
Grain					
Average	4.8	3.4	3.5	3.9	4.6
SE±	1.3	0.8	1.6	1.5	1.2
N	70	27	45	105	65
Stover					
Average	20.9	10.9	11.6	18.9	15.9
SE±	5.7	3.6	5.0	6.7	5.6
N	70	27	41	105	65

[1]Weight ratio of grain to total above-ground crop biomass production.
N = Number of observations.
Source: van Duivenbooden (1992).

Nutrient Replacement

Using Inorganic Fertilizers

Sustainable increases in biomass productivity are fundamental to providing adequate food and feed for expanding human and livestock populations in SAA. The proper use of inorganic fertilizers will be crucial to achieving these increases (Breman, 1990; McIntire and Powell, 1995). However, fertilizers remain costly and unavailable to many farmers. The limited amounts of fertilizer available need to be used judiciously for maximum benefit.

Various approaches have been used to estimate the amount of nutrients needed to replace the nutrients harvested in grain and stover and to achieve desired crop yields. Van Duivenbooden (1992) modelled the nutrient requirements (REQ) for a variety of crops cultivated in semiarid tropical and subtropical environments as follows:

$$REQ = ((CON–FIX)/REC)–NAT1–NAT2–CRR1–CRR2$$

where CON = nutrient content in crop at maturity (kg ha^{-1})
    FIX = N fixation by crop (kg ha^{-1})
    REC = fractional apparent recovery of the nutrient
    NAT1 = natural nutrient soil fertility (kg ha^{-1})
    NAT2 = nutrient availability from additional natural resources (kg ha^{-1})
    CRR1 = nutrient availability from crop roots and stubble of preceding year (kg ha^{-1})
    CRR2 = nutrient availability from crop stover of preceding year (kg ha^{-1})

Only a fraction of the nutrients applied are taken up by crops. Considerable amounts can be lost via such routes as erosion, leaching, as gases (N) or as elements such as phosphorus (P) that recombine with soil components into forms unavailable to plants. Only 30-40% of the N, 10-16% of the P and 25-40% of the potassium (K) applied as fertilizer is actually recovered in the subsequent crop (Table 11.3).

In a review of the effects of fertilizer on crop yields in semiarid West Africa, Williams *et al.* (1995) concluded that the use of light fertilizer applications with crop residues or manure gave better results than heavy fertilizer applications alone. The application of chemical fertilizers alone to poorly buffered soils leads to decreases in soil pH and base saturation, K deficiency and aluminium (Al) toxicity (Pieri, 1989). The addition of even small amounts of crop residues in combination with inorganic fertilizers can counteract these negative effects (de Ridder and van Keulen, 1990) and also substantially decrease the soils' capacity to fix P (Kretzschmar *et al.,* 1991). Applying crop residues also increases soil cation exchange capacity (de Ridder and van Keulen, 1990; Bationo *et al.,* 1995), populations of N$_2$-fixing bacteria, and root length and density, leading to increased P uptake by the crop (Hafner *et al.,* 1993a,b).

Table 11.3. Fertilizer recovery fractions of nitrogen, phosphorus and potassium in five major cereals cultivated in the semiarid tropics.

	Millet and sorghum	Maize	Rice	Deepwater rice	Wheat
Nitrogen:					
25% quartile	0.22	0.21	0.24	0.14	0.26
75% quartile	0.51	0.46	0.52	0.45	0.55
Average	0.37	0.36	0.39	0.29	0.42
SE±	0.18	0.19	0.19	0.21	0.20
N	67	93	123	34	108
Phosphorus:					
25% quartile	0.09	0.02	0.05	–	0.04
75% quartile	0.20	0.18	0.16	–	0.17
Average	0.16	0.13	0.12	–	0.10
N	31	46	61	31	
Potassium:					
25% quartile	0.25	0.17	0.19	–	0.11
75% quartile	0.46	0.50	0.53	–	0.23
Average	0.38	0.34	0.34	–	0.24
N	5	6	39	6	

N = Number of observations.
Source: Adapted from van Duivenbooden (1992).

Animal Manures

Manure is used throughout the semiarid tropics to maintain soil productivity. In SAA it may be transported from corrals and hand-spread on cropland. Alternatively, livestock may be corralled directly on cropland between cropping periods (Powell and Williams, 1993). The amount of manure required to offset nutrient harvests varies with crop yield, nutrient harvests and other losses, the type of animal manure and the way in which it is managed. In Niger, biennial manure requirements to replenish the nutrients removed in millet grain and stover range from 2.1 to 6.9 t ha^{-1} (Williams *et al.*, 1995). Corralling reduces manure requirements since both manure and urine are returned to the soil, resulting in higher crop yields than if manure alone is applied (Powell and Ikpe, 1992; ILCA, 1993; Williams *et al.*, 1995).

Most farmers in semiarid West Africa do not have sufficient livestock to provide enough manure to offset nutrient harvests. In Niger, farmers would need an additional 9 to 21 cattle per hectare of millet (Williams *et al.*, 1995). In Mali, 3 to 7 additional cattle plus 3 to 12 additional small ruminants (or 100 to 200 kg of urea fertilizer) per hectare would be required to offset the deficits listed in Table 11.1. These nutrient balance studies emphasize the need for more information on current rangeland and cropland carrying capacities and stocking rates before advocating any increase in herd size for the purpose of manuring cropland. Moreover, the limitations in animal numbers posed by feed availability in these semiarid locations imply that external inputs, in the form of inorganic fertilizers, are needed for long-term gains in agricultural productivity.

Nutrient cycling can be enhanced by synchronizing the application of organic soil amendments, such as crop residues and manures, to soils so that these decompose and release nutrients in a pattern that coincides with crop nutrient demands (McGill and Myers, 1987; Swift *et al.*, 1989; Ingram and Swift, 1989). Organic materials such as cereal stovers, which have C:N ratios greater than 20 to 30 (Alexander, 1977) and N concentrations below 15 g kg^{-1} (Bartholomew, 1972) temporarily immobilize soil N and release nutrients more slowly than manures that have lower C:N ratios and higher N contents.

Alternative Livestock Feeds

Forage Legumes

The integration of herbaceous forage legumes and shrubs into cereal-based cropping systems can control soil erosion, improve soil moisture conservation, suppress weed growth, accelerate nutrient cycling, enhance soil productivity and provide food, fodder and wood. Forages harvested at their optimum growth stage have higher feed value than the residues of many crops. Therefore, if forage crops with high nutritive value were grown on some land, residues could be retained on the remaining land for conservation purposes. Animal production should not decrease and could even increase, since forage legumes are of much higher quality than crop residues. Production of food crops could be lower initially because of the reduced amount of land devoted to them. However, if the residues were properly managed, soil and water conservation and nutrient cycling would increase, which would raise yields in the medium to longer term (Papendick and Parr, 1988).

While forage legume/cereal intercropping often increases the quantity and feed value of crop residues, it may decrease the grain yield of the companion cereal crop (Waghmare and Singh, 1984; Mohamed-Saleem, 1985; Kouamé *et al.*, 1993). Only when cereals follow short-term fallows of forage legumes are greater grain and stover yields obtained. For example, cereal yields in areas where the cereal was preceded by *Stylosanthes* spp. were always higher and sometimes double those following natural fallow (Tarawali and Mohamed-Saleem, 1995). The superior performance of cereals after legumes is associated with improved soil physical and chemical properties brought about by the legume. Problems associated with introducing forage legumes into cereal-based systems include the shortage of land to devote to forages, the high cost of fencing in areas where livestock are free-range, the management needed to ensure predominance of the legume, soil trampling by livestock, and legume diseases. The low returns in terms of livestock output to the increased labour required are also often a constraint (McIntire *et al.*, 1992).

Alley Cropping

Alley cropping involves growing deep-rooting perennial shrubs or trees in rows spaced far enough apart so that crops can be grown in the interrow area. The shrubs or trees are pruned periodically to reduce competition for light and water, with the leaves and twigs being used as animal feed or as a mulch for the cropped area. Larger twigs and branches can be used as firewood, reducing the need to use crop residues for that purpose. For maximum benefit, shrubs or trees should grow rapidly, fix nitrogen, be multipurpose and have a deep, narrow root system to reduce competition with the companion crop for water and nutrients in the soil. The leguminous species *Leucaena leucocephala* (Lam.) de Wit and *Gliricidia* Kunth spp. performed well under tropical conditions in parts of Nigeria (Wilson *et al.*, 1986; Atta-Krah, 1990), but decreased sorghum, cowpea and castor (*Ricinus communis* L.) yields in India due to serious competition for light and water (Singh *et al.*, 1988).

Utilization of Wasteland

Areas unsuitable for crop production are present on many farms and in or near many villages or communities. These may be next to streams or roads, on rocky outcrops, in low-lying or marshy areas, along property boundaries and in irregular-shaped sites within fields. Plants from such areas often already are used as livestock feed, but greater production could be obtained through improved management, which would further reduce the demand for crop residues as livestock feed. Improved management might also improve erosion control on such areas.

Improved Crop Genotypes and Selective Crop Residue Removal

Many farmers throughout the semiarid tropics cultivate crop varieties that produce adequate stover to satisfy on-farm feed, fuel and construction material requirements. Such diverse crop uses present particular challenges for crop breeders. Varietal improvement programmes have traditionally focused on enhancing grain production at the expense of stover. This results in lower stover yields, which are often of poor feed quality (Reed *et al.*, 1988).

Large differences in cereal morphological and chemical characteristics may offer possibilities for breeding crop genotypes that improve both grain and stover yield and quality. Greater crop residue production may allow sufficient residues to be returned to fields for soil conservation and nutrient cycling while allowing some to be removed. For example, Powell (1989), working in Texas, USA, found that forage sorghums (low grain/very high stover production) and intermediate-type sorghums (moderate grain/high stover production) return sufficient stover to maintain SOM and soil nutrient levels in a continuous grain sorghum cropping system, while allowing large amounts of stover to be harvested for energy conversion.

Cereal stover plant parts vary in nutrient content and carbohydrate and lignin composition, which directly affect their value as soil amendments, animal feeds, alternative energy sources, and so on. Concentrations of N and P are higher in leaves than in stalks and generally decline from upper to lower stover parts (Table 11.4). Selective residue removal can be achieved in both manual-labour and

Table 11.4. Dry matter (DM), nitrogen (N), phosphorus (P), potassium (K), non-structural carbohydrate (NS), cellulose (CL), hemicellulose (HC) and lignin (LG) concentrations in conventional, intermediate-type and forage sorghums.

Stover part	DM (%)	N	P	K	NS	CL	HC	LG
					$(g\ kg^{-1})$			
Conventional grain sorghum								
Total stover	100	6.0	1.5	21.2	14.2	31.3	28.2	7.1
Blades:								
Total	28.4	13.3	2.4	12.3	10.0	24.9	32.9	5.4
Upper	8.3	14.8	2.5	12.3	14.1	25.4	32.5	5.7
Middle	13.2	11.9	2.0	13.2	8.3	26.1	32.0	4.9
Lower	5.8	8.5	1.5	6.6	10.8	29.2	29.3	6.5
Stalks:								
Total	71.6	4.6	1.6	24.3	21.2	31.6	25.5	7.1
Upper	10.5	5.0	1.3	25.7	7.5	33.4	32.8	7.5
Middle	25.0	4.6	1.6	24.3	21.2	31.6	25.5	7.1
Lower	37.2	3.1	1.3	21.8	24.5	33.7	23.4	7.2
LSD all[1]	–	3.2	0.8	5.3	4.1	6.1	4.7	2.4
LSD blades	–	2.8	1.2	4.3	3.5	NS	NS	NS
LSD stalks	–	1.4	NS	NS	3.3	NS	2.9	NS
Forage sorghum								
Total stover	100	4.5	1.4	13.2	27.1	29.1	23.9	7.0
Blades:								
Total	18.0	12.2	2.0	15.0	11.2	29.8	38.7	5.4
Upper	6.9	16.0	2.3	12.5	16.0	26.4	42.1	5.1
Middle	6.3	8.9	1.4	16.6	8.7	33.1	35.5	6.0
Lower	2.7	7.1	1.1	13.6	6.7	33.6	31.1	6.6
Stalks:								
Total	82.0	3.7	1.2	12.9	31.6	27.5	21.3	6.5
Upper	15.9	4.6	1.4	12.2	30.5	26.7	20.4	6.8
Middle	28.1	2.3	1.0	10.7	30.8	27.6	20.4	6.8
Lower	40.0	2.3	1.0	12.9	29.7	30.1	20.5	8.5
LSD all	–	2.7	0.5	4.0	3.2	3.8	4.7	1.5
LSD blades	–	2.8	0.4	2.8	3.6	6.4	1.7	0.5
LSD stalks	–	0.9	NS	NS	2.0	1.4	5.9	1.3

Table 11.4 continued.

Stover part	DM (%)	N	P	K	NS	CL	HC	LG
					(g kg⁻¹)			
Intermediate-type sorghum								
Total stover	100	2.3	0.8	16.9	28.5	25.3	21.7	7.3
Blades:								
Total	13.8	7.4	1.2	12.9	8.9	26.9	33.8	5.5
Upper	5.1	9.6	1.7	9.1	NR	NR	NR	NR
Middle	7.1	5.8	1.0	14.7	NR	NR	NR	NR
Lower	3.6	4.7	0.7	12.4	NR	NR	NR	NR
Stalks:								
Total	86.2	1.9	0.8	16.8	38.9	23.4	20.7	7.6
Upper	21.2	2.5	0.8	15.0	NR	NR	NR	NR
Middle	26.1	1.4	0.7	16.5	NR	NR	NR	NR
Lower	36.9	1.5	0.7	20.6	NR	NR	NR	NR
LSD all	–	1.9	0.5	4.0	5.1	2.3	1.7	0.7
LSD blades	–	2.3	0.9	NS	–	–	–	–
LSD stalks	–	0.8	NS	1.6	–	–	–	–

[1] LSD all indicates LSD (P<0.05) for all comparisons between pairs within columns. LSD blades and LSD stalks indicate LSD for respective upper, middle and lower components. NR indicates no fractionation of carbohydrates corresponding to upper, middle and lower stover components. NS indicates no significant differences.
Source: Adapted from Powell *et al.* (1991).

mechanized farming systems. For example, in semiarid West Africa the manual harvesting of millet and the chemical composition of millet stover parts may allow certain parts to be harvested as high-quality feed while other parts are returned for soil conservation (Powell and Fussell, 1993). In mechanized production systems, selective residue removal can be achieved by: (i) using only the plant parts that pass through the harvester as feed for livestock; (ii) allowing animals to forage on fields, but removing them when remaining residue amounts approach a critical level; and (iii) removing residues only from areas less prone to erosion.

Use of Adapted Tillage Practices

Tillage for crop production involves modifying, usually mechanically, the conditions of the upper soil layer. A tillage system is the combination of operations used in a given situation. Tillage influences water conservation through its effects on soil conditions that retard runoff, enhance infiltration and suppress evaporation, and by controlling weeds that use soil water. Runoff is retarded and infiltration enhanced when: (i) water flow into soil is not restricted by surface conditions; (ii) water is temporarily stored on the surface or in the tillage layer, providing more time for infiltration; and (iii) water movement within the soil profile is not impeded. Under these conditions, the potential for soil erosion by water is reduced. Evaporation is suppressed by insulating and cooling the soil surface, reflecting solar energy, decreasing wind speed at or near the soil surface and providing a barrier against water vapour movement. Timely weed control is highly important because weeds may deplete soil water supplies, in turn reducing crop yields. Where excess water must be removed, conditions generally opposite to the above must be induced and a drainage system may be needed for some soils.

Although conservation tillage, which is based on managing crop residues on or near the soil surface, is most effective for conserving soil and water resources, its use is not possible when adequate residues are not available. No-tillage, defined as planting crops with no preparatory tillage since harvest of the previous crop, is not suitable when surface residues are absent or the soil is severely degraded (Charreau, 1977). Under these conditions a "clean" or reduced tillage system must be relied upon to conserve resources and sustain production.

Clean Tillage

Clean tillage is the process of ploughing and cultivating that incorporates all crop residues and prevents the growth of all plants except the crop being grown (SSSA, 1987). It is often called "conventional" tillage, but this designation has two disadvantages: convention changes with the development and adoption of new practices, and what is conventional in one area may not be so in another.

When tillage is clean, water conservation results primarily from the disruption of soil crusts, providing for temporary water storage, and the control of weeds. Under some conditions, clean tillage may suppress evaporation. Unfortunately, the residue-free surface produced by clean tillage often eventually aggravates the crusting problem.

Tillage-induced surface roughness and cloddiness can reduce runoff velocity and create depressions for temporary water storage on the surface, thereby providing more time for infiltration and reducing the potential for erosion by water. Surface roughness and cloddiness are also highly important for controlling erosion by wind when surface residues are not present (Woodruff and Siddoway, 1973). In addition, tillage-induced soil loosening increases the potential for water storage in the tillage layer (Burwell *et al.*, 1966).

A soil mulch created by shallow tillage reduced evaporation in areas where a distinct dry season followed a wet period that had recharged the soil profile with water (Hammel *et al.*, 1981; Papendick, 1987). Such mulch also reduced evaporation on various soils when evaporativity was low (Gill *et al.*, 1977; Jalota and Prihar, 1979; Gill and Prihar, 1983). However, it was ineffective where precipitation occurred mainly during the summer, when the potential for evaporation was greatest and tillage was needed after each rain to control weeds (Jacks *et al.*, 1955). In addition, much of the water had already evaporated by the time tillage was possible, while the frequent tillage pulverized the soil and so increased the potential for erosion (Jacks *et al.,* 1955).

Weeds compete with crops for water, nutrients and light. Competition for water is generally considered the most important under dryland conditions. Effective weed control before planting and during the growing season is therefore essential if crops are to use available water to produce at their potential in semiarid farming systems. Tillage aids weed control by killing seedlings, burying seeds, delaying the growth of perennials, providing a rough surface that hinders seed germination, ensuring loose soil that permits effective cultivation, creating a clean uniform surface for the efficient action of herbicides, and incorporating herbicides when necessary (Richey *et al.,* 1977). Clean tillage methods such as mouldboard ploughing and disking effectively control some existing weeds. However, they also bury weed seeds that may germinate when they are returned to the surface by subsequent tillage, often a year or more later.

Reduced Tillage

Since weed control is a major reason for tillage, the need for tillage may be reduced if weeds can be controlled by other means, such as herbicides. Herbicides can effectively control weeds during at least part of the crop production cycle in a reduced tillage system. In such systems, tillage may be reduced in intensity or in frequency. Reduced tillage systems are often considered intermediate between clean and conservation tillage as regards their effects on surface residue retention or incorporation.

Some residues often remain on the surface, even where crop residues are used by livestock. In addition, the crowns and root masses of some crop plants, if brought to the surface by an appropriate tillage method, may provide additional surface materials to help protect the soil against erosion. Such materials may even provide some water conservation benefits. Surface residue retention under such conditions is usually best achieved by using a reduced tillage method. Suitable methods include stubble mulch tillage, in which the surface is undercut with sweeps or blades to control weeds and prepare a seedbed, and chisel or tine tillage, which merely loosens the soil. Disking or inversion-type ploughing should be avoided if the goal is to retain some residues on the surface. Even relatively low amounts of surface residues aid soil and water conservation, as pointed out previously. Also, the surface roughness achieved through stubble mulch, chisel or tine tillage provides conservation benefits similar to those resulting from clean tillage, with any residues retained on the surface providing additional benefits.

Complementary Practices

Use of appropriate tillage methods can improve soil and water conservation on lands that have few limitations. However, where constraints are more severe due to surface slope, poor infiltration or other factors, tillage alone may be ineffective and other practices may be needed to complement it. Some of these practices involve relatively minor changes in management, whereas others are more radical.

Contour Tillage and Graded Furrows

Contour tillage is an effective water conservation practice. Although designed primarily to control water erosion, it also helps conserve water because the ridges formed by tillage hold water on the land, providing more time for infiltration. When properly used, contouring promotes uniform water storage over the entire field.

Whereas contour tillage is intended to retain water and is thus suitable for use in semiarid to arid regions, graded furrows are used to remove it and are therefore more common in subhumid and humid areas. However, when furrow gradients are low, water conservation may benefit because runoff is slower and more uniformly distributed over the entire field.

Basin Listing

Basin listing (also called basin tillage, furrow blocking, furrow diking, furrow damming and tied ridging) is the practice of forming small earthen dams in furrows to trap rainfall, thereby preventing runoff and providing more time for infiltration (Jones and Clark, 1987). In some years, basin listing results in greater water storage and higher crop yields than when furrows are not blocked. Lack of response in other years results from: (i) inadequate rain to cause runoff, even with unblocked furrows; (ii) water loss by evaporation before it can be used by a crop; and (iii) abundant rainfall that provides adequate water, even with unblocked furrows.

Although used mainly to conserve water for dryland crops, basin listing is also a component of the low-energy precision application (LEPA) irrigation system (Lyle and Bordovsky, 1981). In LEPA, the basins trap water delivered through drop tubes and orifice-controlled emitters, resulting in irrigation application efficiencies of over 95% (Lyle and Bordovsky, 1981). In contrast, up to 30% evaporative water losses may occur in high-pressure sprinkler systems (Jones and Stewart, 1990).

Land Smoothing

The objective of land smoothing is to move soil from high to low points in a field. When low points are eliminated, water is prevented from concentrating at them. This creates more uniform storage of water in the field for use by the next crop (Singh, 1974). However, land smoothing should not eliminate small-scale surface depressions if these allow water to be stored temporarily rather than running off. Also, land smoothing may not be viable in more arid environments, where yield stability is strongly linked to the favourable soil moisture and fertility conditions associated with micro-high and -low elevations in farmers' fields (Brouwer *et al.*, 1993).

Terraces

Various types of terrace have been developed that provide soil and/or water conservation benefits. These include:

1. Level terraces with open or blocked ends to retain water on the land.
2. Graded terraces to remove excess water at a non-erosive velocity.
3. Conservation bench terraces (CBTs) to retain water and control erosion. These are special terraces in which the adjacent upslope portion of the terrace interval is also levelled. Runoff from the remaining non-levelled part of the terrace interval is captured and spread over the levelled area (Zingg and Hauser, 1959). Jones (1981) developed narrow CBTs (about 10 m wide) to reduce construction costs.
4. Level bench terraces, for which the entire interval is levelled. In some countries such terraces are primarily used on steeply sloping land, but Jones (1981) developed a level bench terrace system to conserve water for dryland crops on gently sloping land (about 1%). When a 5-m terrace interval was used, only a small amount of soil had to be moved, which greatly reduced costs.
5. Specialized terraces, often used on relatively steep slopes. These include: (i) variations of bench terraces, such as outward sloping, inward sloping (or reverse-slope) and step terraces; (ii) intermittent types, including orchard terraces, hillside ditches and lock-and-spill hillside ditches; and (iii) discontinuous parallel terraces (Hudson, 1981; Unger, 1984).

Soil Surface Amendments

Surface sealing of soils with low-stability aggregates can cause major runoff and hence severe soil erosion. If aggregates could be made more stable, the potential for runoff would be reduced, improving soil and water conservation. Runoff was sixfold less when phosphogypsum was applied at 10 t ha^{-1} to a ridged sandy soil than when it was not applied (Agassi *et al.*, 1989). Aggregate breakdown in furrows results in low infiltration and high sediment losses during furrow irrigations under some conditions. When polyacrylamide or starch copolymer solutions were injected into irrigation water at different rates, infiltration increased by 15% and sediment loss decreased by 94% (Lentz and Sojka, 1994). Injection of these materials also improved lateral infiltration and hence the movement of water and nutrients to row crops (Lentz *et al.*, 1992).

Reichert *et al.* (1994) showed that a surface application of ash from power plants reduced runoff from some bare soils. The material was a source of electrolytes and of alkalinity. Runoff was reduced more when it was applied to freshly tilled soil than to dry- or wet-crusted soil.

Conclusions

For many parts of the semiarid tropics, sustainable gains in agricultural productivity can only come from tandem increases in both grain and crop residue production. More grain will be needed for expanding human populations and for livestock feed (Winrock International, 1992). Crop residues will continue to provide vital live-stock feeds in many farming systems where few alternative feeds exist during the long dry season. Most of the plant biomass produced in these systems passes first through the rumen before becoming available as a soil amendment in the form of manure or urine. The sustainability of livestock production will depend, therefore, on increasing the feed supply from croplands, implying both increased crop residue production and better crop residue management.

Although legumes and fertilizers have potentially major roles in increasing grain and crop residue yields and enable more organic matter to be returned to soils, new management strategies will be required that minimize the competition for crop residues between animals and the requirements of soil management. The growing of forages to substitute for crop residues, the partial rather than total removal of crop residues, and an improved balance between feed supplies and animal populations will be crucial to sustainable improvements in agricultural production (Unger *et al.*, 1991).

Developing crop genotypes that produce sufficient grain and crop residues of desirable qualities remains a particular challenge for crop breeders. Many crop improvement programmes need to modify their breeding objectives to reflect farmers' multiple needs. Improved varieties continue to be rejected by farmers because they have lower stover yield and quality than traditionally cultivated geno-types.

Chemical fertilizers will play a crucial role in improving soil fertility and crop yields. However, they continue to be costly and unavailable to most farmers. The small amounts available need to be used judiciously with organic nutrient sources, such as N-fixing legumes, crop residues and manures. These agronomic interventions will be particularly important in regions having high yield potential due to favourable soil water conditions. The necessary increase in fertilizer use may require the granting of credit and/or subsidies to farmers, proper instruction in fertilizer application, the provision of fertilizer-responsive varieties, and policies that give farmers timely access to fertilizer at reasonable cost and attractive prices for the additional output obtained.

Crop residues retained on the soil surface are highly effective in controlling erosion. They also help conserve water. However, when crop residues are used for other purposes, other practices must be relied on to conserve soil and water and sustain crop production. Included are: (i) the use of appropriate tillage methods that reduce runoff, enhance water infiltration, suppress evaporation and control weeds; (ii) the use of complementary practices such as contour tillage, basin listing, land smoothing and terracing to complement tillage; and (iii) the application of amendments to stabilize the soil surface, thereby enhancing infiltration and reducing erosion.

To develop scientifically sound and socially acceptable technologies that improve agricultural productivity while protecting soil and water resources, an understanding of the key biological and socioeconomic interactions that govern crop residue use is needed. The appropriateness and adoption of improved crop residue and soil management systems will depend largely on their profitability and other advantages in the eyes of farmers, who will need to be involved in all stages of technology development and assessment.

References

Agassi, M., Shainberg, I., Warrington, D. and Ben-Hur, M. (1989) Runoff and erosion control in potato fields. *Soil Science* 148: 149-154.

Alexander, M.A. (1977) *Introduction to Soil Microbiology*. John Wiley & Sons, New York, USA.

Atta-Krah, A.N. (1990) Alley farming with leucaena: Effect of short grazed fallows on soil fertility and crop yields. *Experimental Agriculture* 26: 1-10.

Barber, S.A. (1979) Corn residue management and soil organic matter. *Agronomy Journal* 71: 625-627.

Bartholomew, W.V. (1972) *Soil Nitrogen: Supply Process and Crop Requirements*. Soil Fertility Evaluation Improvement Technical Bulletin No. 6. Department of Soil Science, North Carolina State University, Raleigh, USA.

Bationo, A. and Mokwunye, A.U. (1991) Role of manures and crop residues in alleviating soil fertility constraints to crop production, with special reference to the Sahelian and Sudanian zones of West Africa. *Fertilizer Research* 29: 117-125.

Bationo, A., Buerkert, A., Sedogo, M.P., Christianson, B.C. and Mokwunye, A.U. (1995) A critical review of crop residue use as soil amendment in the West African semiarid tropics. In: Powell, J.M., Fernández-Rivera, S., Williams, T.O. and Renard, C. (eds), *Livestock and Sustainable Nutrient Cycles in Mixed Farming Systems of sub-Saharan Africa,* vol. 2: Technical Papers. Proceedings of an International Conference, 22-26 November 1993, International Livestock Centre for Africa (ILCA), Addis Ababa, Ethiopia, pp. 305-322.

Breman, H. (1990) No sustainablity without external inputs. In: *Beyond Adjustment: Sub-Saharan Africa.* Africa Seminar, Maastricht, The Netherlands, pp. 124-134.

Brouwer, J., Burkert, A.C., Stern, R.D., Vandenbeldt, R.J. and Powell, J.M. (1993) Soil and crop growth micro-variability in the Sahel: Boon or bane for farmers and agronomists? In: Kronen, M. (ed.), *Proceedings of the Third Annual*

Conference of the SADC Land and Water Management Programme, 5-7 October 1992, Harare, Zimbabwe. Southern African Development Community, Gaborone, Botswana, pp. 167-176.

Burwell, R.E., Allmaras, R.R. and Sloneker, L.L. (1966) Structural alteration of soil surfaces by tillage and rainfall. *Journal of Soil and Water Conservation* 21: 61-63.

Charreau, C. (1977) Some controversial technical aspects of farming systems in semiarid West Africa. In: Cannell, G.H. (ed.), *Proceedings of the International Symposium on Rainfed Agriculture in Semiarid Regions,* April 1977, University of California, Riverside, USA, pp. 313-360.

Dalal, R.C. (1986) Organic matter dynamics in vertisols of South Queensland. In: *Proceedings of the Thirteenth Congress of the International Society of Soil Science,* 13-20 August 1986, Hamburg, Germany, pp. 710-711.

de Ridder, N. and van Keulen, H. (1990) Some aspects of the role of organic matter in sustainable intensified arable farming systems in the West African semiarid tropics. *Fertilizer Research* 26: 299-310.

Elliott, L.F. and Papendick, R.I. (1986) Crop residue management for improved soil productivity. In: Lopez-Real, J.M. and Hodges, R.D. (eds), *The Role of Microorganisms in a Sustainable Agriculture.* Selected Papers from the Second International Conference on Biological Agriculture, University of London, Wye College, Kent, UK. Academic Publishers, London, pp. 45-56.

Fuller, W.H., Nielsen, D.R. and Miller, R.W. (1956) Some factors influencing the utilization of phosphorus from crop residues. *Soil Science Society of America Proceedings* 20: 218-224.

Geiger, S.C., Manu, A. and Bationo, A. (1992) Changes in a sandy Sahelian soil following crop residues and fertilizer additions. *Soil Science Society of America Journal* 56: 172-176.

Gill, K.S. and Prihar, S.S. (1983) Cultivation and evaporativity effects on the drying patterns of sandy loam soil. *Soil Science* 135: 367-376.

Gill, K.S., Jalota, S.K., Prihar, S.S. and Chaudhary, T.N. (1977) Water conservation by soil mulch in relation to soil type, time of tillage, tilth and evaporativity. *Journal of the Indian Society of Soil Science* 25: 360-366.

Grove, T.L., Lathwell, D.J. and Butler, I.W. (1986) Carbon assimilation in intensively managed agricultural soils. In: *Proceedings of the Thirteenth Congress of the International Society of Soil Science,* 13-20 August 1986, Hamburg, Germany, pp. 756-757.

Hafner, H., Bley, J., Bationo, A., Martin, P. and Marschner, H. (1993a) Long-term nitrogen balance for pearl millet (*Pennisetum glaucum* L.) in an acid sandy soil of Niger. *Zeitschrift für Pflanzenernährung und Bodenkunde* 156: 164-176.

Hafner, H., George, E., Bationo, A. and Marschner, H. (1993b) Effect of crop residues on root growth and nutrient acquisition of pearl millet in an acid sandy soil in Niger. *Plant and Soil* 150: 117-127.

Hammel, J.E., Papendick, R.I. and Campbell, G.S. (1981) Fallow tillage effects on evaporation and seed zone water content in a dry summer climate. *Soil Science Society of America Journal* 45: 1016-1022.

Hudson, N. (1981) *Soil Conservation,* 2nd edn. Cornell University Press, Ithaca, New York, USA.

ILCA (1993) *Annual Report and Programme Highlights.* International Livestock Centre for Africa, Addis Ababa, Ethiopia.

Ingram, J.A. and Swift, M.J. (1989) Sustainability of cereal-legume intercrops in relation to management of organic matter and nutrient cycling. *Farming Systems Bulletin of East and Southern Africa* 4: 7-21. International Maize and Wheat Improvement Center (CIMMYT), Harare, Zimbabwe.

Jacks, G.V., Brind, W.D. and Smith, R. (1955) *Mulching.* Commonwealth Bureau of Soil Science (England) Technical Communication No. 49. Commonwealth Agricultural Bureaux (CAB), Wallingford, UK.

Jalota, S.K. and Prihar, S.S. (1979) Soil water storage and weed growth as affected by shallow tillage and straw mulching with and without herbicide in bare fallow. *Indian Journal of Ecology* 5: 41-48.

Jenkinson, D.S. and Rayner, J.H. (1977) The turnover of soil organic matter in some of the Rothamsted classical experiments. *Soil Science* 123: 298-305.

Jones, O.R. (1981) Land forming effects on dryland sorghum production in the Southern Great Plains. *Soil Science Society of America Journal* 45: 606-611.

Jones, O.R. and Clark, R.N. (1987) Effects of furrow dikes on water conservation and dryland crop yields. *Soil Science Society of America Journal* 51: 1307-1314.

Jones, O.R. and Stewart, B.A. (1990) Basin tillage. *Soil and Tillage Research* 18: 249-265.

Kouamé, C.N., Powell, J.M., Renard, C. and Quesenberry, K.H. (1993) Plant yields and fodder quality related characteristics of millet-stylo intercropping systems in the Sahel. *Agronomy Journal* 85: 601-605.

Kretzschmar, R.M., Hafner, H., Bationo, A. and Marschner, H. (1991) Long- and short-term effects of crop residues on aluminum toxicity, phosphorus availability and growth of pearl millet in an acid sandy soil. *Plant and Soil* 136: 215-223.

Larson, W.E., Clapp, C.E., Pierre, W.H. and Morachan, Y.B. (1972) Effects of increasing amounts of organic residues on continuous corn, 2: Organic carbon, nitrogen, phosphorus and sulfur. *Agronomy Journal* 64: 204-208.

Lentz, R.D. and Sojka, R.E. (1994) Field results using polyacrylamide to manage furrow erosion and infiltration. *Soil Science* 158: 274-282.

Lentz, R.D., Shainberg, I., Sojka, R.E. and Carter, D.L. (1992) Preventing irrigation furrow erosion with small applications of polymers. *Soil Science Society of America Journal* 56: 1926-1932.

Lyle, W.M. and Bordovsky, J.P. (1981) Low energy precision application (LEPA) irrigation system. *Transactions of the ASAE* 24: 1241-1245.

Magdoff, F. (1992) *Building Soils for Better Crops: Organic Matter Management.* University of Nebraska Press, Lincoln, USA.

Mannering, J.V. and Meyer, L.D. (1963) The effects of various rates of surface mulch on infiltration and erosion. *Soil Science Society of America Proceedings* 27: 84-86.

McCown, R.L., Haaland, G. and de Haan, C. (1979) The interaction between cultivation and livestock production in semiarid Africa. In: Hall, A.E., Cannell, G.W. and Lawton, H.W. (eds), *Agriculture in Semiarid Environments*. Springer-Verlag, Berlin, pp. 297-332.

McGill, N.B. and Myers, R.J.K. (1987) Controls on the dynamics of soil and fertilizer nitrogen. In: Follett, R.F., Stewart, J.W.B. and Cole, C.V. (eds), *Soil Fertility and Organic Matter as Critical Components of Production Systems*. Soil Science Society of America Special Publication No. 19, Madison, Wisconsin, USA, pp. 73-99.

McIntire, J. and Powell, J.M. (1995) African semiarid agriculture cannot produce sustainable growth without external inputs. In: Powell, J.M., Fernández-Rivera, S., Williams, T.O. and Renard, C. (eds), *Livestock and Sustainable Nutrient Cycles in Mixed Farming Systems of sub-Saharan Africa*, vol. 2: Technical Papers. Proceedings of an International Conference, 22-26 November 1993, International Livestock Centre for Africa (ILCA), Addis Ababa, Ethiopia, pp. 539-554.

McIntire, J., Bourzat, D. and Pingali, P. (1992) *Crop/Livestock Interactions in sub-Saharan Africa*. World Bank, Washington DC, USA.

Michels, K., Sivakumar, M.V.K. and Allison, B.E. (1993) Wind erosion in the southern Sahelian zone and induced constraints to pearl millet production. *Agricultural and Forest Meteorology* 67: 65-67.

Mohamed-Saleem, M.A. (1985) Effect of sowing time on grain and fodder potential of sorghum undersown with stylo in the subhumid zone of Nigeria. *Tropical Agriculture* (Trinidad) 62: 151-153.

Papendick, R.I. (1987) Tillage and water conservation: Experience in the Pacific Northwest. *Soil Use and Management* 3: 69-74.

Papendick, R.I. and Parr, J.F. (1988) Crop residue management to optimize crop/livestock production and resource conservation in the Near East region. In: Proceedings of a Workshop, 31 January-2 February 1988, Amman, Jordan.

Pichot, J., Sedogo, M.P., Poulain, J.F. and Arrivets, J. (1981) Evolution de la fertilité d'un sol ferrugineux tropical sous l'influence de fumures minérales et organiques. *Agronomie Tropicale* 36: 122-133.

Pieri, C. (1989) Fertilité des terres de savanes. In: *Bilan de Trente Ans de Recherche et de Développement au Sud du Sahara*. Centre de Coopération Internationale en Recherche Agronomique pour le Développement (CIRAD) and Ministère de la Coopération, Paris, France.

Powell, J.M. (1989) Sorghum stover removal and fertilizer nitrogen effects on sorghum yields and nutrient cycling. PhD thesis, Texas A&M University, College Station, USA.

Powell, J.M. and Coulibaly, T. (1995) *The Ecological Sustainability of Red Meat Production in Mali: Nitrogen Balance of Rangeland and Cropland in Four*

Production Systems. Report to the Projet de Gestion des Ressources Naturelles (PGRN), Bamako, Mali.

Powell, J.M. and Fussell, L.K. (1993) Nutrient and carbohydrate partitioning in millet. *Agronomy Journal* 85: 862-866.

Powell, J.M. and Hons, F.M. (1992) Sorghum stover removal effects on soil organic matter content, extractable nutrients and crop yield. *Journal of Sustainable Agriculture* 2: 25-39.

Powell, J.M. and Ikpe, F. (1992) Fertilizer factories. *ILEIA Newsletter* 8: 13-14. Information Service for Low-External-Input Agriculture, Leusden, The Netherlands.

Powell, J.M. and Williams, T.O. (1993) *Livestock, Nutrient Cycling and Sustainable Agriculture in the West African Sahel.* Gatekeeper Series No. 37. International Institute of Environment and Development (IIED), London, UK.

Powell, J.M., Hons, F.M. and McBee, G.G. (1991) Nutrient and carbohydrate partitioning in sorghum stover. *Agronomy Journal* 83: 933-937.

Power, J.F., Doran, J.W. and Wilhelm, W.W. (1986) Uptake of nitrogen from soil, fertilizer and crop residues by no-till corn and soybeans. *Soil Science Society of America Journal* 50: 137-142.

Reed, J.D., Capper, B.S. and Neate, P.J.H. (eds) (1988) *Plant Breeding and the Nutritive Value of Crop Residues.* Proceedings of a Workshop, 7-10 December 1987, International Livestock Centre for Africa (ILCA), Addis Ababa, Ethiopia.

Reichert, J.M., Norton, L.D. and Huang, Chi-hua. (1994) Sealing, amendment and rain intensity effects on erosion of high-clay soils. *Soil Science Society of America Journal* 58: 1199-1205.

Richey, C.B., Griffith, D.R. and Parsons, S.D. (1977) Yields and cultural energy requirements for corn and soybeans with various tillage-planting systems. *Advances in Agronomy* 29: 141-182.

Sandford, S. (1989) Integrated cropping/livestock systems for dryland farming in Africa. In: Unger, P.W., Sneed, T.V., Jordan, W.R. and Jensen, R. (eds), *Challenges in Dryland Agriculture: A Global Perspective.* Proceedings of the International Conference on Dryland Farming, Texas Agricultural Experiment Station, College Station, USA, pp. 861-872.

Sauerbeck, D.R. and Gonzalez, M.A. (1977) Field decomposition of carbon-14-labelled plant residues in various soils of the Federal Republic of Germany and Costa Rica. In: *Soil Organic Matter Studies,* vol.1. Proceedings of a Symposium, 6-10 September 1976, Braunschweig, Austria. International Atomic Energy Agency (IAEA), Food and Agricultural Organization of the United Nations (FAO) and Agrochimica, Vienna, pp.159-170.

Singh, G. (1974) The role of soil and water conservation practices in raising crop yields in dry farming areas of tropical India. FAO/UNDP International Expert Consultation on the Use of Improved Technology for Food Production in Rainfed Areas of Tropical Asia. Mimeo, Food and Agriculture Organization of the United Nations (FAO), Rome.

Singh, R.P., Ong, C.K. and Saharan, N. (1988) Microclimate and growth of sorghum and cowpea in alley cropping in semiarid India. In: Unger, P.W., Jordan, W.R., Sneed, T.V. and Jensen, R.W. (eds), *Challenges in Dryland Agriculture.* Proceedings of the International Conference on Dryland Agriculture, August 1988, Amarillo/Bushland, Texas. Texas Agricultural Experiment Station, College Station, USA, pp. 163-169.

Somda, Z.C., Powell, J.M., Fernández-Rivera, S. and Reed, J.D. (1995) Feed factors affecting nutrient excretion by ruminants and fate of nutrients when applied to soil. In: Powell, J.M., Fernández-Rivera, S., Williams, T.O. and Renard, C. (eds), *Livestock and Sustainable Nutrient Cycles in Mixed Farming Systems of sub-Saharan Africa,* vol. 2: Technical Papers. Proceedings of an International Conference, 22-26 November 1993, International Livestock Centre for Africa (ILCA), Addis Ababa, Ethiopia, pp. 227-246.

SSSA (1987) *Glossary of Soil Science Terms.* Soil Science Society of America, Madison, Wisconsin, USA.

Stoorvogel, J.J. and Smaling, E.M.A. (1990) *Assessment of Soil Nutrient Depletion in sub-Saharan Africa: 1983-2000.* The Winand Staring Centre, Wageningen, The Netherlands.

Swift, M.J., Frost, P.G.H., Campbell, B.M., Hatton, J.C. and Wilson, K. (1989) Nutrient cycling in farming systems derived from savanna: In: Clarholm, M. and Berstrom, D. (eds), *Ecology of Arid Lands: Perspectives and Challenges.* Kluwer Academic Publishers, Amsterdam, The Netherlands, pp. 63-76.

Tarawali, G. and Mohamed-Saleem, M.A. (1995) The role of forage legume fallows in supplying improved feed and recycling nitrogen in subhumid Nigeria. In: Powell, J.M., Fernández-Rivera, S., Williams, T.O. and Renard, C. (eds), *Livestock and Sustainable Nutrient Cycles in Mixed Farming Systems of sub-Saharan Africa,* vol. 2: Technical Papers. Proceedings of an International Conference, 22-26 November 1993, International Livestock Centre for Africa (ILCA), Addis Ababa, Ethiopia, pp. 263-276.

Unger, P.W. (1978) Straw-mulch rate effect on soil water storage and sorghum yield. *Soil Science Society of America Journal* 42: 486-491.

Unger, P.W. (1984) *Tillage Systems for Soil and Water Conservation.* FAO Soils Bulletin No. 54. Food and Agriculture Organization of the United Nations, Rome, Italy.

Unger, P.W., Stewart, B.A., Parr, J.F. and Singh, R.P. (1991) Crop residue management and tillage methods for conserving soil and water in semiarid regions. *Soil and Tillage Research* 20: 219-240.

van Duivenbooden, N. (1992) *Sustainability in Terms of Nutrient Elements with Special Reference to West Africa.* Agrobiological Research Centre (CABO-DLO), Wageningen, The Netherlands.

van Faasen, H.G. and Smilde, K.W. (1985) Organic matter and nitrogen turnover in soils. In: Kang, B.T. and van der Heide, J. (eds), *Nitrogen Management in Farming Systems in Humid and Subhumid Tropics.* Institute for Soil Fertility and International Institute of Tropical Agriculture (IITA), Haren, The Netherlands.

Waghmare, A.B. and Singh, S.P. (1984) Sorghum/legume intercropping and the effects of nitrogen fertilization, 1: Yield and nitrogen uptake by crops. *Experimental Agriculture* 20: 251-259.

Williams, T.O., Powell, J.M. and Fernández-Rivera, S. (1995) Manure utilization, drought cycles and herd dynamics in the Sahel: Implications for cropland productivity. In: Powell, J.M., Fernández-Rivera, S., Williams, T.O. and Renard, C. (eds), *Livestock and Sustainable Nutrient Cycles in Mixed Farming Systems of sub-Saharan Africa,* vol. 2: Technical Papers. Proceedings of an International Conference, 22-26 November 1993, International Livestock Centre for Africa (ILCA), Addis Ababa, Ethiopia, pp. 393-410.

Wilson, G.F., Kang, B.T. and Mulongoy, K. (1986) Alley cropping: Trees as sources of green manure and mulch in the tropics. *Biological Agriculture and Horticulture* 3: 251-267.

Winrock International (1992) *Assessment of Animal Agriculture in Africa.* Morrilton, Arkansas, USA.

Woodruff, N.P. and Siddoway, S.R. (1973) Wind erosion control. In: *Conservation Tillage.* Proceedings of a National Conference, March 1973, Des Moines, Iowa. Soil Conservation Society of America, pp. 156-162.

Zingg, A.W. and Hauser, V.L. (1959) Terrace benching to save potential runoff for semiarid land. *Agronomy Journal* 51: 289-292.

12. Crop Residues for Feeding Animals in Asia: Technology Development and Adoption in Crop/Livestock Systems

C. Devendra
International Livestock Research Institute, PO Box 30709, Nairobi, Kenya
Contact address: 8 Jalan 9/5, 46000, Petaling Jaya, Selangor, Malaysia

Abstract

This paper discusses the importance of crop residues for feeding animals in Asia, with reference to their relevance as the main source of feed in crop/livestock production systems, priorities for their use, the availability of cereal straws in terms of quantity produced and uses, and opportunities for more intensive utilization as feeds. The availability of these feeds depends on type of agroecosystem, cropping patterns and intensity, type and concentration of animal species, and prevailing animal production systems. The paper reviews efforts to develop and transfer technology, with reference to crop morphology and nutritive value, urea-ammonia pre-treatment, supplementation, the use of multinutrient block licks and food-feed intercropping.

Of these, research on pre-treatment and supplementation has dominated overwhelmingly, has often been duplicated, and has had no discernible impact on small farms. The importance of on-farm testing of known technologies now far outweighs the need for basic or strategic research, and a major shift to development efforts is therefore essential. To promote more intensive use of crop residues, projects need to emphasize large-scale on-farm research and development efforts with strong institutional support. Important elements in this approach include a systems orientation, multidisciplinarity, holism, farmer participation, and congruence between productivity and sustainability.

Vigorous intervention of this kind could ensure greater efficiency and economic impact in the use of crop residues as feeds, overcoming nutrition as the major constraint to productivity. The result could be increased livestock production and a better understanding of the importance of animals in sustainable crop/livestock systems.

Introduction

A recent assessment of the major constraints to animal production across the world's developing regions and the opportunities for obtaining increased productivity and more efficient performance from animals concludes that nutrition is by far the most important factor (ILRI, 1995). This conclusion reaffirms the significance of what has long been an acute problem throughout Asia. In this region as elsewhere, in seeking to increase the productivity of animals it is especially important to make full use of available feeds, to aim for a realistic potential level of production and to identify objectives clearly in terms of production and profitability. In this context, crop residues provide enormous opportunities for improving the current level of production from ruminants.

It is important to keep the concept of crop residues in perspective. While the term generally refers to various by-products from crop cultivation, we should avoid lumping all residues together. Each residue, and even each fraction of certain residues, is different in terms of its availability, its nutritive value and its potential impact in relation to the overall feeding system. We can, however, distinguish one major group of residues, the fibrous crop residues (FCRs), which have in common their high biomass, and their low crude protein and high crude fibre contents, of approximately 3-4% and 35-48% respectively. These FCRs form the base of feeding systems for ruminants throughout the developing countries, and include all cereal straws, sugarcane tops, bagasse, cocoa pod husks, pineapple waste and coffee seed pulp. Complementary to FCRs are those crop residues that are more nutritious and can therefore be used judiciously to improve the overall diet. This category includes a variety of oilseed cakes and meals, such as coconut cake, palm kernel cake, cottonseed cake and sweet potato vines, which are often used as dietary supplements.

Among crop residues, cereal straws have been at the forefront of research and development efforts. Their place was assured by two factors. First, large quantities of rice and wheat straws were produced in multiple cropping systems as a result of the Green Revolution, which spread through South and Southeast Asia from the late 1960s onwards. Second, the potential of alkali treatment, notably urea, to increase intake and nutritive value generated considerable interest throughout the developing world. The fact remains, however, that evidence for the delivery of this technology to farmers and its adoption by them is extremely limited (Owen and Jayasuriya, 1989; Devendra, 1991; Dolberg, 1992).

The purposes of this paper are to review past efforts to improve the utilization of crop residues, to identify the reasons why technology application and adoption have been poor, to spell out the lessons that can be (but are not invariably) learnt, and to contribute to the development and transfer of technology that is truly appropriate to mixed crop/livestock production systems. Cereal straws will be given special emphasis because of their near universal availability and widespread use, but I will also refer to other important residues. I will start with a discussion of the feed base.

The Feed Base

The principal determinants of the types of crop grown and animals reared in any particular location are the prevailing agroecological conditions (Duckham and Masefield, 1970; Spedding, 1975; Ruthenberg, 1980; Seré and Steinfeld, 1995). Climate especially, and to a lesser extent soil, affect the natural vegetation and influence farmers' choice of crops. The crops in turn determine the feed base and its quantity, quality and dispersion. The feed resources provide a direct link between crops and animals, and the interaction of the two, together with the disease challenge, largely dictate the development of mixed systems. The genesis of mixed crop/livestock systems is illustrated in Fig. 12.1.

The important influence of the agroecosystem on the feed base is reflected, for example, in the amount of rice straw produced. In the more favourable lowland rainfed and irrigated areas, multiple cropping involving two to three crops of rice a year generates about two to three times the amount of straw available in upland areas, where only one rice crop is grown. This is the case throughout humid Southeast Asia. In India, Singh *et al.* (1995) have reported that the proportions of rice grown in lowland irrigated, lowland rainfed and upland areas were 42%,

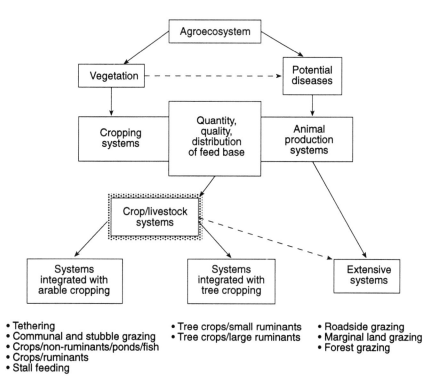

Fig.12.1. Genesis of mixed farming systems in Asia.

42 and 10% respectively, again with implications for the amount of rice straw produced. The availability of such feeds, together with the animal species raised, will determine the type and intensity of the crop/livestock system that will develop. The challenge for research is to ensure that such systems are demonstrably sustainable and have positive economic and environmental impacts. Examples of successes in this area have recently been reviewed (Devendra, 1995a).

The positive economic and environmental impacts potentially resulting from crop/livestock interactions are numerous and include:

1. Diverse and efficient resource use.
2. Reduced risk.
3. Better use of farm labour for higher productivity and increased income.
4. Efficient use of biological and chemical energy in the system and less dependence on external inputs.
5. Improved tillage and soil conservation through the use of draught animal power compared to mechanical power.
6. The supply of dung and urine to improve soil fertility.
7. Reduced carbon dioxide and carbon monoxide emissions from the non-renewable fossil fuels used by tractors.
8. Development of sustainable production systems that recycle nutrients.
9. Development of food-feed systems, resulting in increased (and more stable) feed availability for animals, more fertile soils and hence higher food crop yields.
10. Higher economic output.
11. Greater security for farm households.

Priorities for Use

It is important to identify the priorities for using crop residues, given the range of residues available, the prevailing animal production systems and the objectives of production. Table 12.1 summarizes the three categories of crop residue, their nutrient potential and the animal species that make the best use of them.

Good-quality crop residues have a high nutrient potential because of their high energy, protein and mineral contents. Thus they should be used mainly in production rations for non-ruminants (pigs, poultry and ducks) and also for ruminants producing milk and meat. Good examples of this type of residue are high-quality oilseed cakes and meals, such as soyabean meal, groundnut cake and cassava leaf meal. Medium-quality crop residues are also useful for promoting production in both ruminants and non-ruminants. Examples are coconut cake, palm kernel cake and sweet potato vines.

The third category consists of low-quality residues, synonymous with FRCs as defined above, among which the cereal straws are quantitatively the most important. Other FCRs include cereal stalks, legume haulms and cereal stovers. These bulky fibrous crop residues provide the main source of energy for the

Table 12.1. Priorities for crop residue use by animals in Asia.

Type of residue	Nutrient potential	Species (product/service)[1]
Good-quality (e.g. oilseed cakes and meals, cassava leaves)	High-protein, high-energy supplement, minerals	Pigs, chickens, ducks, ruminants (milk, meat)
Medium-quality (e.g. coconut cake, palm kernel cake, sweet potato vines)	Medium-protein	Pigs, chickens, ruminants (meat, milk)
Low-quality (e.g. cereal straws, palm press fibre, stovers)	Low-protein, very fibrous	Ruminants (meat, draught), camels, donkeys, horses (draught)

[1] Ruminants refers to buffaloes, cattle, goats and sheep.

maintenance of adult animals, including those used for draught and haulage such as swamp buffalo, cattle, horses, camels and donkeys. However, it is the ruminant species (cattle, sheep and goats) that make the most efficient use of FCRs for maintenance and some production.

Egan (1989) further subdivides FCRs into three classes: (i) those relatively low in residual cell contents and high in fibre, typically more highly lignified, of low in vitro digestibility (30-40%) and low intake, usually not greatly improved by urea treatment; (ii) those relatively high in cell wall content, not as extremely lignified, supporting high in vitro organic matter digestibilities (IVOMD) (50-60%) and intake; and (iii) those in the intermediate digestibility range (40-50%), with low cell wall contents, but with cell walls of variable digestibility only broadly related to lignin content, and capable of variable improvement by urea treatment.

It is equally important, therefore, that in defining priorities for the use of crop residues we should take into account not merely the differences between groups but also within them, and especially within the FCRs. Priorities for the use of FCRs will depend on the quantities available, the relative nutritive values, the potential value to individual ruminant species, the state of knowledge regarding their use to enhance animal production, and the potential for technology transfer and application.

Availability of Cereal Straws

Quantity Produced

Table 12.2 represents an attempt to quantify the amounts of cereal straw produced from the major cereals grown in the Asian region. These have been calculated

Table 12.2. Estimated quantities (million t) of cereal straws produced in Asia and availability per ruminant livestock unit[1].

| Country | Cereal | | | | | | Total | Contribution (%) | Total RLUs | Availability per RLU (million t) |
	Maize	Millet	Rice	Sorghum	Wheat					
Bangladesh	-	-	26.6	-	0.9		27.3	3.1	22.4	1.2
Bhutan	0.2	-	0.1	-	-		0.4	0.1	10.3	-
China	151.7	5.7	179.4	12.2	91.0		490.0	49.5	108.2	4.1
India	15.6	10.0	107.5	23.0	54.0		210.1	23.6	231.2	0.9
Indonesia	12.6	-	43.6	-	-		56.2	6.3	11.7	4.8
Japan	-	-	12.9	-	1.0		13.8	1.6	3.8	3.6
Kampuchea	0.2	-	2.1	-	-		2.3	0.3	21.9	0.1
Korea DPR	6.0	-	6.4	0.4	0.9		13.7	1.5	1.0	13.7
Korea Rep.	0.2	-	8.2	-	-		8.4	0.5	1.6	1.0
Laos	0.1	-	1.2	-	-		1.3	0.1	1.7	0.8

Table 12.2 continued.

Country	Cereal					Total	Contribution (%)	Total RLUs	Availability per RLU (million t)
	Maize	Millet	Rice	Sorghum	Wheat				
Malaysia	0.1	-	1.7	-	-	1.8	0.2	0.7	2.6
Myanmar	0.5	0.2	13.6	-	0.2	14.5	1.6	-	-
Nepal	1.8	0.2	2.9	-	0.8	5.7	0.7	8.6	1.7
Pakistan	2.2	0.4	4.8	0.4	14.4	22.2	2.5	34.4	0.6
Philippines	9.0	-	9.5	-	-	18.5	2.1	24.2	0.8
Sri Lanka	0.1	-	1.9	-	-	2.0	0.2	2.5	0.8
Thailand	8.9	-	21.3	0.4	-	30.6	3.4	10.9	2.8
Vietnam	1.8	-	18.1	0.1	-	20.0	2.3	5.3	3.8
Total	211.0	16.5	461.8	36.5	163.2	938.8	100.0	509.0	-
Contribution (%)	23.7	1.8	52.0	4.1	18.4	100.0	-	-	-

[1] Calculated from FAO (1989).
[2] Calculated using grain:straw ratios of 1:1 for millet, rice and wheat, and 1:2 for maize and sorghum.
[3] Buffalo = 1.0, cattle = 0.8, goats = 0.1 and sheep = 0.1.

using data from the Food and Agriculture Organization of the United Nations (FAO, 1989) and, in particular, those for grain yield. The following grain:straw ratios were used: millet, rice and wheat, 1:1; maize and sorghum, 1:2. These ratios represent average values; specific cases will of course differ due to variety, crop management and location. Singh *et al.* (1995), for example, have reported that, in India, the grain:straw ratio for rice varied from 1:1.3 to 1:3. The extent of such differences is often not clear, however, such that further field studies, documentation and a standardized approach may be needed before a more accurate quantitative assessment can be made.

The table shows that about 939 million t of cereal straws are the principal sources of FCRs in the region. This figure probably accounts for about 94% of the total supply of cereal straws. Rice, grown in irrigated, rainfed lowland and upland conditions, contributes about half the total supply.

The last column of the table shows cereal straw availability per ruminant livestock unit (RLU). For the region as a whole, this is 1.8 million t RLU^{-1}. This seems excessive and suggests a large surplus, but actual utilization is far below this level. In some countries, such as Korea and Indonesia, availability is quite high— suggestive again of the high potential value for ruminant production—whereas in other countries, such as India and Pakistan, the value is much lower, reflecting a higher level of utilization in a feed-deficit situation, especially in the more marginal lands.

Uses of Rice Straw

Data on the different uses of rice straw are extremely sparse (Table 12.3). One of the few studies on this subject is from South Korea (Im and Park, 1983). It indicated that feed was the third most important use, at 15% of the total, after fertilizer (46%) and fuel (20%). In Thailand, it was reported that 75% of the rice straw from rainfed upland farms and 82% in the lowland farms is collected by farmers for use as feed (Wanapat, 1990). In Bangladesh, Saadullah *et al.* (1991) found that 47% of rice straw was used as animal feed.

For most countries in Asia, use of rice straw for feed is likely to be very much higher than the value from South Korea. The reason is that probably all buffalo, a large proportion of draught and beef cattle and small numbers of small ruminants subsist mainly on cereal straws for maintenance. Rice straw is especially important during periods when other feeds are scarce. These periods are of two kinds: one occurring regularly from time of planting through to harvesting, when animals have no or limited access to grazing, and the other arising whenever drought or some other cause of crop losses strikes. The maximum intake of straw per 100 kg liveweight is about 1.0 to 1.2 kg.

For Southeast Asia, utilization, which is mainly by swamp buffalo and cattle, was determined as follows:

1. Adult liveweights of swamp buffaloes and cattle are 350 and 200 kg respectively.
2. Average intake is 1 kg of rice straw per 100 kg liveweight. This amounts to a total intake of 1.28 and 0.73 t annually for buffaloes and cattle respectively.
3. The requirements for existing populations of swamp buffaloes and cattle are therefore 51.2 and 89.4 million t respectively, giving a total requirement of 140.6 million t.
4. This represents 30.4% of the total availability of rice straw produced annually.

The figure is somewhat underestimated as it precludes use by goats, sheep and other herbivores. However, it is clear that, in this as in other subregions, rice straw availability is well in excess of the current needs of ruminants, reflecting considerable underutilization and hence major opportunities for promoting more intensive use.

Available Technologies

Increasing the efficiency with which crop residues are utilized by livestock has been a major theme of technology development, transfer and adoption since the 1970s. In the case of cereal straws, the central objective of technology development is to overcome their inherent nutritional limitations: low digestibility, low crude protein (cell and cell wall contents), poor palatability and sheer bulk.

There are several complementary means to this end, of which the following five are the most relevent: (i) interventions associated with morphology and nutritive value; (ii) pre-treatment with urea-ammonia; (iii) supplementation with other feeds (energy, nitrogen and minerals); (iv) use of multinutrient block licks; and (v) introduction of food-feed intercropping.

Morphology and Nutritive Value

The feed value of a crop residue depends on the biomass produced, its intake by animals and its digestibility. It is generally represented in the equation $I = D \times E$, where:

I = voluntary feed intake,
D = digestibility of the feed eaten, and
E = efficiency of extraction of nutrients during digestion.

Maximizing intake is the first prerequisite. This is linked to the potential of a feed to supply, through microbial degradation, quantities of nutrients which, together with those provided through supplementation, can produce a balanced diet.

The main determinants of intake and digestibility in FCRs are the morphological characteristics of the plant. These are: (i) the proportions of different plant parts (leaves, stems and stubble); (ii) the composition and proportions of

Table 12.3. Utilization of rice straw in Asia.

Country/region	Purpose	% of total availability	Reference
Bangladesh	Feed	74.4	Saadullah *et al.* (1991)
Korea	Feed Fertilizer Fuel	15.0 46.0 20.0	Im and Park (1983)
Thailand	Feed	75-82	Wanapat (1990)
China	Feed	25.0	Wang (1996)
Southeast Asia[1]	Feed	30.4	Present study (1996)

[1]The countries involved are Cambodia, China, Indonesia, Laos, Malaysia, Mongolia, Myanmar, The Philippines, Thailand and Vietnam.

different cell types in the various plant parts; (iii) the relative amounts of cell contents and cell walls in those tissues; and (iv) the physical and chemical nature of the cell walls. These factors influence the animal's chewing behaviour and the extent of fragmentation in the reticulorumen.

There is considerable variability in the morphological characteristics within and between crop residues. The proportions of leaves and stems vary, which explains why there is no consistency in the grain:straw ratios of cereals. Pearce *et al.* (1988) reported that the IVOMD of rice straw stem varies from 40 to 75%, that of leaf sheath from 38 to 56%, and that of leaf from 45 to 60%.

Analysing the chemical composition of a residue is a first step in identifying the residual amounts of cell wall contents—a major factor determining variability in nutritive value. In cereal straws, in vitro digestibilities of dry matter of between 35 and 60% have been reported (Doyle *et al.*, 1987), with intakes ranging from 1.2 to 1.9 kg per 100 kg liveweight in cattle and buffaloes respectively (Pearce, 1986). In some rice plants, the stubble has been reported to be of higher nutritive value than the straw, yet intake of the straw by cattle still exceeds that of stubble (Wanapat and Kongpiroon, 1988); the leaf fraction of the straw is more acceptable than the more highly digestible stem fraction of the stubble. In short, digestibility is affected by palatability of the material, as well as by its nutrient content. Among the trace elements affecting digestibility is silica, the content of which in rice straw varies between genotypes and according to growing conditions (Juliano, 1988).

To sum up, it is clear that there exists considerable variability in the morphological characteristics of cereal straws. This variability influences such parameters as intake and retention time in the reticulorumen, which in turn determine the supply of nutrients for maintenance and production by animals. If the nutritive value of cereal straws can be enhanced through genetic programmes, the impact on animal production could be immense. Chesson *et al.* (1995) have recently suggested that, since pre-treatments are rarely cost-effective, the changes they introduce are better engineered into the forage plant. The extent to which improvement of the quality of cereal straws is built into the breeding programmes of the international crop research centres remains unclear. Considerable opportunities evidently exist in this area.

Pre-treatment

The alkali treatment of cereal straws has received a great deal of attention from researchers. Han and Garret (1986) have listed 26 chemical treatments that can be used to improve the quality of cereal straws. These have been variously tried in different countries (Sundstøl and Owen, 1984; Doyle *et al.,* 1986). Urea-ammonia treatment has emerged as the most effective, practical and relevant one (Schiere and Ibrahim, 1989).

The justification for this conclusion is the increased digestibility and higher intake resulting from this treatment (Chesson and Ørskov, 1984; Ghebrehiwet *et al.,* 1988). The main variables affecting the efficiency of urea-ammonia treatment are (i) level of urea; (ii) straw in long or chopped form; (iii) method of application (spraying or impregnation); (iv) moisture content of the straw; (v) storage (open or closed); (vi) method of feeding (with or without additional ingredients); (vii) species of animal; and (viii) objective of production.

The value of individual cereal straws in effective feeding systems can only be defined in terms of their ability to promote a consistently good response in animals. Response will, however, vary according to species and function, with services such as the provision of manure and traction having a value in addition to the more obvious (and measurable) products, milk and meat. In some systems, particularly pastoral ones, the mere maintenance of animals through the dry season has a value, at least in the eyes of their owners. Moreover, beyond the evaluation and testing phase at station level, the final value of the straw can only be realized through the adoption of new technologies at farm level. On-farm trials to determine and demonstrate responses among the farming community are therefore especially important, but unfortunately such experiments are few and far between compared to the numerous studies that have focused mainly on evaluation on the research station.

One example of a large growth response study is shown in Table 12.4, which reproduces data from Sri Lanka. A total of 36 heifers, 12 each of Sahiwal, Sahiwal x local and Jersey x local breeds were given one of two treatments: either rice straw upgraded with urea, or rice straw supplemented with urea. The upgrading consisted of impregnating a solution of 4 kg urea 100 l^{-1} of water into 100 kg of air-dried straw, which was then stored in sealed polythene bags for 9-11 days. In the supplemented treatment the urea was simply added in the feed trough at the rate of 2 kg urea 100 kg^{-1} of air-dried straw.

Growth rates were better when straw was upgraded rather than merely supplemented (217 g head^{-1} day^{-1}, compared with 718 g head^{-1} day^{-1}). The improvement was associated with a higher intake of upgraded straw (2.4 kg 100 kg^{-1} liveweight compared with 1.8 kg). No genotypic differences were found between the different animal groups.

These results are similar to those reported previously by Jaiswal *et al.* (1983) and Perdok *et al.* (1984). They imply higher dry-matter digestibility and nitrogen content obtained through upgrading. The economic implications, including whether urea treatment is indeed justified, depend on the cost of inputs and the price of beef.

In sum, a review of past research on urea-ammonia treatment leads to the following conclusions:

1. There has been far too much attention to the treatment of rice straw per se.
2. This has been at the expense of adequate attention to supplementation strategies.
3. Much of the work has been confined to the research station, with few attempts to transfer the technology through large-scale on-farm testing.
4. The failure to demonstrate cost-effective results has discouraged farmer adoption.

Table 12.4. Effect of animal breed and straw treatment on feed intake and liveweight gain in heifers receiving urea-treated or urea-supplemented straw.

	Sahiwal		Sahiwal x local		Jersey x local	
	U[1]	S	U	S	U	S
Liveweight gain[2] (g day^{-1})	282a	105bc	185b	70c	-	39c
DM intake (kg 100 kg^{-1} BW):[3]						
Straw	2.33a	1.89b	2.49a	1.83b		1.70b
Grass	0.13	0.14	0.25	0.26		0.25
Rice bran	0.29	0.31	0.54	0.56	-	0.54
Total	2.75	2.34	3.27	2.65	-	2.49

[1] U = upgraded; S = supplemented.
[2] Values followed by the same letter (a, b, c) are not significantly different.
[3] Estimated.
Source: Schiere and Wieringa (1988).

Supplementation

Supplementation is justified because the use of FCRs demands it, and because of the feed shortages that commonly occur in so many Asian countries. Feeding on cereal straws (and natural grazing) alone results in reduced liveweight and perpetual low productivity in most animals. It also leads to delayed age at first parturition, increased interval between parturitions, increased non-productive life of the animal, and high mortality.

One solution to these problems is strategic supplementation with feeds that provide additional energy, proteins and minerals. Several alternative supplementation strategies have been pursued, the commonest being the use of purchased protein supplements such as cotton- or oilseed cakes, and the installation of multinutrient liquids or block licks, often consisting of molasses and minerals. Also gaining in importance is the introduction of leguminous fodder trees into cropping systems, with species such as *Leucaena leucocephala* proving especially popular. These supplements serve two essential functions: (i) to promote efficient microbial growth in the rumen; and (ii) to increase protein supply for digestion in the small intestines (bypass or rumen non-degradable proteins).

One impressive example of research on supplementation comes from Henan Province, China, where cottonseed cake (CSC) was used to develop beef production. Dolberg and Finlayson (1995) report the following results of this work:

1. The projected project output of 3000 t of treated straw (with 3% anhydrous ammonia or 5% urea) was exceeded by an estimated 4000% in the counties covered by the project.
2. Cattle were finished quicker, reaching a slaughter weight of 450 kg faster due to CSC supplementation.
3. A supplement of 2 kg CSC day^{-1} gave daily growth rates of 600-700 g in 2-year-old crossbred cattle. The feeding period was halved from 36 months to 18 months when 0.5 kg CSC day^{-1} was fed. At 4 kg CSC day^{-1}, the feeding period was further reduced to 10.2 months.
4. Economic analysis indicated considerable financial benefits to farmers. Feeding urea-treated straw alone gave an additional profit of US$ 26 head^{-1}. With CSC supplementation, this increased tenfold.
5. The most profitable rate of supplementation was 2 kg CSC day^{-1} (Table 12.5).
6. Cottonseed cake supplementation significantly reduced the dependence on more expensive purchased concentrates.
7. Increased beef production making optimum use of local feeds is clearly a viable activity on small-scale farms with good market access.

Table 12.6 illustrates the value of block licks in recent studies conducted at village level in India. Even when buffalo and cattle were fed considerable amounts of green forage, millet straw and a concentrate, supplementation with a urea-molasses nutrient block improved milk yield by an average of 30%.

Table 12.5. Returns to feeding urea-treated straw and cottonseed cake in Henan Province, China.

Variable	Amount of cottonseed fed (kg day^{-1})[1]					
	0	0.5	1	2	3	4
Feeding period (months)	36	18	15	12	10.7	10.2
Daily gain (m)	250	500	600	750	845	883
Costs (US$ head^{-1}):						
Non-feed	161	137	133	129	127	127
Feed	106	69	70	77	88	100
Revenue (US$ head^{-1})	293	293	293	293	293	293
Net returns (Y):						
Head	26	87	90	87	77	66
Day	0.03	0.16	0.20	0.24	0.24	0.21
Marginal rate of return (%)[2]		103	52	36	0	

[1] Protein content = 43%.
[2] Calculated by dividing the change in net returns by the change in costs per day between adjacent treatments.
Source: Dolberg and Finlayson (1995).

The use of leguminous fodder trees as supplements is probably under-estimated, given the wide range and diversity of these species available throughout the tropics. Despite this diversity, research and development activities have tended to focus on a relatively narrow range of species, with *Leucaena leucocephala* and *Gliricidia sepium* in Asia and Africa, and *Erythrina* spp. in Latin America, monopolizing most of the efforts.

The many advantages of leguminous fodder trees have been summarized by Devendra (1988, 1993) and include: (i) availability on farms or in surrounding countryside; (ii) accessibility to farmers; (iii) flexible use (browsing/cut-and-carry); (iv) the provision of variety in the diet; (v) the provision of dietary nitrogen, energy, minerals and vitamins; (vi) their laxative influence on the alimentary system; (vii) the reduced requirement for purchased concentrates resulting from their use; and hence (viii) the reduction in the cost of feeding they permit.

Table 12.6. Average feed intake (kg day[-1]) over a 6-month trial in India with Mehsani buffalo and crossbred cattle fed according to village management, with and without supplementation with a urea-molasses multinutrient block.

	Village (number of animals)				
	Boriavi[1] (36)	Ejipura (28)	Bhesana (34)	Dholasan (48)	Fatehpura (28)
Green fodder:[2]					
Without[3]	25.5	21.5	13.2	16.7	22.4
With	25.9	22.6	13.2	16.8	25.1
Dry fodder:[4]					
Without	6.4	8.6	7.0	7.1	7.7
With	6.7	8.7	7.6	7.3	9.1
Concentrate:[5]					
Without	7.3	3.6	3.4	7.2	6.2
With	7.2	3.6	3.3	7.2	6.9
Total DM:[6]					
Without	22.5	10.6	14.6	19.6	21.5
With	22.9	20.2	15.1	19.8	24.4
Supplement (g day[-1])	204	204	210	135	330

[1] Four cows were included in this group.
[2] Usually sorghum, roadside grass or Napier grass fed at a constant rate.
[3] Without = without supplementation; with = with supplementation.
[4] Mainly millet straw, fed ad libitum.
[5] Local product consisting of agroindustrial by-products (18% crude protein).
[6] Assuming 25% dry matter in green forage and 90% in concentrate and straw.
Source: Manget-Ram *et al.* (1992).

The benefits of including leguminous tree forages in the diet have not been adequately demonstrated to farmers, and efforts in this area should therefore be vigorously intensified. Table 12.7 summarizes the benefits of supplementing various FCRs, notably rice straw, with leucaena. This research was conducted over a number of years by various authors in The Philippines and Thailand.

Table 12.7. Benefits of supplementing cattle diets with leucaena.

Feeding regime	Forage supplement	Country	Significant response[1]	Result	Reference
Rice straw	Leucaena leaf meal	Philippines	LW gain	Reduced costs	Sevilla *et al.* (1976)
Sugarcane tops	Leucaena	Philippines	LW gain	-	Lopez *et al.* (1981)
Rice straw	Leucaena	Philippines	LW gain	-	Le Trung *et al.* (1983)
Urea-treated rice straw	Leucaena	Thailand	Milk	Reduced costs	Promma *et al.* (1984)
Rice straw	Leucaena	Thailand	LW gain	Reduced costs	Cheva-Isarakul and Potikanond (1985)
Rice straw + dried poultry litter + concentrate	Leucaena leaf meal	Philippines	LW gain	Reduced costs	Le Trung *et al.* (1987)

[1] LW = liveweight.

Food-feed Intercropping

Food-feed intercropping is not a new concept, but rather the extension and intensification of the traditional intercropping systems long practised in the region's more favoured environments. New technology, in the form of shorter-duration crops and new or improved irrigation schemes, has contributed greatly to this process. At first, scientists tended to view such systems solely from the point of view of their potential contribution to human food supplies in the short term. Nowadays the perspective also includes the system's contribution to animal feeds, and its sustainability over the long term.

The strategy is to integrate other food- and feed-producing crops within the rice cropping system, principally through intercropping or relay cropping. Figure 12.2 shows a system of this kind in The Philippines. The crops included are often food or dual-purpose legumes, such as mungbean, pigeonpea or cowpea, while other food crops such as sorghum may be included alongside rice or maize. In South Asia, similar rice-based systems involve lentil, chickpea, groundnut and lathyrus. The aim in each case is to diversify the diet and spread the risk of crop failure, notably by warding off pests and diseases and avoiding losses caused by

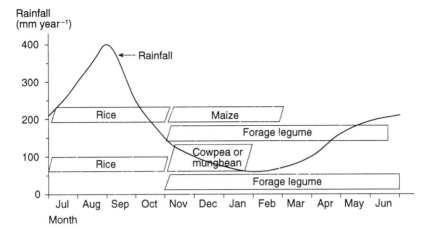

Fig. 12.2. Rice-based intercropping system in The Philippines.

drought. Other criteria for the choice of intercrops include their compatibility in terms of competition for light, nutrients and moisture, their contribution to soil fertility, their potential forage or crop residue biomass yield, the type of animals to be fed, the ease with which they can be eradicated and the labour requirements for planting, weeding and harvesting.

Until the new technology became available, the tendency throughout much of the region's rainfed lowlands and uplands was to grow only a single rice crop. Recent studies in the rainfed lowlands of Pangasinan in The Philippines have shown that rice-mungbean has now largely replaced rice-fallow (Devendra, 1995b). Intercropping with Siratro and use of the last cutting as green manure resulted in higher yields of the succeeding crop, in addition to the improved animal feed supplies derived from earlier cuttings (Sevilla *et al.*, 1995).

The Need for On-farm Testing

The Current Situation

Research on nutrition has been at the forefront of animal science in Asia for several decades—justifiably so because it remains the major constraint to increased animal production in the region. Considerable research has gone into the identification and/or further development of suitable feeds, especially cereal straws, and there has been no lack of on-station feeding trials to evaluate nutritive value and animal responses. The results, which have often been impressive, have been reported at the many meetings held on this subject in different countries throughout the region (Table 12.8).

Table 12.8. Meetings on animal feed resources and feeding systems held in Asia, 1977-1995[1].

Country	Year
Bangladesh	1981, 1982,1983
China	1993, 1995
India	1986, 1987, 1988a, 1988b, 1993, 1994a, 1994b
Indonesia	1981, 1984, 1985, 1987, 1992
Japan	1990, 1991
Malaysia	1977, 1982, 1987
Pakistan	1983
Philippines	1983, 1987, 1989, 1995
Sri Lanka	1984, 1985, 1986
Thailand	1984, 1985, 1988
Vietnam	1993

[1] See reference list, under country and date, for details of proceedings.

Let us look more closely at this table. Over the period 1977 to 1995, no less than 34 meetings, mostly regional and international, have been held in Asia. That works out at 1.7 meetings annually! It is interesting to speculate as to the impact of these meetings on production at the farm level, since this is ultimately more important than the fund of knowledge that has been accumulated and disseminated among the scientists.

As regards cereal straws, which have been the main focus of attention at virtually all these meetings, I conclude that there has been no discernible impact whatsoever at the farm level. Technologies such as urea-treated straw have simply not been adopted by farmers.

The key factors associated with this situation are:

1. Too much emphasis on the characterization and pre-treatment of cereal straws. This research is often duplicated, even within countries.
2. Work is generally confined to the research station, with few or no on-farm efforts.
3. Insensitivity to the broader need for on-farm work, with its strong client orientation, obliging scientists to conduct research in response to farmers' real needs.
4. Inadequate project design, lacking in a systems approach. When properly conducted, a systems approach to research can ensure effective priority setting in

accordance with farmers' needs, useful component research, farmer participation in technology design, on-farm evaluation with farmers, and appropriate mechanisms for technology transfer. Such elements have been almost wholly absent from research on the pre-treatment of cereal straws.

5. No clear demonstration of the economic benefits of this technology to farmers. Without such benefits, farmer acceptance and adoption are likely to remain nil.

A review of the potential of rice straw as animal feed, conducted over a decade ago by Doyle *et al.* (1986), concluded that the need for on-farm testing and demonstration far outweighed the need for further documentation of the effects of supplementation and pre-treatments. That situation has not changed today. In other words, the review's conclusion has apparently been ignored. Limited field testing with rice straw has been attempted in India, Sri Lanka and China, but farmer adoption still remains very low. No on-farm work with FCRs from sorghum, maize or finger millet has been reported.

The Way Forward

On-farm Research and Development

The above assessment leads inescapably to the conclusion that the overriding need now is for more concerted large-scale testing of technologies for improving crop residue utilization on the farm. This push needs to take into consideration three fundamentals: (i) a systems orientation; (ii) a participatory approach; and (iii) congruence between productivity and sustainability.

A systems approach is especially important. Eponou (1993) suggests the following prerequisites for the successful integration of agricultural research and technology transfer:

1. Shared goals. If all partners in the system have the same goals, this will promote unity and commonality of approach.
2. Synergy. By definition, this will result in outputs which will be more than the sum of the outputs of individual components. Creating synergy requires a high level of managerial skill and dynamism.
3. Strong leadership. A motivating, inspiring and unifying leader capable of allocating resources judiciously and rationally is essential. Setting priorities, clarifying responsibilities and ensuring the accountability of all partners are the main tasks of the leader.
4. Decision making by consensus. To avoid conflict and confusion, it is always best if possible to make decisions through consultation among the partners, keeping the process transparent.
5. Accountability. Each partner should be held accountable to clients and policy makers for the tasks he or she has agreed to perform. Appropriate evaluation and

monitoring mechanisms should be built into the research process. The criteria for resource allocations and rewards to individuals and institutions should be consistent with the goals of the research.

6. Farmers as partners. As the ultimate users of new technologies, farmers should be equal partners in the research process and should be involved in decision making at each phase, from problem diagnosis through to technology transfer.

In this context it is worth noting that the International Service for National Agricultural Research (ISNAR) reported that, in 20 case studies in seven developing countries, links between research and technology transfer were generally inefficient or non-existent in the public sector. In most of the cases studied, one or more of the above six key elements were missing (Eponou, 1993).

Figure 12.3 shows an approach to intensifying the use of crop residues as animal feeds. The approach integrates on-station and on-farm efforts, involving participation by farmers. Research is thus brought to bear on development. The

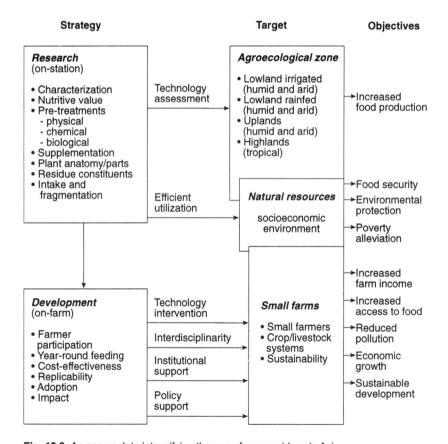

Fig. 12.3. An approach to intensifying the use of crop residues in Asia.

approach should allow us to identify the critical points of intervention necessary for the development of sustainable year-round feeding systems, including the identification of what crop residues are to be fed, how and when to feed them, how to overcome critical periods of feed shortage, and the need for strategic supplementation. The benefits, in target agroecological zones, are associated with several objectives: increased food production and food security, poverty alleviation, increased farm income, reduced pollution, and so on.

In the case of the pre-treatment of cereal straws, there must be a concerted effort to define, marshall and apply all the elements necessary for successful on-farm implementation, including monitoring and evaluation, the introduction of feedback mechanisms and the assessment of impact. If the push to transfer this technology through on-farm research fails, so that we are forced to acknowledge that the technology is unlikely to be adopted by farmers in the foreseeable future, we must then have the courage and honesty to give up this line of research and pursue other projects that have higher chances of success.

References

Bangladesh (1981) *Proceedings of an International Conference on Maximum Livestock Production from Minimum Land* (eds, Jackson, M.G., Dolberg, F., Davis, C.H., Haque, M. and Saadullah, M.), 2-5 February 1981, Bangaladesh Agricultural University, Mymensingh.

Bangladesh (1982) *Proceedings of an International Conference on Maximum Livestock Production from Minimum Land* (eds, Preston, T.R., Davis, C.H., Dolberg, F., Haque, M. and Saadullah, M.), 13-18 February 1982, Bangaladesh Agricultural University, Mymensingh.

Bangladesh (1983) *Proceedings of an International Conference on Maximum Livestock Production from Minimum Land* (eds, Davis, C.H., Preston, T.R., Haque, M. and Saadullah, M.), 2-4 May 1983, Bangaladesh Agricultural University, Mymensingh.

Chesson, A. and Ørskov, E.R. (1984) Microbial degradation in the digestive tract. *Development in Animal and Veterinary Science* 14: 305-339.

Chesson, A., Forsberg, C.W. and Grenet, E. (1995) Improving the digestion of plant cell walls and fibrous feeds. In: Journet, M., Grenet, E., Farce, M.H., Theriez, M. and Demarquilly, C. (eds), *Proceedings of the Fourth International Symposium on the Nutrition of Herbivores,* 11-15 September 1995, Clermont-Ferrand, France, pp. 271-278.

Cheva-Isarakul, B. and Potikanond, B. (1985) Performance of growing cattle fed diets containing treated rice straw and leucaena leaves compared to urea-treated rice straw. In: Doyle, P. T. (ed.), *Proceedings of a Workshop on the Utilization of Fibrous Agricultural Residues as Animal Feeds.* International Development Program of Australian Universities and Colleges (IDP), Canberra, pp. 140-144.

China (1993) *Proceedings of the International Conference on Increasing Animal*

Production with Local Resources, 18-23 October 1993, Ministry of Agriculture, Beijing.

China (1995) *Proceedings of the International Conference on Increasing Animal Production with Local Resources,* 27-30 October 1995, Ministry of Agriculture, Beijing.

Devendra, C. (1988) Forage supplements: Nutritional significance and utilization for draught, meat and milk production. In: *Proceedings of the Second World Buffalo Congress,* vol. 2, 12-17 December 1988, Indian Council of Agricultural Research (ICAR), New Delhi, India, pp. 409-423.

Devendra, C. (1991) Technologies currently used for the improvement of straw utilization in ruminant feeding systems in Asia. In: *Proceedings of the NRI/ MARDI Workshop on the Utilization of Straw in Ruminant Production Systems,* 7-11 October 1991, Malaysian Agricultural Research and Development Institute, Kuala Lumpur, pp. 1-8.

Devendra, C. (1993) Trees and shrubs as sustainable feed resources. In: *Proceedings of the Seventh World Conference on Animal Production,* vol. 1, 28 June-2 July 1993, University of Alberta, Edmonton, Canada, pp.119-138.

Devendra, C. (1995a) Mixed farming and intensification of animal production in Asia. In: *Proceedings of the ILRI/FAO Round Table on Livestock Development in Low-income Countries,* 22 February-2 March 1995, International Livestock Research Institute (ILRI), Addis Ababa, Ethiopia, pp. 133-144.

Devendra, C. (1995b) Environmental characterization of crop/animal systems in rainfed areas. In: Devendra, C. and Sevilla, C. (eds), *Crop/Animal Interaction.* Proceedings of a Workshop, 27 September-1 October 1993, University of The Phillipines, International Rice Research Institute (IRRI) and International Development Research Centre (IDRC), Los Baños, The Philippines, pp. 43-64.

Dolberg, F. (1992) Progress in the utilization of urea-ammonia treated crop residues: Biological and socio-economic aspects of animal production and application of the technology on small farms. *Livestock Research for Rural Development* 4: 20-31.

Dolberg, F. and Finlayson, P. (1995) Treated straw for beef production in China. *World Animal Review* 82: 14-24.

Doyle, P.T., Devendra, C. and Pearce, G.R. (1986) *Rice Straw as a Feed for Ruminants.* International Development Program of Australian Universities and Colleges (IDP), Canberra, Australia.

Doyle, P.T, Pearce, G.R. and Djajanegara, A. (1987) Intake and digestion of cereal straws. In: *Proceedings of the Fourth AAAP Animal Science Congress,* Department of Scientific and Industrial Research, Hamilton, New Zealand, pp. 59-62.

Duckham, A.M. and Masefield, O.D. (1970) *Farming Systems of the World.* Chatto & Windus, London, UK.

Egan, A. (1989) Living with, and overcoming limits to, feeding value of high-fibre roughages. In: Hoffman, D., Nari, J. and Pethrem, R.J. (eds), *Draught Animals in Rural Development.* ACIAR Proceedings No. 27, Australian Center for International Agricultural Research, Canberra, pp. 176-180.

Eponou, T. (1993) Integrating agricultural research and technology transfer. Special Issue on Managing Agricultural Research, *Public Administration and Development* 13: 307-318.

FAO (1989) *Production Yearbook*, vol. 43. Food and Agriculture Organization of the United Nations, Rome, Italy.

Ghebrehiwet, T., Ibrahim, M.N.M. and Schiere, J. B. (1988) Response of growing bulls to diets containing untreated or urea-treated rice straw and rice bran supplemantation. *Biological Wastes* 25: 269-280.

Han, I.K. and Garret, W.N. (1986) Improving the dry-matter digestibility and voluntary intake of low-quality roughage by various treatments: A review. *Korean Journal of Animal Science* 28: 89-96.

ILRI (1995) *Gobal Agenda for Livestock Research*. Proceedings of a Consultation (eds, Gardiner, P. R. and Devendra, C.), 18-20 January 1995, Nairobi, Kenya.

Im, K.S. and Park, Y.I. (1983) *Animal Agriculture in Korea*. Department of Agriculture, Seoul National University, Korea.

India (1986) Crop residues as livestock feeds: Factors limiting their utilization with special reference to anti-nutritional factors. In: Gupta, B.N., Prasad, T. and Walli, T.K. (eds), *Proceedings of the Fifth Conference of the Animal Nutrition Society of India*, 14-17 July 1986, Udaipur.

India (1987) *Proceedings of a Workshop on the Biological, Chemical and Physical Treatment of Crop Residues* (eds, Kiran Singh, Flegel, T.W. and Schiere, J.B.), 20-21 January 1987, Indian Council of Agricultural Research (ICAR), New Delhi.

India (1988a) *Proceedings of an International Workshop on Fibrous Crop Residues as Animal Feeds* (eds, Kiran Singh and Schiere, J.B.), 27-28 October 1988, Indian Council of Agricultural Research (ICAR), Bangalore.

India (1988b) *Proceedings of a Workshop on Expanding the Utilization of Non-conventional Feeds and Fibrous Agricultural Residues* (ed., Devendra, C.), 21-29 March 1988, Hisar. International Development Research Centre (IDRC) and Indian Council of Agricultural Research (ICAR), Singapore.

India (1993) *Feeding of Ruminants on Fibrous Crop Residues*. Proceedings of an International Workshop (eds, Kiran Singh and Schiere, J.B.), 4-8 February 1991, Indian Council of Agricultural Research (ICAR), New Delhi.

India (1994a) *National Seminar on Variation in the Quantity and Quality of Fibrous Crop Residues* (eds, Joshi, A.L., Kiran Singh and Oosting, S.J.), 8-9 February 1991, Indian Council of Agricultural Research (ICAR), New Delhi.

India (1994b) *National Seminar on Farming Systems Research for Improving Livestock Production and Crop Residue Utilization* (eds, Singh, B., Rao, S.V.N. and Jain, D.K.), 24-26 November 1994, Indian Council of Agricultural Research (ICAR), New Delhi.

Indonesia (1981) *Proceedings of the First ASEAN Workshop on the Technology of Animal Feed Production Utilizing Food Waste Materials*, 26-28 August 1981, Bandung.

Indonesia (1984) *Proceedings of a Workshop on Biological, Chemical and*

Physical Evaluation of Lignocellulosic Residues, 22-27 October 1994, University of Gadja Mada, Yokjakarta.

Indonesia (1985) *Proceedings of a Workshop on Ruminant Feeding Systems Utilizing Fibrous Agricultural Residues* (eds, Dixon, R.M. and Pearce, G.R.). International Development Program of Australian Universities and Colleges (IDP), Canberra.

Indonesia (1987) *Bioconversion of Fibrous Crop Residues for Animal Feed and Other Purposes.* Proceedings of a Workshop (eds, Soejono, M., Musofie, M., Utomo, R., Wardhapi, M. and Schiere, J.B.), 16-17 November 1987, Institute for Animal Production (IAP), Grati.

Indonesia (1992) *International Conference on Livestock and Feed Development in the Tropics* (eds, Ibrahim, M.N.M., de Jong, R., van Bruchem, R. and Purnomo, H.), 21-25 October 1991, Brawijaya University, Malang.

Jaiswal, R.S., Verma, M.L. and Agarwal, I.S. (1983) Effect of urea and protein supplements added to untreated and urea-treated rice straw on digestibility, intake and growth of crossbred steers. In: Davis, C.H., Preston, T.R. , Haque, M. and Saadullah, M. (eds), *Maximum Livestock Production in Minimum Land.* Proceedings of the Fourth Seminar, 2-4 May 1983, Nimaysingh, Bangladesh, pp. 26-31.

Japan (1990) *Proceedings of a Symposium on Animal Feed Resources in Asia and the Pacific,* 14-29 August 1987, Asian Productivity Organization (APO), Tokyo.

Japan (1991) *Proceedings of a Symposium on Farm-level Feeding Systems in Asia and the Pacific,* 23 July-3 August 1990, Asian Productivity Organization (APO), Tokyo.

Juliano, B.Q. (1988) Effect of silica level on some properties of rice plant and grain. In: Dixon, R. M. (ed.), *Proceedings of a Workshop on Ruminant Feeding Systems Utilizing Fibrous Agriculture Residues.* International Development Program of Australian Universities and Colleges (IDP), Canberra, pp.115-122.

Le Trung, T., Abenir, E.E., Palo, L.P., Matios, J.M. and Lapinid, R.R. (1983) Evaluation of rice straw diets supplemented with *Leucaena leucocephala* and/ or dried poultry manure. In: Doyle, P.T. (ed.), *Proceedings of a Workshop on the Utilization of Agricultural Residues as Animal Feeds.* International Development Program of Australian Universities and Colleges (IDP), Canberra, pp. 202-212.

Le Trung, T., Palo, L.P., Matios, J.M., Abenir, E.E., Lapinid, R.R. and Atega, T.R. (1987) Dried poultry manure and leucaena in rice-straw based diets for dairy cattle. In: Dixon, R.M. (ed.), *Proceedings of a Workshop on Ruminant Feeding Systems Utilizing Fibrous Agricultural Residues.* International Development Program of Australian Universities and Colleges (IDP), Canberra, pp. 199-210.

Lopez, P.H., Calub, A.D, Calub, A.C., Alferez, A.C. and Infante, L. (1981) Fresh leucaena leaves and processed sugarcane tops with concentrate supplementation in growing fattening cattle. IDRC-PCARRD Research Project Report. Philippines Council for Agricultural Resources Research and Development, Los Baños, The Philippines.

Malaysia (1977) *Proceedings of a Symposium on Feedingstuffs for Livestock in Southeast Asia* (eds, Devendra, C. and Hutagalung, R.I.), 17-19 October 1977, Malaysian Society of Animal Production, Serdang.

Malaysia (1982) *Proceedings of a Workshop on the Utilization of Fibrous Agricultural Residues as Animal Feeds* (ed., Doyle, P.T.), 3-7 May 1982, University of Melbourne, Australia. International Development Program of Australian Universities and Colleges (IDP), Canberra.

Malaysia (1987) *Advances in Animal Feeds and Feeding in the Tropics* (eds, Hutagalung, R.I., Chen, C.P., Wan Mohamed, W.E., Law, A.T. and Sivarajasingam, S.). Malaysian Society of Animal Production, Serdang.

Manget-Ram, Tripathi, A.K., Dave, A.S., Metha, A.K. and Kurup, M.P.G. (1992) Effect of Urea-molasses Block Supplementation on Buffaloes and Cattle at the Village Level. Mimeo, National Dairy Development Board, Anand, India.

Owen, E. and Jayasuriya, M.C.M. (1989) Use of crop residues as animal feeds in developing countries: A review. *Research for Development in Agriculture* 6: 124-128.

Pakistan (1983) *Proceedings of the Workshop on Least-cost Ration Formulation.* Food and Agriculture Organization of the United Nations (FAO) and Pakistan Agriculture Research Council (PARC), Islamabad.

Pearce, G.R. (1986) Possibilities for improving the nutritive value of rice straw without pre-treatment. In: Dixon, R.M. (ed.), *Proceedings of a Workshop on Ruminant Feeding Systems Utilizing Fibrous Agricultural Residues,* 1-3 April 1986. International Development Program of Australian Universities and Colleges (IDP), Canberra, pp. 101-105.

Pearce, G.R., Lee, J.A., Simpson, J.R. and Doyle, P.T. (1988) Sources of variation in the nutritive value of wheat and rice straws. In: Reed, J.D., Capper, B.S. and Neate, P.J.H. (eds), *Plant Breeding and the Nutritive Value of Crop Residues.* Proceedings of a Conference, 7-10 December, International Livestock Centre for Africa (ILCA), Addis Ababa, Ethiopia, pp. 195-229.

Perdok, H.B, Muttettuwegana, G.S., Kasschieter, G.A., Boon, H.M, van Wagengingen, N., Arumugam, V., Linders, M.G.F.A. and Jayasuriya, M.C.N. (1984) Production responses of lactating or growing ruminants fed urea-ammonia treated paddy straw with or without supplements. In: Doyle, P.T. (ed.), *The Utilization of Fibrous Agricultured Residues as Animal Feeds.* University of Melbourne, Victoria, Australia, pp. 213-230.

Philippines (1983) *Proceedings of a Workshop on the Utilization of Fibrous Agricultural Residues* (ed., Pearce, G.R.), 18-23 May 1981. Australian Development Assistance Bureau (ADAB), Canberra.

Philippines (1987) *Proceedings of a Workshop on Ruminant Feeding Systems Utilizing Fibrous Agricultural Residues* (ed., Dixon, R.M.), 1-3 April 1986. International Development Program of Australian Universities and Colleges (IDP), Canberra.

Philippines (1989) *Non-conventional Feedstuffs for Livestock and Poultry in Asia.* FFTC/SEARCA Seminar on Feed Grains Substitutes, 13-19 October 1989, Food and Fertilizer Technology Centre (FFTC) and South-East Asia Research Centre (SEARC) for Agriculture, Los Baños.

Philippines (1995) *Proceedings of the FFTC/PCARRD Workshop on the Efficient Utilization of Feed Resources,* Food and Fertilizer Technology Centre and South-East Asia Research Centre for Agriculture, Los Baños.

Promma, S., Tuibumpee, S., Ratnavenija, A., Vidhyakorn, N. and Bromert, R.W. (1984) The effects of urea-treated rice straw on growth and milk production of crossbred Holstein-Friesian dairy cattle. In: Doyle, P.T. (ed.), *The Utilization of Fibrous Agricultural Residues as Animal Feeds.* International Development Program of Australian Universities and Colleges (IDP), Canberra, pp. 88-93.

Ruthenberg, H. (1980) *Farming Systems in the Tropics,* 2nd edn. Clarendon Press, Oxford, UK.

Saadullah, M., Haq, M.A., Mandol, M., Wahid, A. and Azizul Haque, M. (1991) *Livestock and Poultry Development in Bangladesh.* Rotary Club of Mymensingh and Bangladesh Agricultural University, Mymensingh.

Schiere, J.B. and Ibrahim, M.N.M. (1989) *Feeding of Urea-ammonia Treated Rice Straw.* Pudoc, Wageningen, The Netherlands.

Schiere, J.B. and Wieringa, J. (1988) Overcoming the nutritional limitations of rice straw for ruminants: Response of growing Sahiwal and local cross heifers to urea-upgraded and urea-supplemented straw. *Asian-Australian Journal of Animal Science* 1: 209-212.

Seré, C. and Steinfeld, H. (1995) *World Livestock Production Systems: Current Status, Issues and Trends.* FAO Animal Production and Health Paper, Food and Agricultural Organization of the United Nations, Rome, Italy.

Sevilla, C.C., Perez, C.B. Jr and Gatmaitan, O.M (1976) The effect of ipil ipil *(Leucaena leucocephala* Lam. de Wit) leaves in rice straw-based rations on growth performance and carcass characteristics. *Philippine Journal of Veterinary and Animal Science* 2: 21-32.

Sevilla, C.C., Carangal, V. and Ranola, R.F. Jr (1995) Development of crop/animal methodology for lowland areas in Pangrinan, Philippines. In: Devendra, C. and Sevilla, C. (eds), *Crop/Animal Interaction.* Proceedings of a Workshop, 27 September-1 October 1993, University of The Philippines, International Rice Research Institute (IRRI) and International Development Research Centre (IDRC), Los Baños, The Philippines, pp. 265-286.

Singh, R.B., Saha, R.C., Singh, M., Dinesh, C., Shukla, S.G., Walli, T.K., Pradhan, P.K. and Kessels, K.P. (1995) Rice straw: Its production and utilization in India. In: Kiran Singh and Schiere, J.B. (eds), *Handbook for Straw Feeding Systems.* Indian Council of Agricultural Research (ICAR), New Delhi, pp. 325-338.

Spedding, C.R.W. (1975) *The Biology of Agricultural Systems.* Academic Press, London, UK.

Sri Lanka (1984) *Proceedings of a Workshop on the Utilization of Fibrous Agricultural Residues as Animal Feeds* (ed., Doyle, P.T.), 17-22 April 1983, University of Melbourne, Australia.

Sri Lanka (1985) *Potential of Rice Straw in Ruminant Feeding: Research and Field*

Application (eds, Ibrahim, M.N.M., Schiere, J.B. and de Siriwardne, J.A.), Proceedings of a Seminar, 7-8 February 1985, Department of Livestock Services, Kandy.

Sri Lanka (1986) *Rice Straw and Related Feed in Ruminant Rations.* Straw Utilization Project, Publication No. 2: International Workshop on Rice Straw and Related Feeds in Ruminant Rations (eds, Ibrahim, M.N.M. and Schiere, J.B.), 24-28 March 1986, Department of Livestock Services, Kandy.

Sundstøl, F. and Owen, E. (eds) (1984) *Straw and Other Fibrous By-products as Feed.* Elsevier Press, Amsterdam, The Netherlands.

Thailand (1984) *Proceedings of an International Workshop on the Relevance of Crop Residues as Animal Feeds in Developing Countries* (eds, Wanapat, M.and Devendra, C.), 29 November-2 December 1984, University of Khon Kaen.

Thailand (1985) *Proceedings of a Workshop on the Utilization of Fibrous Agricultural Residues as Animal Feeds* (ed., Doyle, P.T.), 10-14 April 1984. International Development Program of Australian Universities and Colleges (IDP), Canberra.

Thailand (1988) *Proceedings of a Workshop on Ruminant Feeding Systems Utilizing Fibrous Agricultural Residues* (ed., Dixon, R.M.), 18-22 April 1988, Chiang Mai. International Development Program of Australian Universities and Colleges (IDP), Canberra.

Vietnam (1993) *Increasing Livestock Production by Making Better Use of Local Feed Resources.* Proceedings of a Regional Workshop (eds, Preston, T.R. and Ogle, B.), 25-29 November 1991, Hanoi/Ho Chi Min City. Food and Agriculture Organization of the United Nations (FAO) and Swedish Agency for Research Cooperation with the Developing Countries, Stockholm.

Wanapat, M. (1990) Nutritional Aspects of Ruminant Production in Southeast Asia with Special Reference to Thailand. Mimeo, University of Khon Kaen, Thailand.

Wanapat, M. and Kongpiroon, N. (1988) Intake and digestibility by native cattle of straw and stubble of glutinous and nonglutinous varieties of rice. In: Dixon, R.M. (ed.), *Proceedings of a Workshop on Ruminant Feeding Systems Using Fibrous Agricultural Residues.* International Development Program of Australian Universities and Colleges (IDP), Canberra, pp.133-135.

Wang Jiaqi (1996) Utilization of crop residues in sustainable crop/livestock farming systems in China: Present status and prospects. Mimeo. Country paper presented to the International Workshop on Crop Residues in Sustainable Mixed Crop/Livestock Farming Systems, 22-26 April 1996, International Crops Research Institute for the Semi-Arid Tropics (ICRISAT), Patancheru, Andhra Pradesh, India.

13. The National Perspective: A Synthesis of Country Reports Presented at the Workshop

J.H.H. Maehl

Gustav-Leo Strasse 4, D-20249, Hamburg, Germany

Abstract

This paper synthesizes information on the use of crop residues for feeding ruminants provided by 19 countries present at the workshop. The countries are representative of the world's four major developing regions—sub-Saharan Africa (SSA), Asia, West Asia-North Africa (WANA) and Latin America—and account for about 50% of its cattle, 80% of buffaloes, 20% of camels, 30% of sheep and 60% of goats.

Information is presented on land use patterns, livestock populations and trends over the past two decades. A summary description of the major agricultural systems in each country is given, reflecting considerable variation between countries in the approach taken to the classification of agroecologies. Data are presented on the quantities of crop residues available, their relative importance as livestock feed, and current practices for improving their nutritive value through supplementation and processing. Constraints in utilizing crop residues are indicated and opportunities identified for addressing them through research and development activities.

The nutritive value of crop residues is generally low but shows considerable variability, caused by plant genetic and environmental factors. In all four regions, there is immense potential for making better use of them as livestock feed. In view of the continuing decline of natural grazing, crop residues may be expected to play an increasingly important role in the future, especially in Asia and WANA, where they will be needed to meet the rapidly rising demand for milk and meat.

Introduction

This paper synthesizes the information given in 19 country reports presented at the workshop. Of these reports, six represented sub-Saharan Africa (SSA), four West

Asia-North Africa (WANA), six Asia and three Latin America. For the latter region each report summarized the situation in three subregions. The countries and subregions involved were:

Region	*Country or Subregion*
SSA	Ethiopia, Kenya, Nigeria, Malawi, Mali, Senegal
WANA	Egypt, Iraq, Morocco, Tunisia
Asia	Bangladesh, China, India, Indonesia, Nepal, Thailand
Latin America	Central America and Mexico (Costa Rica, El Salvador, Guatemala, Honduras, Mexico, Nicaragua, Panama) Andean Region (Peru), Southern Cone (Argentina, southern Brazil, Chile, Paraguay, Uruguay).

To ensure that their presentations would be comparable, country representatives were provided, before the workshop, with a common list of the topics their reports should address. This list, drawn up and approved by the Workshop Organizing Committee, was as follows:

1. Livestock population structure and trends.
2. Role of livestock in major agricultural systems.
3. Mechanization and its effects on livestock use.
4. Importance of crop residues as a source of feed in major agricultural systems (quantitative and qualitative aspects).
5. Current practices in the processing, treatment, supplementation and storage of crop residues in major agricultural systems.
6. Nutritive values of crop residues (absolute and relative).
7. Constraints in utilizing crop residues as livestock feeds.
8. Opportunities for collaborative research on developing improved strategies for using crop residues as animal feeds.
9. Role and involvement of agribusiness and research and development (R&D) institutions in promoting crop residues as animal feeds.

Overall, the information provided in the country reports represented a valuable database on the current situation pertaining to most of the above issues. However, given the technical and regional complexity of the workshop's subject matter, it was not always possible to get a complete picture across all countries for the entire set of topics. This paper will therefore focus on those topics for which fairly complete and consistent information was provided. (Hence there is some disparity between the topics listed above and the headings under which I report below.)

Crop residue production varies greatly according to agroecological conditions within a given country as well as across countries and regions. A more systematic characterization of these conditions would have been beneficial for the purposes of comparison and synthesis. As it was, the country reports did not provide information consistently across a range of identically defined agroecological zones (AEZs).

To provide an overview of the AEZs found in the 19 countries, the classification developed by the Technical Advisory Committee (TAC) of the Consultative Group on International Agricultural Research (CGIAR) is presented in Table 13.1. This classification approach was developed for reviewing and setting CGIAR research priorities and is based on the original work on agroecological characterization of the Food and Agriculture Organization of the United Nations (FAO, 1978-1981). Definitions used by TAC for the classification of AEZs are also specified in the table.

Land Use Patterns and Livestock Populations

Feed and fodder resources for ruminant livestock are determined, among other things, by the areas under permanent pasture and cultivation. Table 13.2 presents FAO data on land use. The data indicate the relative importance of different land use patterns in different countries, together with changes that have occurred over the past 20 years (FAO, 1984, 1994). Since most countries have land in more than one AEZ, it would have been preferable to present land use data according to AEZs. However, such data were not readily available in the country reports.

In the Asian countries except China, permanent pastures account for a small proportion of the total land area. The percentage of area cultivated is usually larger than the area under permanent pasture, extreme cases being Bangladesh (74% cultivated, 5% grazing land) and India (57 and 4% respectively). In Africa, Latin America and parts of WANA this situation is reversed, with permanent pasture covering a larger area than cultivated land. The role of pasture is especially pronounced in the countries of the Andean region and the Southern Cone of Latin America, where natural grasslands still provide the major fodder resource for ruminant livestock.

Changes over the past 20 years were calculated as the difference between the relative proportion of land use types in 1993 and the corresponding 3-year average figures of 1974/76. Over the period, Southeast Asian countries showed the biggest increases in the proportion of cultivated land relative to total land area, which rose by 6% or more. In Thailand, for example, the share of cultivated area increased from 33% in 1974/76 to 41% in 1993. Expansion of the cultivated area was also observed in Central America and Mexico and in some countries of SSA and WANA, while a slight decrease in cultivated area occurred in the Southern Cone. The area under forest and woodland declined in about half the countries listed in Table 13.2. This may be attributed mainly to the increasing pressure on land for cultivation.

Table 13.3 gives the populations of ruminant livestock in each country, again following FAO data as the information given in the country reports was incomplete and inconsistent with regard to the period covered. Some countries gave different figures to those of FAO, but without explaining them. The countries represented at the workshop accounted for 52, 79, 23, 30 and 60% of the world populations of cattle, buffalo, camels, sheep and goats respectively. The table also shows trends in

Table 13.1. Classification of countries according to prevailing agroecologies.

Region/country	Agroecological zone[1]								
	1	2	3	4	5	6	7	8	9
Sub-Saharan Africa									
Ethiopia	+	+		+					
Kenya	+			+					
Malawi	+	+							
Mali	+								
Nigeria	+	+	+						
Senegal	+								
West Asia-North Africa									
Egypt									+
Iraq									+
Morocco									+
Tunisia									+
Asia									
Bangladesh			+						
China					+	+	+	+	
India	+	+			+	+		+	
Indonesia			+						
Nepal								+	
Thailand	+	+	+						
Latin America									
Costa Rica		+	+	+					
El Salvador		+		+					
Guatemala			+	+					
Honduras			+	+					
Mexico	+	+	+	+	+				
Nicaragua			+						
Panama			+						
Peru			+	+					
Argentina					+	+	+	+	+
Brazil	+	+	+				+		
Chile									+
Paraguay		+					+		
Uruguay								+	

[1] Numbered zones identified as follows: 1: warm arid and semiarid tropics; 2: warm subhumid tropics; 3: warm humid tropics; 4: cool tropics; 5: warm arid and semiarid subtropics with summer rainfall; 6: warm subhumid subtropics with summer rainfall; 7: warm/cool humid subtropics with summer rainfall; 8: cool subtropics with summer rainfall; 9: cool subtropics with winter rainfall. *Definitions:* tropics: all months with monthly mean temperature (corrected to sea level) above 18°C; subtropics: 1 or more months with monthly mean temperature (corrected to sea level) below 18°C; temperate: 1 or more months with monthly mean temperature (corrected to sea level) below 5°C; length of growing period (LGP): period (in days) during the year when rainfed available soil moisture supply is greater than half potential evapotranspiration

Notes to Table 13.1 *(continued)*

(PET) (including the period required to evapotranspire up to 100 mm of available soil moisture stored in the soil profile, but excluding any interval when the daily mean temperature is less than 5°C); warm: daily mean temperature during the growing period greater than 20°C; cool: daily mean temperature during the growing period in the range 5-20°C (includes the moderately cool range 15-20°C); cold: daily mean temperature less than 5°C; warm/cool: daily mean temperature during part of the growing period greater than 20°C and during another part less than 20°C; arid: LGP less than 75 days; semiarid: LGP in the range 75-180 days; subhumid: LGP in the range 180-270 days; humid: LGP greater than 270 days.
Source: TAC (1992).

livestock numbers from 1974/76 to 1993. Following the same order of species, average numbers for all the countries listed rose over this period by 21, 26, 8, 6 and 47%.

Livestock numbers in a given country often fluctuate dramatically over short periods. Annual changes of up to 50% have been reported in some countries. The trend figures in Table 13.3 might look quite different if the reference period were altered by only a few years, or if data from national reports had been used instead of FAO figures. Factors influencing numbers over time vary greatly from country to country. The main factors reported in the country papers were as follows:

1. Drought appears to be the single most important factor causing decreases in livestock numbers, especially in the arid and semiarid zones of SSA and WANA. Its long-term effect is often reported to be greater in cattle than in small ruminant populations, due to the different feeding behaviour of these species. Because of their shorter reproduction cycles, small ruminant populations also tend to recover more rapidly after a decline in numbers caused by either drought or disease.
2. Increasing pressure on land to meet the rising demand for human food reduces the supply of natural grazing land for livestock in general and for large ruminants in particular. This leads to a decline in animal productivity and ultimately to falling animal numbers.
3. Political instability and war were reported to have substantially reduced livestock numbers in countries such as Iraq and Peru and in parts of Central America.
4. Growing demand for livestock products such as milk and meat, coupled with favourable cost:price ratios, improve the profitability of keeping livestock and hence may lead to an increase in livestock numbers. This is exemplified by the data for sheep populations in Tunisia, for female cattle and buffalo populations in India (where, during the same period, the bullock population remained constant), and for specialized dairy cattle operations in a number of favourable environments.
5. Improved animal health management leads to more stable or increased livestock numbers, either in general or for certain classes of stock, as in the case of dairy cattle in Kenya and Central America.
6. Livestock sector policy was reported to influence livestock populations in many ways: for example, through the liberalization of producer prices, through

Table 13.2. Land use patterns in 1993 and changes in relative proportions of each land use type since 1974/76.

Country/region	Land use patterns in 1993[1]									Changes since 1974/76 (%)		
	('000 ha)						(% of total area)					
	Area	AL	PC	PP	F/W	Other	A + PC	PP	F/W	A + PC	PP	F/W
Sub-Saharan Africa												
Ethiopia	100 000	12 000	650	40 000	25 000	22 350	13	40	25	0	-1	0
Kenya	56 914	4 000	520	21 300	16 800	14 294	8	37	30	4	31	25
Malawi	9 408	1 640	30	1 840	3 700	2 168	18	20	39	-6	0	-14
Mali	122 019	2 500	3	30 000	6 900	82 616	2	25	6	0	0	-2
Nigeria	91 077	29 850	2 535	40 000	11 300	7 392	36	44	12	3	21	-6
Senegal	19 253	2 330	20	3 100	10 450	3 353	12	16	54	-14	-14	22
Asia												
Bangladesh	13 017	9 450	244	600	1 900	823	74	5	15	6	0	-2
China	932 640	92 708	3 267	400 000	130 496	306 169	10	43	14	-1	12	2
India	297 319	166 100	3 550	11 400	68 500	47 769	57	4	23	1	0	1
Indonesia	181 157	18 900	12 087	11 800	111 774	26 596	17	7	62	6	0	-6
Nepal	13 680	2 325	29	2 000	5 750	3 576	17	15	42	0	2	9
Thailand	51 089	17 600	3 200	800	13 500	15 989	41	2	26	8	1	-10
West Asia-North Africa												
Egypt	102 345	2 800	2 450	350	31	96 714	5	0	0	2	0	0
Iraq	43 737	5 250	200	4 000	192	34 095	12	9	0	0	0	-3
Morocco	44 630	9 256	664	20 900	8 970	4 840	22	47	20	5	19	8
Tunisia	15 626	2 987	1 965	3 103	676	6 895	32	20	4	0	2	1

[1] AL = arable land; PC = permanent crops; PP = permanent pasture; F/W = forest and woodland; A+PC = arable land and permanent crops.

Table 13.2 continued.

Country/region	Area	Land use patterns in 1993								Changes since 1974/76 (%)		
		('000 ha)					(% of total area)					
		AL	PC	PP	F/W	Other	A + PC	PP	F/W	A + PC	PP	F/W
Central America/Mexico												
Costa Rica	5 106	285	245	2 340	1 570	666	10	46	31	0	14	-13
El Salvador	2 072	565	165	610	104	628	35	29	5	4	0	-3
Guatemala	10 843	1 324	556	2 602	5 813	548	17	24	54	2	13	8
Honduras	11 189	1 683	332	1 533	6 000	1 641	18	14	54	3	-17	14
Mexico	190 869	23 150	1 580	74 499	48 700	42 940	13	39	26	1	0	-1
Nicaragua	11 875	1 100	170	5 500	3 200	1 905	11	46	27	0	8	-16
Panama	7 443	500	160	1 490	3 260	2 033	9	20	44	2	5	-13
Andean region												
Peru	127 700	3 400	30	27 120	84 800	12 350	3	21	66	0	0	10
Southern Cone												
Argentina	273 669	25 000	2 200	142 000	50 900	53 569	10	52	19	-3	-1	-3
Brazil (total)	845 651	42 000	6 955	185 000	488 000	123 696	6	22	58	-1	3	-12
Chile	74 880	3 984	273	13 600	16 500	40 523	6	18	22	-1	3	2
Paraguay	39 730	2 190	80	21 700	12 850	2 910	6	55	32	3	17	-20
Uruguay	17 481	1 260	44	13 520	930	1 727	7	77	5	-1	-1	2

Source: FAO (1984, 1994).

Table 13.3. Ruminant livestock populations in 1993 and percentage change since 1974/76.

Country/ region	1993 populations ('000 head)					% change since 1974/76				
	Cattle	Buffalo	Camels	Sheep	Goats	Cattle	Buffalo	Camels	Sheep	Goats
Sub-Saharan Africa										
Ethiopia	29 450		1 000	21 700	16 700	13		4	-6	-4
Kenya	11 000		810	5 500	7 300	14		44	79	75
Malawi	970			195	888	47			138	26
Mali	5 380		232	4 926	7 029	38		29	1	44
Nigeria	16 316		18	14 000	24 500	47		6	38	3
Senegal	2 750		15	4 400	3 118	19		150	162	271
Asia										
Bangladesh	23 923	866		989	25 967	-6	-21		-7	238
China	85 781	22 217	401	109 720	97 812	52	23	-22	16	48
India	192 700	78 555	1 510	44 608	117 547	7	31	40	12	62
Indonesia	11 300	3 450		6 300	11 800	82	45		94	63
Nepal	6 237	3 073		4 911	5 452	-5	-20		-61	134
Thailand	7 190	4 747		136	151	71	-17		300	403
West Asia-North Africa										
Egypt	2 977	3 466	110	3 707	3 017	42	57	5	93	128
Iraq	1 120	103	14	6 300	1 050	-40	-53	-94	-43	-69
Morocco	2 924		36	16 302	4 773	-18		-82	9	-27
Tunisia	659		231	7 110	1 417	-25		28	24	39

Table 13.3 continued.

Country/ region	1993 populations ('000 head)					% change since 1974/76				
	Cattle	Buffalo	Camels	Sheep	Goats	Cattle	Buffalo	Camels	Sheep	Goats
Central America/Mexico										
Costa Rica	2 122			3	2	18			50	100
El Salvador	1 197			5	15	13			25	36
Guatemala	2 236			440	78	48			-15	3
Honduras	2 077			13	27	14			160	23
Mexico	30 649			5 876	11 300	7			-7	25
Nicaragua	1 645			4	6	-36			100	-14
Panama	1 437				5	7				-17
Andean region										
Peru	3 924			11 915	1 780	-5			-22	-13
Southern Cone										
Argentina	50 856			18 468	3 370	-10			-47	-20
Brazil (total)	152 300	1 430		20 000	12 180	61	444		13	68
Chile	3 557			4 629	600	2			-18	-19
Paraguay	8 000			378	119	55			4	11
Uruguay	10 093			24 414	15	-8			60	25
World	1 281 606	148 475	18 632	1 096 049	592 874	7	29	15	4	42
Total for represented countries	670 770	117 907	4 377	332 949	358 018	21	26	8	6	47
% of world livestock	52	79	23	30	60					

Source: FAO (1984, 1994).

programmes aimed at increasing the incomes of smallholders, or through the regional stratification of production according to economic comparative advantage, as reported for China and the Southern Cone of Latin America.

The factors mentioned above are indicative only—many others were reported in the country papers. The information available did not, however, allow more detailed analysis.

Major Agricultural Systems and the Role of Livestock

Table 13.4 shows the major agricultural systems reported by the different countries. The criteria used for classification varied between countries. Besides agroecological zone, other criteria used included geographical position (lowland, upland, coastal, hillsides, etc), ecozone (forest, savanna, etc), intensity of livestock production (extensive, semi-intensive, intensive, etc), or a combination of these criteria. The systems given in the table are further characterized according to their main crops and ruminant livestock species. The predominant feeding systems and the main livestock products are also specified, where this information was given.

The relative importance of different agricultural systems within a country could not be quantified, as relevant data were not available in most reports.

The role of livestock in mixed crop/livestock farming systems is manifold. Besides producing food (milk, meat) and non-food (wool, hair, hide, fuel, manure, draught power) items, livestock have several social or socioeconomic functions (conferring status, providing security, storing wealth, raising cash, etc). Through the utilization of crop residues and agricultural by-products, ruminant livestock also contribute to the recycling of nutrients and so to soil fertility and structure, particularly in highly integrated farming systems. The sale of livestock generates cash for the purchase of agricultural inputs and thus contributes to the intensification of crop production. These roles, and the important contribution of livestock to the development of productive and sustainable mixed farming systems, were generally acknowledged by most countries participating in the workshop.

Utilization of Crop Residues as Livestock Feed

"There is enormous potential for a better utilization of crop residues as livestock feed!" This statement, quoted (with its exclamation mark) directly from one of the country reports, epitomizes the opinions expressed in most of the reports and reiterated by participants in many sessions of the workshop itself. At the same time, it reflects one of the main areas of concern expressed: that is, crop residues are generally much underutilized as livestock feed at present, and this even in countries with a feed deficit, such as Tunisia.

Table 13.4. Main agricultural systems in different countries.

Country report[1]	Classification of zone (mm annual rainfall, m altitude)	Main agricultural systems			
		Main crops	Livestock[2]	Main livestock products	Predominant feeding systems
Sub-Saharan Africa					
Senegal	Sahelian (200-500 mm)	Rice, sugarcane, cassava, vegetables, melon	G, C, S	Meat, milk	Pasture grazing By-products
	Sahelo-Sudanian (500-700 mm)	Groundnut, millet, sorghum, maize, vegetables, fruits	C, S, G	Meat, milk	Pasture grazing Groundnut haulms, cereal straws Commercial concentrates
	Sudano-Sahelian (700-900 mm)	Groundnut, millet, sorghum, maize, cotton	C, S, G	Meat, milk	Pasture grazing Groundnut haulms, cereal straws Cottonseed
	Sudanian (900-1200 mm)	Cotton, groundnut, millet, sorghum, maize	C, S, G	Meat	Pasture grazing Groundnut haulms, cereal straws Cottonseed
	Sudano-Guinean (1200-1800 mm)	Rice, groundnut, millet, sorghum, palm, vegetables, cassava, fruits	C, G, S	Meat	Pasture grazing Crop residues Browsing
Ethiopia	Highlands (>1200 mm, >1500 m)	Wheat, barley, oats, pulses, ensete	C, S, G	Cropping areas: Draught, meat, milk, manure	Grazing communal lands and fallows Stubble grazing Crop residues
	Mid-altitude (700-1200 mm, <1500 m)	Teff, maize, sorghum, millet, pulses, root crops, ensete	C, G, S	Pastoral areas: Milk, animals, meat	As above

[1] No information received from Mali, Nepal and Morocco. NA = information not available in country report.
[2] B = buffalo; C = cattle; DC = dairy cattle; BC = beef cattle; DPC = dual-purpose cattle; S = sheep; G = goats.

Table 13.4 continued.

Country report	Classification of zone (mm annual rainfall, m altitude)	Main agricultural systems			Predominant feeding systems
		Main crops	Livestock	Main livestock products	
Kenya	Tropical alpine (>2300 m)	Natural forest, bog	C, S	NA	Grazing
	Highlands (1700–2300 m)	Coffee, tea, pyrethrum, wheat, sorghum, barley, maize, vegetables	DC, G, S	NA	Grazing Maize silage, hay Chopped stover Concentrates, vegetables
	Upper midlands (1600–2000 m)	Maize, sugarcane, sorghum, banana	C, G	NA	Grazing Crop residues
	Lower midlands (1200–1500 m)	Rice, cotton, groundnut, millet	C, G	NA	Grazing Crop residues Browsing tree forages
	Coastal lowlands (<1200 m)	Coconut, sugarcane, cashew, cassava	C, S, G	NA	Free-range
Malawi	Plateau (1000–1500 m)	Maize, tobacco, groundnut, pulses, cassava, potato	C, G, S	Meat, milk, manure, draught	Grazing natural grasses/crop residues
	Lake shore (500–750 m)	Sugarcane, maize, cassava, potato, pulses, groundnut	C, G	Meat, milk, manure	Grazing natural grasses/crop residues
	Shire Valley (200–300 m)	Sorghum, millet, cotton, groundnut, pulses	C, G	Meat, milk	Grazing natural grasses/crop residues

Table 13.4 continued.

Country report	Main agricultural systems			Main livestock products	Predominant feeding systems
	Classification of zone (mm annual rainfall, m altitude)	Main crops	Livestock		
Nigeria	Humid forest	Vegetables, yams, cassava, cocoa, oil palm, sugarcane	G, S	Meat	Tree forage, grasses, legumes Cereal straws, tuber peels, sugarcane tops
	Mid-altitude and derived coastal savanna	Vegetables, yams, maize, cassava, oil palm, coffee	C, G, S	Meat	Grazing all year Cereal straws, legumes, tuber peels, tree foliage
	Southern Guinea savanna	Sorghum, millet, maize, cotton, fruits, rice	DPC, G, S	Milk, meat	Transhumant grazing Cereal straws, legumes Browsing tree foliage and fruits Molasses
	Northern Guinea savanna	Sorghum, millet, maize, cotton, fruits, rice	DPC, G, S	Milk, meat	Transhumant grazing Cereal straws, legumes Browsing tree foliage and fruits Molasses
	Sudano-Sahelian savanna	Sorghum, cotton, rice, groundnut	DPC, G, S	Meat, milk	Transhumant grazing Cereal straws, legumes Molasses
Asia					
Thailand	Subhumid/dry (lowlands)	Rice, soybean, groundnut	B, BC, DC	Draught, meat, milk	Stall feeding, tethering, grazing of straw, concentrates, crop residues
	Subhumid/dry (uplands)	Rice, sugarcane, cassava	B, BC, DC	Draught, meat, milk	Grazing rice straw, Crop residues, concentrates
	Subhumid/dry (highlands)	Tree plantations	S	Meat	Grazing and cut-and-carry Tree fodder

Table 13.4 continued.

Country report	Main agricultural systems		Livestock	Main livestock products	Predominant feeding systems
	Classification of zone (mm annual rainfall, m altitude)	Main crops			
Thailand (continued)	Subhumid lowlands	Rice, maize	BC, DC	Meat, milk	Grazing and cut-and-carry Rice straw, concentrates, crop residues
	Humid lowlands	Fruit tree plantations	BC, S, G	Meat	Grazing and cut-and-carry Tree fodder
Indonesia	Lowland (<100 m)	Rice, maize, cassava, vegetables, sugarcane, oilpalm, coconut, rubber	B, C, S	Draught, meat, calves	Cut-and-carry, limited grazing of crop residues, tree fodder Tree-livestock system
	Upland (>500 m)	Maize, rice, potato, sweet potato	C, G	Draught, meat, milk	Cut-and-carry, limited grazing of crop residues, Crop-livestock system
	Pasture (100-500 m)	Rainfed rice, maize	C, B, G	Trampling, meat, calves	Mainly grazing natural grasses, crop residues Pasture-tree system
Bangladesh	Humid tropical	Rice, wheat, jute, potato, sweet potato, sugarcane, oilseeds, pulses, fruit trees	C, G, S, B	Draught, meat, milk, manure	Mainly cut-and-carry of crop residues, weeds, roadside grass, green fodder
China	Pasturing area (northern China)	NA	C, S, G, Camel	Meat, milk, wool, transport, fuel	Grazing
	Cropping-pasturing area	NA	C, S, G	Transport, draught, manure, fuel, meat, milk, wool	NA
	Cropping area (southern China)	Rice, wheat, maize, soyabean, rapeseed, groundnut, sugarcane, sugarbeet, linseed	C, S, G, B	Meat, milk, draught, manure	NA

Table 13.4 continued.

Country report	Classification of zone (mm annual rainfall, m altitude)	Main agricultural systems			
		Main crops	Livestock	Main livestock products	Predominant feeding systems
India	Arid (Rajasthan and surroundings)	Millet, pulses, cotton, oilseed, wheat	DC, S, G, camel	Milk, wool, hair, meat, draught	Grazing Feeding chopped millet, tree/shrub leaves
	Semiarid (south and central)	Sorghum, millet, wheat, pulses, maize, sugarcane, cotton	DC, B, S, G, camel	Milk, wool, hair, meat, draught	Grazing Feeding chopped stover, straws and grasses
	Humid/subhumid tracts (north and east)	Rice, wheat, sugarcane, pulses, oilseeds, sorghum, cotton	DC, B, S, G	Milk, wool, hair, draught	Grazing Stall feeding of straws, chopped green stover, grasses
	Subtropical/temperate (Himalayan foothills)	Rice, wheat, maize, pulses (forest)	S, G, DC	Wool, meat, milk	Mainly grazing Stall feeding of stover and straws
	Highlands (alpine/subalpine tracts)	Rice, buckwheat, small millets, vegetables (forest)	S, G, mithun	Wool, meat, milk	Grazing Partly stall feeding of chopped fodders
	Coastal region	Rice, coconut, cassava, small millet (forest)	C, S, G	Milk, wool, meat	Grazing Partly stall feeding with chopped straws and other fodders
Latin America					
Central America/ Mexico	Highlands	Vegetables, fruits, potato	DC	Milk, cheese	Grazing Chopped grass and conserved forages Commercial concentrates, molasses, banana fruits

Table 13.4 continued.

Country report	Main agricultural systems				
	Classification of zone (mm annual rainfall, m altitude)	Main crops	Livestock	Main livestock products	Predominant feeding systems
Central America/ Mexico (continued)	Humid hillsides (600-1200 m)	Coffee, sugarcane	DPC or DC	Milk, cheese, weaners	Limited grazing Chopped grasses Residues from plantains, bananas, legumes Sugarcane tops
		Slush-mulch maize or sorghum, legume cover crops, beans, cassava	DPC	Cheese, milk, weaners or mature animals	Grazing Maize/ sorghum straw and legumes Browsing tree foliage
	Dry hillsides (600-1200 m)	Maize, sorghum or upland rice, beans, cassava	DPC	Cheese/milk (in rainy season), weaned or mature stock	Grazing all year Dry season: straws and stover, *guatera* and conserved forages, browsing tree foliage and fruit
	Humid lowlands (<600 m)	Maize, rice, cassava, yams, banana	DPC, DC, BC	Milk/cheese, weaned (in DPC systems) and mature (mainly in BC system) stock	Grazing Molasses, banana fruits, concentrates Legume tree foliage
	Subhumid/dry lowlands (<600 m)	Maize or sorghum, rice, cassava, cowpea, sugarcane, citrus fruits	DPC or BC	Cheese/milk (mainly in rainy season), weaned or mature stock	Grazing almost all year Dry-season cereal legume straws and stover, *guatera* and other conserved forages, browsing tree foliage and fruits
Andean region (Peru)	Extensive livestock system (>3800 m)		S, camelids	Wool, meat	Native rangelands
	Mixed system (3000-4500 m)	Barley, wheat, maize, oats	S, G, C	Wool, milk, meat	Native rangelands Crop residues

Table 13.4 continued.

	Main agricultural systems				
Country report	Classification of zone (mm annual rainfall, m altitude)	Main crops	Livestock	Main livestock products	Predominant feeding systems
Andean region (Peru) (continued)	Semi-intensive (3700–4000 m)	NA	S, DPC, DC	Wool, milk, meat	Native rangelands Cultivated pasture Crop residues
	Intensive (2000–2800 m)	NA	DC, DPC	Milk, meat	Cultivated pasture Agroindustrial by-products
Southern Cone	Temperate (beef: cow/calf system)	For whole region: Cereals: maize, wheat, rice, sorghum, oats, barley, rye	BC	Meat, weaners	Grazing natural grassland, cereal stover, perennial pasture, pasture hay
	Temperate (beef: fattening)	Oilseeds: soyabean, sunflower	BC	Meat	Perennial pasture, maize stover, annual forage crops (oats, rye), maize/oat grain, wheat, rice bran
	Temperate (dairy cattle: exotic breeds)	Horticulture and fruits: limited importance	DC	Milk	Perennial pasture grazing, green fodder (oats, ryegrass), maize silage, grass silage, grain (maize, barley, oat), wheat bran, whole cottonseed, sunflower meal, commercial concentrates
	Subtropical (beef production)		BC	Meat, animals	Natural grassland, tropical pasture, sugarcane, molasses, urea, cassava
	Subtropical (dairy cattle: crossbreds)		DC	Milk	Tropical pasture, sugarcane, molasses, urea, cassava, commercial concentrates
	Semiarid (sheep production)		S	Meat, wool	Natural grassland

Table 13.4 continued.

Country report	Main agricultural systems		Livestock	Main livestock products	Predominant feeding systems
	Classification of zone (mm annual rainfall, m altitude)	Main crops			
West Asia-North Africa					
Tunisia	Humid hillsides	(Forest) wheat, fruits	C	Milk, meat	Forest grazing, straws
	Semiarid hillsides	Fruits, rangeland, cereals	S, G	Meat	Rangeland grazing / Oats-vetch hay, straw / Concentrates
	Subhumid lowlands	Wheat, sugarbeet, fodder crops	DC	Milk/cheese	Limited grazing / Silage, hay, beet pulp / Cereal staws
	Semiarid lowlands	Cereals, forage crops, rangeland, olive	S	Meat, wool	Commercial concentrates / Grazing, hay / Cereal and olive tree residues
	Arid lowlands	Rangeland, barley, olive	S, G	Meat	Barley grain, straw / Grazing, shrubs, olive tree residues
	Desert	Rangeland	G, camels	Meat, milk	Grazing, forage from oases, date palm by-products
Egypt	Irrigation-based systems	Wheat, cotton, maize, rice, sugarcane	B, C, S	Meat, milk, draught	Green fodder (berseem) / Crop residues
Iraq	Arid/rainfed (north)	Wheat, barley, pulses	S, G, C	Milk and dairy products, weaner and mature stock	Grazing natural rangeland / Stubble and fallow grazing, chopped straw feeding / Barley grain
	Arid/irrigated (central and south)	Wheat, barley, rice, vegetable, maize, cotton, sunflower	S, C, B, G	Milk/cream, meat, animals	Stubble grazing (sheep/goats) / Hand-feeding of long straws, irrigated green fodder (cattle, buffalo)

Before coming to a more qualitative discussion of this topic, let us first have a look at some figures that will illustrate the above statements. In the country reports it was often not clear whether the figures given for available crop residues referred to fresh weight or dry matter (DM) weight. Nor did most reports describe how residue yields were estimated (harvest index, multiplication factor, etc). For the purposes of this paper it was therefore felt appropriate to limit the information on the quantitative availability of crop residues to the most important crops world-wide and to use a comparable database—FAO statistics again—for the estimation of crop residue yields.

Cereal straws and stovers are by far the most important residues. Their availability is shown in Table 13.5, along with data on other important crops such as all roots and tubers, all pulses, soyabean, groundnut and sugarcane. The estimated residue yields are based on FAO figures for 1993 for the production of food commodities, multiplied by different factors as proposed by Kossila (1988). These factors are shown in Table 13.6, where the factors used by other authors are also given for comparison. Countries unhappy with the multipliers used in this paper can always recalculate their crop residue figures using factors considered more applicable to their situation.

The topic of crop residue utilization can be discussed from two angles: (i) what is the relative importance of crop residues in the total diet composition of livestock? and (ii) what proportion of residues is actually utilized to feed animals? Details on both these angles are shown in Table 13.7. As quantitative data were relatively scarce in the country reports, the table also includes qualitative information. Some information was obtained from the papers by Kiran Singh *et al.* and by Nordblom *et al.* presented at the workshop (see pp. 113 and 131).

The total feed resource base for ruminant livestock may be divided into three categories: permanent natural rangeland, crop residues (including cultivated fodder), and energy feeds in the form of grain, concentrates or agroindustrial by-products. Among these, the relative importance of crop residues obviously varies between regions, depending directly or indirectly on such factors as agroecological conditions, the extent of cultivated land and the intensity of crop and livestock production.

Rangeland or grassland, also called natural pasture, traditionally forms the mainstay of ruminant diets. However, as we have seen, important changes in land use have taken place in several parts of the developing world, with pasture declining in extent as cultivated land expands in response to rising demand for food. Data from the country reports suggest that, in the humid and subhumid zones of Asia (including Bangladesh, Thailand, Indonesia, southern China and parts of India), crop residues have gained in importance in animal diets at the expense of roughage from natural grazing areas, which have decreased in productivity as well as in area.

In contrast, in the subhumid and humid tropics of Central America, Mexico and the Andean region, the availability of natural grazing land is still higher than in Asia, such that crop residues are of less importance. Here grazing is practised throughout most of the year, with crop residues playing a supportive role, especially during the dry season.

Table 13.5. Estimated crop residue availability ('000 t fresh weight) in 1993[1].

Region/country	Wheat	Rice	Barley	Maize	Rye	Oats	Millet	Sorghum	Roots and tubers (total)	Pulses (total)	Soyabean	Groundnut (in shell)	Sugar-cane
Asia													
Bangladesh	1 529	35 003	10	9	0	0	252	0	364	2 068	0	160	1 877
China	138 314	233 968	4 550	309 138	1 400	910	15 844	22 448	28 990	23 828	61 292	33 984	17 249
India	73 791	152 168	1 958	28 959	0	0	34 576	47 212	4 449	52 580	18 488	30 504	56 963
Indonesia	0	62 635	0	19 380	0	0	0	0	3 971	2 016	6 836	4 272	8 250
Nepal	995	4 550	36	3 600	0	0	920	0	175	824	48	0	342
Thailand	1	25 892	3	9 984	0	0	0	832	4 084	1 760	2 052	0	9 456
Latin America													
Andean region	130	1 258	146	2 316	2	12	0	68	482	404	4	20	1 250
Central America	5 423	1 430	658	43 164	0	65	36	18 280	360	7 296	1 932	512	16 872
Southern Cone	16 567	15 139	1 031	127 113	144	1 427	236	12 956	6 302	11 744	141 608	2 096	66 116
Sub-Saharan Africa													
Ethiopia	1 794	0	1 494	4 932	0	105	1 360	5 395	393	3 208	88	216	425
Kenya	300	66	56	4 848	0	6	290	450	336	800	0	56	1 088
Malawi	2	85	0	6 102	0	0	75	110	117	1 056	0	220	275
Mali	4	556	0	849	0	0	3 540	3 470	30	256	0	524	71
Nigeria	60	4 887	0	6 900	0	0	19 000	20 000	8 483	6 600	640	4 680	226
Senegal	0	251	0	414	0	0	3 270	495	12	176	0	2 512	213
West Asia-North Africa													
Egypt	6 283	5 409	173	15 117	34	0	0	3 116	393	1 284	200	120	2 945
Iraq	1 184	280	1 157	870	0	0	12	8	81	160	8	0	16
Morocco	2 045	70	1 335	606	4	0	24	84	179	1 108	16	136	237
Tunisia	1 837	0	621	0	0	0	0	4	40	400	0	0	0

[1] Based on conversion factors in Kossila, 1988 (see Table 13.6).
Source: FAO (1994).

Table 13.6. Multipliers used by different authors to estimate crop residue production from food commodity yields.

Commodity	Reference[1]	Africa (a)	Central America (a)	South America (a)	Asia (a)	World (a)	WANA (b)	India (c)	Not specified (d)	Ethiopia (e)	Thailand (f)	Southeast Asia (g)
Wheat		2	1.5	1.2	1.3		0.8	1	1	1.5		
Rice		1.5	1.2	1.3	1.3		0.6	1.3	1	1.5	1	1
Barley		1.5	1.2	1.3	1.3		1.2	1.3		1.5		
Maize		3	2	3	3		2	3		2	0.1	3
Rye					2							
Oats		1.5	1.3	1.3	1.3					1.5		
Millet		5	4	4	4		3	4	3	2.5		
Sorghum		5	4	4	4		3	4	1.5	2.5		3
Roots and tubers (total)						0.2		0.2				
Pulses (total)						4		4		1.2		
Soybean						4		2				
Groundnut (in shell)						4		2	2			
Sugarcane						0.25	0.1	0.25			0.2	0.2
Groundnut (waste)											0.57	
Cottonseed (meal)											0.45	
Cassava (waste)											0.2	
Cassava (leaves)											0.08	0.06
Rice (hulls)												0.15
Sugarcane (bagasse)												0.12

[1] References: (a) Kossila (1988); (b) Nordblom (1988); (c) Kiran Singh *et al.* (this proceedings, p.113); (d) ICRISAT (personal communication); (e) FAO (1987); (f) Devendra (1992); (g) Roxas *et al.* (this proceedings, p.101), adapted from Ranjhan (1986).

Table 13.7. Utilization of crop residues as livestock feed[1].

Country report	Importance of CRs relative to other feed resources	Extent of utilization relative to availability of CRs
Sub-Saharan Africa		
Senegal	CRs and AIBPs perceived as alternative feeds in view of deteriorating natural vegetation cover and increasing cropping area	Rice straw often burnt; maize residue losses due to early harvest; AIBPs exported
Mali	No information available	
Ethiopia	Feed composition on DM basis (Nordblom et al., p. 131): 4% range grazing, 7% grains/concentrates, 39% CRs/forage crops	In cereal cropping areas: 63% livestock feed, 20% fuel, 10% construction, 7% bedding
Kenya	CRs mainly used in dry season when rangeland productivity is lower	
Malawi	Greater in dry season	<50% when grazed in field
Nigeria	Greater in dry season	Despite abundant availability and cheapness, utilization very low (most CRs rot or are burnt)
Asia		
Thailand	CRs and AIBPs abundantly available CRs important due to lack of natural grazing, especially during dry season	CRs are far from fully utilized, mainly due to collection and storage problems
Indonesia	Rice straw (single most important CR) serves as survival feed during dry season in areas where no green fodder available	CRs and especially rice straw are largely underutilized
Bangladesh	85% of total feed comes from CRs, AIBPs, cropland weeds, roadside grass, tree leaves 15% of total DM requirements from green fodder	Rice straw: 41% used as feed, 20% spoilt in field (drying problem), 18% fuel, 21% packing/building materials, handling and storage losses
China	No information reported	25% of 570 million t CRs used as feed, rest is mostly burnt

[1] Abbreviations: CRs = crop residues; AIBPs = agroindustrial by-products; DM = dry matter; TDN = total digestible nutrients.

Table 13.7 continued.

Country report	Importance of CRs relative to other feed resources	Extent of utilization relative to availability of CRs
Asia (continued)		
India	Estimates (on DM basis) from Kiran Singh *et al.*, p. 113 Total feed: 530 million t available, 848 million t required; deficit = 318 million t	No information reported
	Dry fodder (CRs): 66% available, 69% required Green fodder: 27% available, 23% required Concentrates: 7% available, 8% required	
	Relative importance of CRs greater in agriculturally developed areas with lower grazing and vice-versa	
	Deterioration of natural grasslands due to overgrazing increases contribution of CRs in diet	
Nepal	Sources of TDN supply for ruminants: 51% cereal straw, 30% green fodder (range), 12% tree leaves, 7% others 30% deficit in feed supply	Despite overall shortage of 30%, most available feed resources are not fully used
Latin America		
Central America/ Mexico	CRs more important under drier conditions and/or in dual-purpose and goat production systems	No information reported
	Grazing of cereal CRs is traditional during drier months in humid and dry hillsides, subhumid/dry lowlands; complemented with pasture, tree foliage	
Andean region (Peru)	Native rangelands are main fodder resource CRs are important for maintaining production levels during seasonal feed shortages	Full utilization can be assumed, as cereal harvest occurs during period of least pasture availability

Table 13.7 continued.

Country report	Importance of CRs relative to other feed resources	Extent of utilization relative to availability of CRs
Latin America (continued)		
Southern Cone	With change from extensive to more intensive livestock production systems, AIBPs become more important. Natural grasslands are main component in cattle diet; supplementation with hay, silage, cereal-based concentrates. CRs could be an alternative in marginal semiarid areas where natural pastures are scarce	Cereal and oilseed CR availability is 290 million t, but utilization as livestock feed is limited as cost per unit energy/protein is higher than in high-quality natural pastures. Situation is different in marginal/semiarid areas with less grazing land
West Asia - North Africa		
Tunisia	Feed composition on DM basis (Nordblom *et al.*, p. 131): 46% range grazing, 15% grains/concentrates, 39% CRs/forage crops	Cereal straws: 50-100%, higher during droughts, lower when forage crops are available. AIBPs: 100%, except for a few of lower quality
Egypt	Feed composition on DM basis (Nordblom *et al.*, p. 131): 4% range grazing, 16% grains/concentrates, 80% CRs/forage crops; 60% of total diet for cattle/buffalo is green fodder (berseem, hay, silage)	Available CRs are underutilized as feed, partly due to high opportunity costs
Morocco	Feed composition on DM basis (Nordblom *et al*, p. 131): 33% in range grazing, 18% grains/concentrates, 49% CRs/forage crops	No information reported
Iraq	Feed composition on DM basis (Nordblom *et al.*, p.131): 32% range grazing, 15% grains/concentrates, 53% CRs/forage crops; . 7.5 million t DM = total feed resources available, of which: 48% rangeland, 32% CRs, 7% irrigated pasture, 13% grains (mainly barley) and AIBPs	There is shortage of feed and hence relatively low livestock productivity (milk, meat). Certain wastage/loss can be assumed due to constraints in postharvest management

Turning to the Southern Cone of Latin America, the situation is again different: vast areas are covered with natural grassland. In the temperate humid zone of Argentina, this grassland is of relatively high quality. The importance of crop residues in such areas is limited.

The warm arid and semiarid tropics and subtropics cover large parts of sub-Saharan Africa, India, China, Mexico and South America. Farming systems in these zones are generally more extensive in view of the rainfall pattern, which limits the potential for cropping. Nevertheless, gradations in the utilization of crop residues can be detected. In Senegal and Mali as well as in northern China, for example, crop residues are less used than in semiarid India. Such differences are explained by the greater availability of natural rangelands in the former countries, as a result of which crop residues are considered more as a dry-season fodder, whereas in India, which has fewer natural grazing reserves, they tend to be used more intensively.

Most of the WANA region is characterized by a cool subtropical climate with winter rainfall. In countries like Tunisia or Iraq, rangeland grazing accounts for about 50% of available feed resources, while crop residues account for 30 to 40%. Climatic conditions strongly influence utilization patterns, however. In Tunisia, poor-quality crop residues and by-products are fully used in drought years but neglected in years when forage crops are available. In the marginal arid to semiarid parts of the country, which have mixed barley-sheep farming systems, cereal stubble, fallow and straw constitute about 20% of the total available feed and are the most important feed resource.

The information provided in the country reports does not allow more detailed discussion of this topic. Nevertheless, it is sufficient to obtain a "bird's eye view" of the relative importance of crop residues in different zones and countries, as we have done above. There are, of course, other factors influencing the diet composition of livestock. For instance, high prices for livestock products promote more inten-sive production, which in turn increases the demand for higher-quality feeds. Milk and meat production in the peri-urban areas of most countries and the barley-based sheep production systems of some WANA countries are examples reported during the workshop. The country reports on China and the Southern Cone mention strategies to optimize the utilization of different feed resources through the regional stratification of livestock production:

In China, the northern pastoral areas used to be the traditional production base for herbivores. Overgrazing led to a fall in annual grass yields of about 30-50%. At the same time, in the so-called "cropping areas" of southern China, the per capita production of grain over the past decade has declined due to population increase and a reduction in the area of arable land. This area produces grain not only for human food but also for the country's pig subsector, which accounts for 75% of total meat production. Since 1990, China has been trying to introduce a new grain-saving livestock production system so as to release the pressure on grain produc-tion. Greater emphasis is being placed on the utilization of crop residues, with the aim of meeting the rising demand for meat through ruminant livestock rather than pig production. Production figures from recent years indicate a promising shift from pork towards beef and mutton, due in part to the rising price of feed grain. To reduce grazing pressure and protect the fragile ecology of the northern grazing lands, it was proposed to use them only for animal breeding and to transfer animals for fattening to the cropping belt.

In the Southern Cone of Latin America, scenarios are being developed for the re-allocation of production to specific areas according to their capacity. Livestock production would be encouraged in areas with an ecological comparative advantage. In these areas, animal products could be produced at lower cost and supplied to other, lower-potential areas, which for years have been producing at relatively low efficiency and high cost in order to maintain self-sufficiency. Greater cost-effectiveness would bring intensification, leading to increased utilization of higher-quality feed resources such as hay and silage or concentrates, including agroindustrial by-products.

Supplementation and Processing

Supplementation

Crop residues are generally of low nutritive value. One way of improving the quality of the animal's diet is to supplement them with other feed resources which are richer in energy and protein and/or superior in digestibility or intake.

Table 13.8 presents the information on this topic available in the country reports. In broad terms, four types of supplementary feed can be distinguished. They are: (i) other crop residues, including those of cultivated fodders; (ii) cereal and/or legume grains; (iii) by-products from agriculture or agroindustry; and (iv) other feeds, such as fodder tree crops (fruit, leaves, pods, wastes).

Once again, several countries did not provide sufficient information on this topic, so the following conclusions are indicative only. Supplementing low-quality crop residues such as grain straws and stovers with residues from legume or oilseed crops or with green fodder appears fairly widespread among farmers in general. In contrast, the use of cereal or legume grain as a feed supplement for ruminants is still comparatively rare in most countries. Exceptions are countries such as Tunisia and Iraq, where grains are sometimes fed in the barley-based sheep production system of the arid and semiarid zones, and the temperate zone of the Southern Cone of South America, where they frequently serve as an energy supplement to the natural grasslands.

Agroindustrial by-products are more commonly used as energy feeds than grain. In intensive production systems, such as peri-urban dairying or sheep fattening operations in the WANA region, they may contribute up to 50% of the animal feed ration.

The use of other supplements, such as browse from legume trees and shrubs, appears fairly widespread in sub-Saharan Africa and Asia, especially during the

Table 13.8. Supplementation and processing practices.

Country report[1]	CR feeding system	Supplementation				Processing	
		Other CRs	Grains	By-products	Others	Physical	Chemical
Sub-Saharan Africa							
Senegal	Cereals grazed in field. Legumes fed at homestead	Yes	Not common	Oilseed cake and brans	Tree crops (green or dry leaves) Browse		Urea (experimental)
Ethiopia	Collected and fed at homestead. Grazing of aftermath and crop weeds	Yes, e.g. teff with wheat/barley straw				Chopping in some areas	Not common
Kenya	Maize stover mainly grazed directly, but also stored for dry season. Green sorghum stover ensiled. Wheat/barley straw baled, also stubble grazing. Rice straw and sugarcane tops mainly used for mulching in coffee plantations	Ground and mixed with molasses and alkalis			Cereal crop by-products supplemented with legume tree fodder (*Leucaena, Sesbania, Gliricidia*)	Chopped and soaked in water	Marginal: NaOH and ammonia on CRs
Malawi	Mostly grazed in field after harvest (maize, groundnut). Some stall feeding in intensive systems	Stover with groundnut haulms, maize bran, legume leaves, sweet potato vines, banana leaves	Not done	Cottonseed/groundnut cake, urea-molasses, bean meal	Cereal crop by-products supplemented with legume fodder tree foliage (*Leucaena, Sesbania, Gliricidia*)	Chopping and soaking of stover in water (traditional)	NaOH, ammonia, urea; however, limited adoption by farmers

[1]No information from Mali or Morocco.

Table 13.8 continued.

Country report	CR feeding system	Supplementation				Processing	
		Other CRs	Grains	By-products	Others	Physical	Chemical
Sub-Saharan Africa (continued)							
Nigeria	Grazed in field after harvest Most residues rot or are burnt before next season	Not mentioned		Oilseed cake, palm oilseed cake, brans, molasses, brewer's grain	Tree legume fodder	Grinding and pelleting	NaOH, urea
Asia							
Thailand	Rice straw, maize stover, legume CRs, grazed directly or cut-and-carried	Information only refers to trials	Not relevant	Information only refers to trials		Marginal use of baling, chopping	Treatment of rice straw with urea
Indonesia	Most important CRs: rice straw, maize stover, groundnut haulms, tuber crop wastes	Yes, very desirable	Not done	Done, but not specified	Tree legumes (*Leucaena, Gliricidia, Sesbania, Calliandria*) are very desirable	Hay making Spraying with salt solution if diet is very fibrous Chopping	Urea, urea-molasses block, but low adoption
Bangladesh	CRs constitute 70% of animal feed: straws (rice, wheat, pulses); potato stems/ leaves; sugarcane tops By-products: bran; oilcakes (mustard, sesame, coconut, cottonseed); molasses	90% of farmers mix straws with green grass, fodder, water hyacinth or sugarcane tops	Not done	Mustard/ sesame oilcake, rice bran are widespread; other by-products of minor importance	Wastes and leaves of fruit trees (mango, jackfruit, pineapple) Aquatic plants and weeds	Chopping and soaking for cereal straws Green forage chopped only	Urea shows good effect, but low adoption

Table 13.8 continued.

Country report	CR feeding system	Supplementation				Processing	
		Other CRs	Grains	By-products	Others	Physical	Chemical
Asia (continued)							
China	Only 25% of 570 million t CRs are used as feed. Most important: rice straw, maize stover, wheat straw, rape, other coarse grains, soyabean, sugarcane tops, groundnut haulms	Not stated	Not common	Oilseed cakes (cotton, rape)	Legume tree leaves, shrubs, animal manure expected to increase in importance as oilseed cakes rise in price	Chopping and soaking Ensiling	Treatment with urea (successful), lime, ammonium bicarbonate, NaOH
India	CRs become increasingly important. Main CRs from wheat/rice, sorghum/millets, maize, pulses, oilseeds	Legume/ non-legume fodders Hay and silage Mixed cereal and legume CRs	Different cereal grains, wheat flour, cottonseed	Range of different agricultural and agroindustrial by-products	Legume fodder trees Horticulture wastes and leaves Forest products	Chaffing, grinding, pelleting, soaking, boiling	Urea-generated ammoniation is sound and profitable, but adoption is low
Nepal	Main CRs are cereal straws Despite shortage, CRs not fully utilized	Not common	Not common	Not common	Non-conventional by-products are available, but are not used systematically Green leaves of fodder trees in winter/scarcity periods	Chaffing only done in some urban and Terai areas	Urea treatment not adopted by farmers

Table 13.8 continued.

Country report	CR feeding system	Supplementation					Processing	
		Other CRs	Grains	By-products	Others		Physical	Chemical
Latin America								
Central America/ Mexico	Use of CRs is greater in drier conditions, more common in dual-purpose cattle and goat systems; concentrates and chopped grass more common in dairy systems Grazing of CRs is traditional in dry months in humid/dry hillsides and subhumid/dry lowlands; complemented with grazing of pasture, browsing of tree foliage, fruits *Guatera* system mainly in Guatemala, Honduras, El Salvador, Nicaragua: sorghum/maize planted as forage late in rainy season, harvested at vegetative state, sun-cured, conserved for feeding in late dry season	Not specified	Not specified	Not specified	Grazing of pasture Browsing of tree foliage		Chopping of straw	Ammonification technology is available but not widely adopted
Andean region (Peru)	Stubble grazing of cereal CRs and cut-and-carry of cultivated fodder Barley, wheat, maize, oats are main crops	Not specified	Not done	Rice and wheat bran, cottonseed cake are used in intensive systems	Alfalfa, ryegrass, trifolium in semi-intensive systems		Chopping and soaking in water	Not done

Table 13.8 continued.

Country report	CR feeding system	Supplementation				Processing	
		Other CRs	Grains	By-products	Others	Physical	Chemical
Latin America (continued)							
Southern Cone	Livestock systems mainly based on natural grasslands and pastures In mixed crop/livestock systems, grazing of cereal (maize, wheat, rice) and oilseed (soyabean, sunflower) CRs is common	Temperate zone: green oats, rye, pasture hay, maize and grass silage Subtropical zone: chopped sugarcane	Temperate zone: maize, oats, rye, whole cottonseed	Temperate zone: brans (wheat, rice); sunflower meal Subtropical zone: molasses and urea	Concentrates (temperate/sub-tropical zones), cassava; horticulture by-products only in intensive systems close to production area (Chile)	Chopped sugarcane stall-fed in subtropical zone	Only marginal importance due to high costs
West Asia - North Africa							
Tunisia	Cereal (wheat, barley) straw and stubble Olive tree prunings (leaves, twigs) are most important CR CRs underutilized when green fodder abundant	Not specified	Barley grain (sheep); commercial concentrates	Brans, olive cake, molasses, sugarbeet, and tomato pulp silage	Spineless cactus (*Opuntia*), acacia, *Atriplex*	Sieving of olive cake Improved olive tree pruning	Straw with ammonia and urea Olive cake with NaOH, ammonia, but little adopted Ensiling of sugarbeet and tomato pulp

Table 13.8 continued.

Country report	CR feeding system	Supplementation				Processing	
		Other CRs	Grains	By-products	Others	Physical	Chemical
West Asia - North Africa (continued)							
Egypt	Maize, green leaves Straws: rice, sorghum, maize, wheat, barley	Not common	Not done			Chopping of straws and sugarcane tops Bailing of rice straw for sale	Ammonia on rice straw (medium/large farms) Urea suitable for small farms, but little adopted
Iraq	CRs 32% of total feed resources for ruminants Wheat straw (stubble grazing or collected) is main CR, followed by rice, barley, maize	Not specified	Barley grain to small ruminants only	Wheat and rice bran, oilseed meal, date stone/pulp	Horticulture by-products	Chopped wheat and barley straw	Urea-treated straw, but little adopted

dry season. Leaves and wastes from horticulture are less important, but may be important locally in some fruit- and vegetable-growing areas, especially those adjacent to areas in which feed resources are scarce.

Processing

Processing of crop residues consists of the physical and/or chemical treatment of mainly fibrous cereal straws and stovers to make them more easily digestible by animals. Physical processing covers a range of different methods, including chopping, soaking in water and ensiling. Chemical processing mostly refers to the treatment of straws and stovers with various alkalines, of which treatment with urea appears to be by far the most common.

The country reports suggest that, with the exception of a few countries, physical processing is a fairly common practice among farmers, who thereby improve the feed intake of their animals and reduce fodder wastage at feeding. With regard to chemical treatment, the picture is quite different. Although these technologies may, in some cases, have become locally important, overall adoption by farmers is extremely low, especially considering the very considerable amount of scientific information available in most country reports and elsewhere on the positive effects of different treatment methods on the nutritive value of straw. Adoption is severely hampered for several reasons, including the high costs of this technology, its additional labour requirements and the attendant health risks for people and animals. Among the country reports, only China reported a degree of success in securing farmer adoption.

Most country reports emphasized the need to adapt existing or even to develop new technologies for both physical and chemical treatment, but especially the latter. The adoption of such technologies by farmers must remain the principal criterion of success.

Nutritive Value

The nutrititive value of feeds, including crop residues, is usually expressed in terms of digestibility, DM intake and energetic efficiency. Straws and stovers are relatively high in fibre compared to better quality forage crops. After protein, minerals and vitamins, the fibrous carbohydrates form the most important source of nutrients for ruminants (van Soest, 1988).

The quality of crop residues is greatly influenced by environmental conditions. As these are highly variable, so also is the quality of crop residues. Other factors contributing to variability in quality include genetic differences between and within crop species, crop management practices, and the post-harvest handling of crop residues.

In the country reports a considerable amount of information was presented on the chemical composition (nutrient content) and nutritive value of a range of crop residues and by-products. The number and type of parameters analysed varied greatly between countries. Background information on the fodder material analysed or the feeding trials conducted was specified in a few country reports, but not in most. With regard to the methodology used for fibre analysis, most countries reported results based on the proximate analysis system, while seven had used the more recently developed detergent system. These inconsistencies in reporting make it difficult to synthesize the results on the chemical composition and nutritive value of crop residues across countries. However, the three papers received from Indonesia, India and Tunisia gave a range of values obtained from analyses of different crop residues.

Constraints and Research Opportunities

Constraints

Table 13.9 provides an overview of the constraints faced in the utilization of crop residues. The country reports from Mali, Ethiopia, Morocco and Central America did not discuss this topic at all, while in the other reports the specification of contraints is probably by no means exhaustive. However, taking the available information as a whole, a fairly conclusive picture emerges.

The constraints described can be regrouped into the following three major types, with the key issues stated under each:

Type 1: Quality of crop residues
1. Poor nutritive value, associated with low contents of available energy, protein, minerals and vitamins.
2. Poor feeding value, usually associated with low palatability, intake and digestibility.

Type 2: Postharvest management and feeding
3. Suboptimal utilization of available crop residues.
4. Poor storage and conservation techniques, incurring considerable losses (including those associated with the transportation of bulky material).
5. Insufficient use of processing and supplementation technologies.

Type 3: Socioeconomic and other aspects
6. Limited labour availability for the additional operations required for more effective crop residue utilization (e.g. the tasks of collection, transport, storage and processing).
7. Unfamiliarity with existing or new utilization and processing technologies, especially in more remote areas.

Table 13.9. Major constraints to the utilization of crop residues as animal feed.

Country report [1]	Quality	Postharvest management/feeding	Socioeconomic and other aspects
		Type of constraint	
Sub-Saharan Africa			
Senegal		Transport from field to farm/other region	Lack of labour for proper storage
		Lack of machines for physical processing	Inadequate use of available technology
Kenya	Low digestibility and intake		High cost of chemical treatment
	Low organic matter		High transportation cost (if CRs used away from production sites)
	Imbalanced minerals		
Malawi	Low nitrogen, energy and protein contents	Poor processing and storage	Lack of technical know-how
			Lack of crop/livestock integration
			Extension service gives little attention to livestock
Nigeria	Poor nutritive value	Lack of proper storage techniques	Labour requirements for storing conflict with harvesting
		Bulkiness	Lack of farmer awareness
			Production sites in areas with lower livestock density
			Poor adoption of treatment techniques due to: water scarcity in dry areas, costly ensilage/covering materials, costly machinery for chopping, competition with industry for molasses
Asia			
Thailand		Lack of quick drying techniques	Lack of labour
		Scattered nature of feed	Cost of transport
Indonesia	Deterioration higher where hot and humid climate stimulates fungi growth	Scattered nature of feed	Sites of production and utilization are often distant and transport costs are high
		Lack of labour and drying technology means large amounts are burnt	

[1] No information was received from Mali, Ethiopia, Morocco, Mexico and Central America.

Table 13.9 continued.

Country report	Type of constraint		
	Quality	Postharvest management/feeding	Socioeconomic and other aspects
Asia (continued)			
Bangladesh	Low digestibility	Lack of supplementation technology acceptable to farmers Bulkiness only allows storage outside house, resulting in straw losses	Poor adoption of chemical/biological treatment methods by farmers
China	Low intake/digestibility	Present system of supplementation depends on oilseed cakes, availability of which is highly limited Lack of farmer awareness of nutrient requirements of animals at specific physiological stages	Increasing cost of urea and plastic limit extension of chemical treatments
India	Highly fibrous and lignified, low in available energy, protein, vitamins, minerals Low palatability, intake, digestibility		
Nepal	Low nutrient content/digestibility, having adverse effects on performance of productive animals in particular	Lack of straw treatment and supplementation results in inefficient utilization in conventional feeding systems	In livestock areas with high populations (mid-hills), insufficient availability to meet animal requirements
West Asia-North Africa			
Tunisia			High opportunity cost for certain CRs/by-products used outside agricultural sector (e.g. molasses in yeast industry)

Table 13.9 continued.

Country report	Type of constraint		
	Quality	Postharvest management/feeding	Socioeconomic and other aspects
West Asia-North Africa (continued)			
Egypt		Lack of suitable feed processing infrastructure in villages (e.g. for cotton stalks)	High opportuniy cost of wheat straw used in poultry industry Lack of technologies adapted to specific farming systems
Iraq		Collection, handling, transportation and storage	
Latin America			
Andean region		Poor adoption of physical/chemical treatment techniques due to lack of technical know-how and disadvantages of technology	Lack of effective extension service
Southern Cone			Due to low nutritive value CRs more expensive to use than high-quality pastures in temperate zones High opportunity cost, e.g. of exported soyabean meal, reduces its utilization as animal feed

8. Insufficient integration of crop and livestock production in many farming systems.

9. Mismatch between surplus crop residue production in one area and demand for feed in another area.

10. Economic aspects, including high cost of processing and conservation, high opportunity cost of alternative uses forgone, cost per unit of energy or protein higher than that of high-quality natural grassland.

Research Opportunities

All country reports except for Mali, Nepal, Egypt and Morocco addressed this topic. The research themes identified by each country are compiled and listed in Table 13.10. They can be classified under five main headings:

1. Crop genetic enhancement (CGE).
2. Crop management (CM).
3. Postharvest handling and storage (PHS).
4. Feeding systems and supplementation (FSS).
5. Animal nutrition (AN).

To some extent the table represents a "wish list" that will have to be thinned down by further priority-setting exercises. There is, however, a reasonably high degree of unanimity regarding the main lines of research that should be addressed. These are not radically different to those of 25 years ago, but there are some interesting shifts in emphasis.

Several countries appear to have shared the perception of the international research centres that genetic enhancement offers one of the more promising avenues for future research. Work on this hitherto neglected topic is being led by the International Centre for Agricultural Research in the Dry Areas (ICARDA) in the WANA region and by the International Livestock Research Institute in other regions. In addition, supplementation of crop residues with other, more nourishing feeds continues to be a popular line of enquiry, perhaps receiving increased emphasis now that hopes for securing the rapid adoption of chemical processing treatments have receded. Several African and Asian countries retain an interest in the intercropping or rotation of forage legumes with cereals, despite the difficulties, evident as a result of research over the past two decades, of fitting such legumes into systems where there is high pressure on land and a high demand for subsistence food crops. This is encouraging, since such crops clearly do have a potential in the more intensive smallholder dairying systems found in peri-urban areas. Increased interest in fodder trees is another welcome development, reflecting increased awareness of the multiple functions of trees in mixed smallholder farming systems and the emergence of the International Centre for Research in Agroforestry (ICRAF) as an energetic partner in this field.

Table 13.10. Opportunities for collaborative research on the use of crop residues as animal feeds.

Country/ region	Opportunities
Andean region	Range evaluation and management (FSS) Conservation and management of CRs and techniques for enhancing their digestibility (PHS)
Bangladesh	Development of technologies for straw treatment and supplementation which are based on farmers' own resources and implements, and which are simple to adopt (PHS) Integration of fodder legume production with existing cropping systems, to supplement straw and other CRs (CM)
Central America/ Mexico [1]	Effect of form of CR utilization (cut-and-carry versus grazing) on soil-related sustainability indicators (physical, biological, nutrient cycling) (FSS) Effect of defoliation on crop yields and forage quantity and quality in monoculture and multiple cropping systems (CM) Practical methods for improving the quality of CRs (e.g. intercropping with legumes, ammonification) (CM) Synergism between CRs and pastures, tree foliages and/or other supplements (FSS) Crop-pasture rotations as a way of restoring the productivity of degraded land (CM; FSS)
China	Selection of crops for better feed value (CGE) Development of cost-effective treatment methods to improve nitrogen utilization (PHS) Exploitation of new resources for supplementation (legumes, tree leaves, brewer's grain, animal manure) (FSS)
Ethiopia	Physical or chemical treatment of CRs with very low inputs (PHS) Intercropping of cereals with legumes (CM) Development of cultivars with high straw quality (CGE)
India	Investigate variations in nutrient characteristics and phenology of crops and breed for yield and nutritive value of CRs without affecting yield and quality of grain (CGE) Study effects of specific plant components, crop management and agroecological factors on quality and quantity of CRs (CM) Study relationship between measured nutritive characteristics and actual digestion of CRs fed to livestock (AN) Study effect of crop management and postharvest handling on quantity and quality of CRs (CM, PHS) Identify physical and chemical factors inhibiting utilization of CRs and develop technologies to enhance intake and rumen fermentation (AN) Develop efficient and economical feeding systems based on fibrous CRs for different agroecological regions (FSS) Conduct on-farm research to test appropriate technologies in mixed crop/livestock production systems in various agroecological zones Promote exchange of information on use of CRs in animal production

[1] Research to be concentrated in the most fragile agroecological zones and in systems with strongest crop-livestock interactions (hillsides, subhumid/dry lowlands).

Table 13.10 continued.

Country/region	Opportunities
Indonesia	How to integrate animal production in feed-producing areas such as crop estates and forests (FSS)
Iraq	Carrying capacity of cereal stubble (PHS) Nutritive value of cereal stubble (AN) Methods for detecting mineral deficiencies under local conditions (AN) Enrichment of feed blocks with bypass protein and trace elements (FSS)
Kenya	Simplified chemical treatment methods (PHS) Supplementation with legume fodder and other crops produced on farm (FSS)
Malawi	Improved storage (PHS) Nutritive value, including toxic constituents (AN) Varietal differences in quality (CGE) Acceptability to farmers of introduced improvement methods (PHS) Better integration of livestock with crops and trees (FSS)
Nigeria	Exploitation of new supplementation resources (FSS) Selection for better feed value (CGE) Adoption of enhancement techniques (PHS) Development of cost-effective and socially feasible treatments (PHS)
Senegal	Increased forage production in sylvo-pastoral areas through study of agronomic characteristics such as fertilizer requirements, cutting frequency and height, grass/legume mixtures, other factors affecting yield and nutritional value (CM) Utilization of by-products in beef and sheep fattening (FSS) Integration of crop and animal production in mixed farming systems in different agroecological zones (FSS) [2] Inventory of type and availability of CRs (CGE) Feed value and toxicity (AN) Technologies of conversion (PHS) Improvement of nutritive value (All) Use of fodder trees in different systems/locations (FSS) Feed value and utilization of household and kitchen wastes as supplementation (FSS)
Southern Cone	Treatment of by-products such as oilseed cake and meal to improve bypass protein (PHS)
Thailand	Seasonal availability of protein-rich CRs (FSS) Use of high-quality concentrate mixtures (FSS) Levels of supplementation of CR-based diets (e.g. for dairy cattle) (FSS) Processing of rice straw to reduce bulkiness and so increase storage capacity, facilitate transportation, increase efficient utilization (PHS)
Tunisia	Breed for more biomass (grains and straw) of better quality (CGE) Develop low-cost methods for improving feed value of CRs (PHS) Integrate CRs and by-products with other feed resources (FSS) Focus on supplementation using protein-rich fodder trees and shrubs, by-products and urea (feed blocks) (FSS)

[2] Focus on systems where crop production is farmers' main occupation.

With regard to chemical processing, there is a noticeable increase in the emphasis on the acceptability of this technology to farmers and in particular on creating new, low-cost options. Studies on the actual conversion of feeds within the animal appear comparatively underemphasized, however, perhaps reflecting their high cost and complexity. Both the African countries listing research needs in this area mention anti-nutritional factors, perhaps reflecting the capacity in this field developed by the International Livestock Centre for Africa (ILCA) during the 1980s.

References

Devendra, C. (1992) *Non-conventional Feed Resources in Asia and the Pacific: Strategies for Expanding Utilization at the Small Farm Level.* Fourth revised edition. Regional Office for Asia and the Pacific (RAPA), Food and Agriculture Organization of the United Nations (FAO), Bangkok, Thailand.

FAO (1978-1981) *Reports of the Agroecological Zones Project.* World Soil Resources Report No. 48, Food and Agriculture Organization of the United Nations, Rome, Italy.

FAO (1984) *FAO Production Yearbook*, vol. 38. Food and Agriculture Organization of the United Nations, Rome, Italy.

FAO (1987) *Assistance to Land Use Planning in Ethiopia: Economic Analysis of Land Use.* UNDP/FAO Technical Report No. 8, Food and Agriculture Organization of the United Nations, Rome, Italy.

FAO (1994) *FAO Production Yearbook*, vol. 48. Food and Agriculture Organization of the United Nations, Rome, Italy.

Kossila, V. (1988) The availability of crop residues in developing countries in relation to livestock populations. In: Reed, J.D., Capper, B.S. and Neate, P.J.H. (eds), *Plant Breeding and the Nutritive Value of Crop Residues.* Proceedings of a Workshop, 7-10 December 1987, International Livestock Centre for Africa (ILCA), Addis Ababa, Ethiopia, pp. 29-39.

Nordblom, T.L. (1988) The importance of crop residues as feed resources in West Asia and North Africa. In: Reed, J.D., Capper, B.S. and Neate, P.J.H. (eds), *Plant Breeding and the Nutritive Value of Crop Residues.* Proceedings of a Workshop, 7-10 December 1987, International Livestock Centre for Africa (ILCA), Addis Ababa, Ethiopia, pp. 41-64.

Ranjhan, S.K. (1986) Sources of feed for ruminant production in Southeast Asia. In: Blair, G.J., Ivory, D.A. and Evans, T.R. (eds), *Forages in Southeast Asian and South Pacific Agriculture.* ACIAR Proceedings No. 12, Australian Centre for International Agricultural Research, Canberra, Australia, pp. 24-28.

TAC (1992) *Review of CGIAR Priorities and Strategies,* Part 1. Technical Advisory Committee of the Consultative Group on International Agricultural Research (CGIAR), Rome, Italy.

van Soest, P.J. (1988) Effect of environment and quality of fibre on the nutritive value of crop residues. In: Reed, J.D., Capper, B.S. and Neate, P.J.H. (eds),

Plant Breeding and the Nutritive Value of Crop Residues. Proceedings of a Workshop, 7-10 December 1987, International Livestock Centre for Africa (ILCA), Addis Ababa, Ethiopia, pp. 71-96.

Acknowledgements

I would like to thank the following authors of the country reports for their contributions to the workshop and to this paper: M.A. Akbar (Bangladesh), G. Ashiono (Kenya), M. Cissé (Senegal), A. Douiyssi (Morocco), I. Gomaa (Egypt), Z. Kouyate (Mali), G.P. Lodhi and V.S. Upadhyaya (India), A. Nefzaoui (Tunisia), N.E. Nyirenda (Malawi), O.L. Oludimu (Nigeria), S.B. Panday (Nepal), D. A. Pezo (Central America and Mexico), D. Rearte (Southern Cone of Latin America), A. D. Salman (Iraq), F. San Martín (Andean region), A. Tukue (Ethiopia), M. Wanapat (Thailand), Wang Jiaqi (China) and M. Winugroho (Indonesia). Lastly, I would like to thank Charles Renard for his help and advice.

Concluding Remarks

C. Renard
International Crops Research Institute for the Semi-Arid Tropics, Patancheru 502 324, Andhra Pradesh, India

This workshop has brought together 72 participants drawn from 33 countries and six continents: Asia, Africa, North America, South America, Europe and Australia. It has provided a unique opportunity to pursue a multidisciplinary approach to the theme of crop residues in sustainable mixed crop/livestock farming systems in the tropics and subtropics.

Scientists from different institutional settings—international agricultural research centres, national agricultural research systems, universities, advanced institutions, development agencies—and different disciplines—agronomy, plant breeding, animal production, animal nutrition, economics—have had, for four-and-a-half days, stimulating exchanges about crop residues across continents, ecozones, disciplines.

We have shared a common concept—the "crop/livestock system"—and we all agree that crop residues play a major role in the four ecoregions we have retained as priorities for research: the highlands, the subhumid and humid tropics, the arid tropics and subtropics, and the semiarid tropics. These four ecoregions have many common features, but also major differences, in terms of the crops and livestock raised in them, the traditions of their peoples, and the status of the technology available.

So we have asked ourselves, to what extent can we transfer tools and techniques from one region or system to another?

We have agreed on the following priorities or recommendations:

Genetic enhancement
1. Scientists and others should be encouraged to breed and cultivate crops for more and better straw or haulm. Both the quantity and quality of residues are important.
2. When a new variety of cereal or legume ispublicized, information on potential biomass production and the quality of the straw or the haulm produced should be given alongside the usual data on grain characteristics and yields.

Crop residues and soil management

3. There is major concern over losses in soil organic matter and nutrients and the increased risk of erosion incurred by removing crop residues from farmers' fields to feed animals.

4. The soil/crops/livestock interface must be looked at from the perspective of overall system sustainability.

Feeding systems

5. For the purpose of improving feed supplies, the issue of interaction between irrigated and rainfed production systems is crucial. Irrigated rice systems produce huge amounts of straw which are used for feeding livestock in rainfed systems.

6. We have a powerful knowledge base in this area, and technology exchange is most important.

7. More information or research is needed on the response profile to supplementation. We know enough about the chemical treatment of straw, but this technology has not been successfully transferred. We should investigate the constraints to transfer.

Modelling

8. Models must be developed according to specific needs and not simply borrowed from elsewhere and then, with minimal adaptation, be expected to address real needs.

9. Models are made of modules. These can be taken from different models and assembled, after adaptation, to form a new model suited to real needs. Modelling is a part of the systems approach.

Technology design and transfer

10. Our target is the small-scale farmer in different ecoregions. To get our technologies to him or her, we must ensure that these are adoptable, acceptable, very simple and—most important—low-cost.

One further outcome of this workshop, besides the proceedings, will be a global proposal for research on crop residues, divided into projects according to ecoregions. This will be available from the Livestock Policy Group at ILRI, based in Nairobi.

Let me conclude by encouraging all researchers to maintain the multidisciplinary approach that has been adopted at this workshop. I would also like to stress the complementarity between different ecoregions: much can be gained by exchanges—of information, expertise, materials and methods—between them. Such exchanges, already so much more extensive than they were a quarter of a century ago, should be further promoted through the active participation of national researchers in the new gobal research initiative being coordinated by ILRI. I wish that initiative every success.

Index